BIOINFORMATICS OF GENOME REGULATION AND STRUCTURE

BIOINFORMATICS OF GENOME REGULATION AND STRUCTURE

Edited by

NIKOLAY KOLCHANOV

Institute of Cytology & Genetics, Siberian Branch
of the Russian Academy of Sciences, Novosibirsk, Russia

RALF HOFESTAEDT

Bielefeld University, Bielefeld, Germany

Kluwer Academic Publishers
Boston/Dordrecht/London

Distributors for North, Central and South America:
Kluwer Academic Publishers
101 Philip Drive
Assinippi Park
Norwell, Massachusetts 02061 USA
Telephone (781) 871-6600
Fax (781) 681-9045
E-Mail: kluwer@wkap.com

Distributors for all other countries:
Kluwer Academic Publishers Group
Post Office Box 322
3300 AH Dordrecht, THE NETHERLANDS
Telephone 31 786 576 000
Fax 31 786 576 254
E-Mail: services@wkap.nl

Electronic Services <http://www.wkap.nl>

Library of Congress Cataloging-in-Publication Data

A C.I.P. Catalogue record for this book is available
from the Library of Congress.

Title: Bioinformatics of Genome Regulation and Structure
Editor: Ralf Hofestaedt and Nikolay Kolchanov
ISBN: 1-4020-7735-1

Printed on acid-free paper.
Printed in the United States of America.

The Publisher offers discounts on this book for course use and bulk purchases. For further information, send email to <Laura.Walsh@wkap.com>.

Contents

Contributing Authors

D.A. Afonnikov
Y.M. Akimov
E.A. Ananko
H.E. Assmus
I.V. Avdeeva
J. Błażewicz
M.B. Chaley
V.F. Chekhun
S.F. Chekmarev
N.A. Chuzhanova
D.N. Cooper
V.A. Debelov
G.N. Erokhin
S.I. Fadeev
S. Fischer
A. Freier
F.E. Frenkel
D.P. Furman
A.V. Galimzyanov
A.V. Gavrilov
D.A. Grigorovich
L.I. Gunderina
V.D. Gusev
P.L. Hammer
H. Herzel
R. Hofestaedt

E.V. Ignatieva
A.G. Istomina
V.A. Ivanisenko
L.N. Ivanisenko
H. Jönsson
A.V. Katokhin
T.M. Khlebodarova
S.M. Kielbasa
I.I. Kiknadze
A.V. Kochetov
N.A. Kolchanov
Yu.V. Kondrakhin
H. Kono
J.O. Korbel
E.V. Korotkov
N.A. Kudryashov
M. Lange
A.A. Laskin
V.G. Levitsky
V.A. Likhoshvai
K.A. Loktev
P. Łukasiak
A.M. Matsokin
Yu.G. Matushkin
T.I. Merkulova
E.M. Meyerowitz

S.V. Mikhalap
L. Milanesi
E. Mjolsness
A.N. Naumochkin
T.M. Naykova
E.A. Nedosekina
L.A. Nemytikova
N.A. Omelyanchuk
D.Yu. Oshchepkov
A.Yu. Palyanov
A.G. Pichueva
S.S. Pintus
N.L. Podkolodny
O.A. Podkolodnaya
M.A. Pozdnyakov
A.L. Proscura
A.V. Ratushny
M. Régnier
I.B. Rogozin
A.G. Romashchenko
A. Sarai
S.M. Schwarzl
B.E. Shapiro
L.M. Shlapatska
S.P. Sidorenko
K.G. Skryabin
J.C. Smith
I.L. Stepanenko
R.N. Tchuraev
I.I. Titov
I.I. Turnaev
O.V. Vishnevsky
E.E. Vityaev
M.I. Voevoda
D.G. Vorobiev
Yu.N. Vorobjev
N.S. Yudin
M.Y. Yurchenko

Preface

Completion of the human genome sequencing could be undoubtedly viewed as the greatest scientific event of the last three years. This event marked the beginning of the post-genome era in biology. This era is characterized by a drastic elevation in the research scale in the fields of transcriptomics, proteomics, and systemic biology (gene interactions, gene network operation, and signal transduction pathways) without losing the basic interest to studying structure–function genome organization.

The structure and regulation of genome are the counterparts of life at molecular level; that is why the understanding of basic principles underlying operation of the regulatory genomic machinery is impossible unless their detailed structural organization is known, and vice versa.

The huge volume of experimental data that has been acquired on the genome structure, functions, and gene expression regulation is growing by leaps and bounds. Development of informational and computational technologies of the new generation is a challenging problem of bioinformatics. Bioinformatics has entered that very phase of development when solutions for the challenging problems determine realization of large-scale experimental research projects directed to studying genome structure, function, and evolution.

One of the international scientific forums presenting and discussing topical problems of bioinformatics related to genome regulation and structure is the international conference Bioinformatics of Genome Regulation and Structure–BGRS, organized biennially by the Laboratory of Theoretical Genetics with the Institute of Cytology and Genetics, Siberian Branch of the Russian Academy of Sciences, Novosibirsk, Russia. BGRS'2002 was the third event in the series.

Participants of BGRS'2002 concentrated their attention on the following hottest items in bioinformatics: (i) regulatory genomic sequences: databases, knowledge bases, computer analysis, modeling, and recognition; (ii) large-scale genome analysis and functional annotation; (iii) gene structure determination and prediction; (iv) comparative and evolutionary genomics; (v) computer analysis of genome polymorphism and evolution; computer analysis and modeling of transcription, splicing, and translation; structural computational biology: genomic DNA, RNA, and protein structure–function organization; (vi) gene networks, signal transduction pathways, and genetically controlled metabolic pathways: databases, knowledge bases, computer analysis, and modeling; principles of organization, operation, and evolution; (vii) data warehousing, knowledge discovery, and data mining; and (viii) analysis of basic patterns of genome operation, organization, and evolution.

This book comprises selected reviewed papers prepared on the basis of results presented at BGRS'2002.

This book may be interesting to the specialists working on development of databases on genome regulation and structure, algorithms and software for analysis and simulation of molecular genetic systems of various organisms at various levels of their function and organization as well as to a wide range of biologists using bioinformatics methods and approaches in their work.

All the papers of this volume passed double reviewing: first, when submitted as presentation to BGRS'2002; second, when preparing the book.

Professor Nikolay Kolchanov
Head of Laboratory of Theoretical Genetics
Institute of Cytology & Genetics, Siberian Branch
of the Russian Academy of Sciences, Novosibirsk, Russia
Chairman of the Conference

Professor Ralf Hofestaedt
Faculty of Technology Bioinformatics Department
University of Bielefeld, Germany
Co-Chairman of the Conference

Foreword

After deciphering the total genomes of hundreds of living organisms from viruses and bacteria to mammals, fishes, etc., including humans, the structural and functional genomics became a key avenue in converting this new knowledge into better understanding of the living matter. The only reasonable way to digest this immeasurable mass of data is to develop new strategies for genome analyses, and this is the major goal of bioinformatics.

This book is composed of contributions from countries all over the world including France, Germany, Japan, Poland, Russia, United Kingdom, and USA. The book comprises a set of preselected reviewed papers based on the presentations delivered at the Third International Conference on Bioinformatics of Genome Regulation and Structure–BGRS (Novosibirsk, Russia, 2002). When compiling this volume, we focused not only on the computer technologies for designing the relevant software or databases on various aspects of genome regulation and structure, but rather on application of these technologies to studying a variety of features of the structure–function organization of molecular genetic systems operating in living organisms. The levels corresponding to genomic DNA, genes, gene regulatory regions, RNA, proteins, gene networks, and metabolic pathways were considered. The book is of interest to the researchers working in the fields of both bioinformatics and experimental biology who use bioinformatics methods for analyzing the data generated in the course of experiments.

Lev Kisselev, member of the Russian Academy of Science, EMBO, and Academia Europaea, Head of the Scientific Committee on Russian Human Genome Project

PART 1. COMPUTATIONAL GENOMICS

NUCLEOSOMAL DNA ORGANIZATION: AN INTEGRATED INFORMATION SYSTEM

V.G. LEVITSKY[1,2]*, A.V. KATOKHIN[1,2], O.A. PODKOLODNAYA[1,2], D.P. FURMAN[1,2]
[1]*Institute of Cytology & Genetics, Siberian Branch of the Russian Academy of Sciences, prosp. Lavrentieva 10, Novosibirsk, 630090 Russia, e-mail: levitsky@bionet.nsc.ru;*
[2]*Novosibirsk State University, ul. Pirogova 2, Novosibirsk, 630090 Russia*
* *Corresponding author*

Abstract An integrated information system Nucleosomal DNA Organization has been designed and is being further developed. This system comprises (1) Nucleosome Positioning Region Database, compiling the available experimental data on locations and characteristics of nucleosome formation sites as well as detailed information on other factors that influence nucleosome positioning, and (2) NucleoMeter software package for recognition of these sites. The system allows prognostic estimates of both the ability of arbitrary genomic sequences to position nucleosomes and functional significance of such positioning to be obtained.

Key words: nucleosome positioning, computer analysis, nucleosome formation site recognition, integrated information system

1. INTRODUCTION

Nucleosomes are the major structural element of chromatin. Each nucleosome is formed by a 147-bp DNA fragment wrapped around an octamer comprised by four pairs of histone molecules of the four types. The neighboring nucleosomes are connected by linker DNA with a length ranging from 20 to 80 bp (Luger *et al.*, 1997; Kornberg and Lorch, 1999).

In addition to DNA compacting, the most important function of nucleosomes is their interaction with the molecular components of the nucleus machinery involved in DNA replication, reparation, and recombination. The key role in gene transcription is also assigned to nucleosomes (Aalfs and Kingston, 2000).

Chromatin from various genomic regions is, as a rule, represented by regular arrays of nucleosomes. Among the factors determining the regularity of nucleosome location *in vivo* (Blank and Becker, 1996; Widom, 1998; Aalfs and Kingston, 2000; Becker, 2002), the sequence-directed nucleosome positioning plays an important functional role in providing a proper interaction of DNA functional sites with non-histone proteins, which is especially important in regulatory regions of genes, in particular, promoters (Richmond and Widom, 2000).The mechanisms of sequence-directed nucleosome positioning have been studied in numerous experiments both *in vivo* and *in vitro*; the results obtained suggest the existence of a specialized chromatin (nucleosome) code determining such positioning through multiple histone–DNA interactions (Trifonov, 1997).

Despite an ambiguity of sequence–function relationships, various research teams have succeeded in discovering a number of periodic contextual and conformational signals and rules regulating nucleosome positioning and ordering. The achievements of this laborious "hunt for specific signals or mechanisms involved in nucleosome positioning" are summarized in the review by Kiyama and Trifonov (2002).

Although this field is intensely studied, the mechanisms underlying nucleosome positioning are yet far from being clearly understood (Lowary and Widom, 1998; Fitzgerald and Anderson, 1999). Development of bioinformatics technologies and their application to systematization and analysis of large volumes of experimental data, which have been accumulated and are still being accumulated, provide new approaches to discovering characteristic DNA features in the nucleosome formation sites.

We have designed and are further developing an integrated information system Nucleosomal DNA Organization, comprising (1) Nucleosome Positioning Region Database, which compiles the experimental data on locations and characteristics of nucleosome formation sites (NFS) and detailed information on the other factors influencing nucleosome positioning, and (2) NucleoMeter software package for NFS recognition (Levitsky *et al.*, 1999; 2001a).

This system allows for realizing complex queries for detection of nucleosome positioning characteristics in DNA sequences analyzed taking into account their location in the genome, species of the organism, particular type of experiment, etc., to obtain prognostic estimates of putative nucleosome formation in genomic DNA sequences yet not studied experimentally.

2. SYSTEM AND METHODS

2.1 Basic scheme of the information system Nucleosomal DNA Organization

The Nucleosome Positioning Region Database (NPRD) is the basis of the system Nucleosomal DNA Organization (Figure 1). Thorough annotation of the information inputted allows various NFS training samples to be formed. These samples are used for designing NucleoMeter software package (Levitsky *et al.*, 2001a), intended for analysis of contextual NFS properties and prediction of the ability of arbitrary DNA sequences to form nucleosomes. The samples of NFSs are also used for detailed analysis of their conformational properties by the tools with the analytical system BDNA Sites Video, section Nucleosome Sites (Levitsky *et al.*, 1999; Ponomarenko *et al.*, 1999).

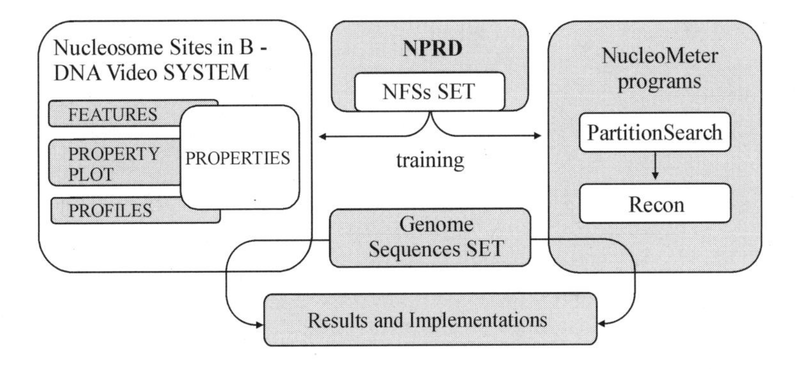

Figure 1. The basic components of the system Nucleosomal DNA Organization.

Within the program section NucleoMeter, we are developing a set of original analytical methods and methods for accounting the basic characteristics of the code underlying a regular nucleosome positioning. We used the sample of NFSs determined experimentally (Ioshikhes and Trifonov, 1993) for analysis of local characteristics of nucleosomal DNA dinucleotide context and designing of the program PartitionSearch (Levitsky *et al.*, 2001a). The results obtained formed the basis for the program Recon intended to calculate the nucleosome formation potential for an arbitrary DNA fragment (Levitsky *et al.*, 2001a). The section Nucleosome Sites in BDNA Sites Video contains the Nucleosomal DNA Property Database. This section compiles profiles of conformational and physicochemical properties of nucleosomal DNA, information on most significant conformational and physicochemical properties of nucleosomal DNA as well as the programs for recognition of potential NFSs developed taking these properties into account (Levitsky *et al.*, 1999; Ponomarenko *et al.*, 1999).

The section Genome Sequence Set contains samples of various structural and functional genomic units of various organisms—telomeric and centromeric repeats, transposable elements, gene regions and their constituents, such as enhancers, promoters, 5'UTRs, 3'UTRs, exons, introns, etc. Each sample is examined with reference to their generalized and individual context-dependent conformational DNA properties using the analytical tools of BDNA Sites Video and NucleoMeter to estimate their capability of forming nucleosomes. Examples of such analysis are described below in the section Results and Implementation.

2.2 Nucleosome Positioning Region Database (NPRD)

The specific feature of this database is a formalized description of NFSs, which we are annotating using published scientific sources. An example of the NPRD entry is shown in Figure 2.

```
AC N00077
RI Levitsky V.G.; 04.09.03
IV in vivo
OS Mus musculus
OC Eukaryota; … Mus.
BI GenBank; MUSENT01; L17322;
DE nidogen 1 gene, entactin 1 gene,
DE MGI:97342; LocusLink:18073; DoTS:DT.94280402; GEA U74A:100120_at
DE NID1 gene, NID gene, ENT gene;
SS gene
SR 5' region
BM L17322; ST: 1227; -600 to -400;();
CN The positions of nucleosomes in the proximal 5' regions ...
CN ... new and/or the modification of existing trans-acting factors.
SQ tccattcccaaactgttttattccttggactccaagcaaatctgactttgaattgtgtc ...
SQ ... acacacatttgtcaggctttt
EP murine F9 stem, ... [MPE X Fe(II)]) treatment;
KW transcriptional activation; basal and high level of gene expression;
AU Chen Y, Keller JM.
TI Transcriptional state and chromatin structure of ...
SO J Cell Biochem.
YR 2001
VL 82(2): 225-233 PM 11527148
```

Figure 2. An example of the Nucleosome Positioning Region Database entry: NFS of murine *nidogen 1* gene 5'-region.

When annotating results of NSF experimental mapping and inputting their nucleotide sequences into the database, we pay a special attention to several important characteristics, such as the location of nucleosome relative to functional components of the genome—within genes (5'- or 3'-regions, enhancers, etc.) or outside genes (repetitive DNA: satellite, centromeric, etc.)—as well as to the type of gene activity related to nucleosome position, influence of nonhistone proteins, occurrence of translational or rotational nucleosome positioning, characteristics of tissue types and states of cell activity, detailed characterization of experimental methods used and accuracy of determining the nucleosome position, and results of applying theoretical and computer methods to analysis of contextual and conformational DNA properties.

2.3 NFS Sets

To develop programs for detecting and recognizing nucleosomal context and to evaluate their prognostic capability, several samples of NFSs—DNA sequences with experimentally demonstrated ability to form nucleosomes—are used. They include (1) 141 sequences extracted from the GenBank/EMBL databank according to the accession numbers and positions indicated in the database Nucleosomal DNA (Ioshikhes, Trifonov, 1993) and used as the training sample for the program Recon; (2) fragments of mouse genomic DNA identified in SELEX experiments for an increased capability of forming nucleosomes (Widlund *et al.*, 1997; Thåström *et al.*, 1999; http://www.molbiotech.chalmers.se/research/mk/nuc/selex.txt); (3) artificially synthesized DNA sequences obtained in SELEX experiments for an increased capability of forming nucleosomes (Lowary, Widom, 1998; Thåström *et al.*, 1999; (http://widomlab1.biochem.northwestern.edu/~wdmgrp/sequences.php); (4) antiNFSs, artificially synthesized DNA sequences isolated in SELEX experiments for a decreased capability of forming nucleosomes (Cao *et al.*, 1998; Thåström *et al.*, 1999; http://www.molbiotech.chalmers.se/research/mk/nuc/antiselex.txt); and (5) the NFSs compiled in NPRD (Nucleosome Positioning Region Database).

2.4 Programs PartitionSearch and Recon

The program PartitionSearch realizes a genetic algorithm based on iterative application of discriminant analysis. PartitionSearch allows locally positioned contextual dinucleotide signals in the sample of NFS sequences to be detected (Levitsky *et al.*, 2001a). By Monte Carlo technique, the alignment of NFS sequences is subjected to numerous partitions into several nonoverlapping blocks of different widths that contain fragments of the alignement. Then, the rules of genetic algorithm (recombinations between regions within blocks and shifting of the boundaries between them) are applied to the partitioned sequences followed by choosing the regions that differ most pronouncedly in the dinucleotide context from the randomized sequences by calculating the distance R^2 (Mahalanobis, 1936). As a result, the program PartitionSearch generates an optimal set of parameters: the widths and arrangement order of regions and dinucleotide frequencies in each region. These parameters are used to train the program Recon. Recon performs a comparative estimation of an arbitrary DNA sequence for its ability to form nucleosomes, that is, the so-called nucleosome formation potential (NFP), which is characterized by the value $\varphi(X)$. When analyzing an arbitrary sequence, the value $\varphi(X)$ is calculated at each position of the sliding window with a length of about 160 bp $\varphi(X)$ is calculated according to the following equation:

$$\varphi(X)=\frac{1}{R^2}\cdot\sum_{n=1}^{N}\sum_{k=1}^{N}\left\{[f_n(X)-(1/2)\cdot[f_n^{(2)}+f_n^{(1)}]\cdot S_{n,k}^{-1}\cdot[f_k^{(2)}-f_k^{(1)}]\right\},$$

$$\text{where } R^2=\sum_{k=1}^{N}\sum_{n=1}^{N}\left\{[f_n^{(2)}-f_n^{(1)}]\cdot S_{n,k}^{-1}\cdot[f_k^{(2)}-f_k^{(1)}]\right\}.$$

Here, $f_n(X)$ is a vector of dinucleotide frequencies, which was constructed by accounting partitioning of the fragment X under study into local regions; $f_n^{(1)}$ is the vectors of average dinucleotide frequencies, which were constructed by accounting the partitioning; $f_n^{(2)}$ is the corresponding vector in the set of random sequences; S^{-1}, inverse matrix for the united covariation matrix, $S = S(1) + S(2)$; and $S(1)$ and $S(2)$, are covariation matrices of vectors $f_n^{(1)}$ and $f_n^{(2)}$.

2.5 Description of WWW interface for Recon

The WWW interface of the program Recon allows the user to construct NFP profile for the sequence of interest (http://wwwmgs.bionet.nsc.ru/mgs/programs/recon/). The output data are in either graphical representation or numerical form (option Graphic Mode). Under Standardization by Dispersion option the original NFP profile is transformed into the profile $\varphi_\alpha(X) \leq +1$, so that the value $\varphi_\alpha(X) = +1$ corresponds to the best prediction; the interval $\varphi_\alpha(X) \geq 0$, to reliable predictions ($P_\alpha > 0.95$); and the interval $\varphi_\alpha(X) \leq 0$, to unreliable predictions ($P_\alpha < 0.95$) (Levitsky *et al.*, 2001a).

2.6 Section Nucleosome Sites in the system BDNA Sites Video

The section Nucleosome Sites in the system BDNA Sites Video comprises the following components (Levitsky *et al.*, 1999): (1) PROPERTIES, description of 38 context-dependent conformational and physicochemical DNA properties; (2) DNA PROPERTY PLOT, a program for calculating profiles of these conformational and physicochemical properties over an arbitrary nucleotide sequence; (3) PROFILES, a database integrating the profiles of conformational and physicochemical DNA properties for samples of nucleosomal sites; and (4) FEATURES, a knowledge base on the conformational and physicochemical DNA properties that are most significant for detecting nucleosomal sites in nucleotide sequences. The knowledge base is linked to the programs for recognizing NFS designed basing on the properties in question.

3. IMPLEMENTATION AND RESULTS

3.1 Computer study of DNA sequences with characterized NFSs

The Recon was used to construct NFP profiles for the DNA sequences whose certain regions were characterized experimentally as NFSs (therefore, these sequences were annotated and stored with NPRD). These sequences are not in the list of Nucleosomal DNA (Ioshikhes and Trifonov, 1993).

NFP profile of the mouse genome DNA region spanning the promoter and upstream region of nidogen1 (Nid1) gene (MGI:97342; LocusLink:18073; DoTS:DT.94280402) (Figure 3). Experiments detected three NFSs in this region (Chen and Keller, 2001). It is evident that the values of NFP for these three NFSs are rather high (0.513, 0.663, and 0.599, respectively), whereas the mean NFP values in linker regions are essentially lower (0.161 and 0.199, respectively). Note also rather low NFP values in the promoter region [–200; +1] (the mean value over this fragment amounts to –0.155). Earlier, we demonstrated that the NFP values of promoters of human housekeeping genes, displaying a constitutive expression, equal on the average -0.609 ± 0.012, whereas the corresponding values for the promoters with tissue-specific expression are 0.481 ± 0.003 (Levitsky *et al.*, 2001a). The calculated mean NFP value for the *Nid1* gene promoter complies with the data on expression of this gene in many tissue types (Chen and Keller, 2001).

NFP profile of human tandem alpha-satellite dimeric repeat of D1-D2 type (Figure 4). For this analysis, a sample of fragments of D1-D2 alpha-satellites from the following contigs was formed: M64779.1 (HUMD8Z2AA), M65181.1 (HUM18ASA), J04773.1 (HUMSATCH2), and D29750.1 (HUMASFA). Location of nucleosomes and CENP-B box, responsible for binding of the protein CENP-B, with reference to the repeated units is given according to the scheme by Yoda *et al.* (1998).

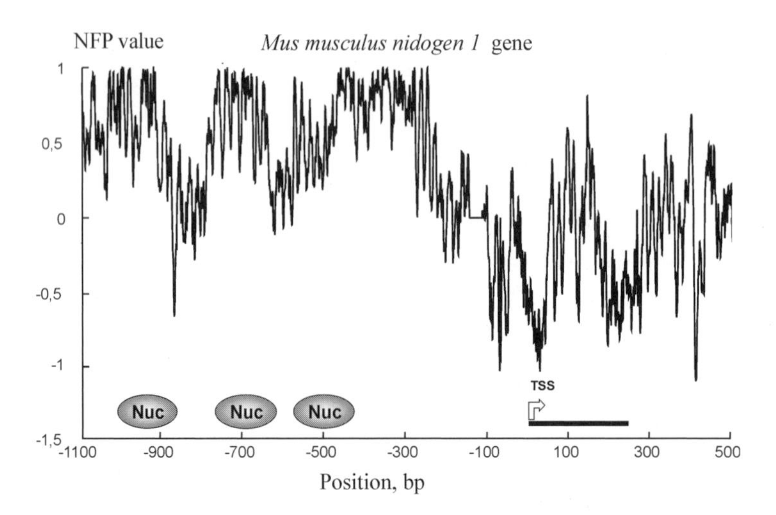

Figure 3. Nucleosome formation potential profiles for sequence L17322 of mouse *nidogen 1* gene: TSS, transcription start site, is designated by arrow; location of the first exon is shown below as solid bar; ovals, nucleosome formation site regions detected *in vivo* (Chen and Keller, 2001).

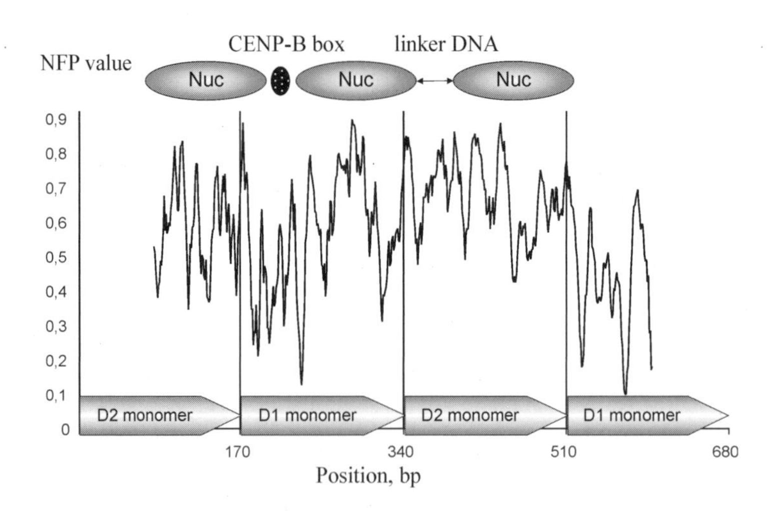

Figure 4. Nucleosome formation potential profiles for sequence set of human alpha-satellite of D1-D2 types aligned as tetrameres: location of monomers is shown below; ovals, the NFSs detected *in vivo* (Yoda *et al.*, 1998); dark circle, CENP-B box; and two-direction arrow, linker DNA.

As is evident from Figure 4, the regions of nucleosomes display highest NFP values, whereas the region of CENP-B box has lower NFP values. This result suggests that the program Recon is capable of detecting efficiently enough the nucleosome positioning signals in highly repetitive fractions of genomic DNA.

3.2 Computer study of arbitrary DNA sequences whose nucleosomal context have not been characterized experimentally

Earlier, Recon was applied to study various regulatory regions (enhancers, the *Not*I sites associated with CpG islands), coding and noncoding gene regions (exons and introns), regions of splicing sites, and regions of transposable element insertions (Levitsky *et al.*, 2001a, b; Podkolodnaya *et al.*, 2001; Kutsenko *et al.*, 2002, Furman *et al.*, this issue). Figure 5 shows NFP profiles for insertion regions of two *Drosophila melanogaster* transposable elements—*P* element (DNA transposon) and *roo* (LTR retrotransposon). The sample of *P* element insertion regions was extracted from the sample constructed by Liao *et al.* (2000); the sample of *roo* insertion regions is characterized by Furman *et al.* (this issue).

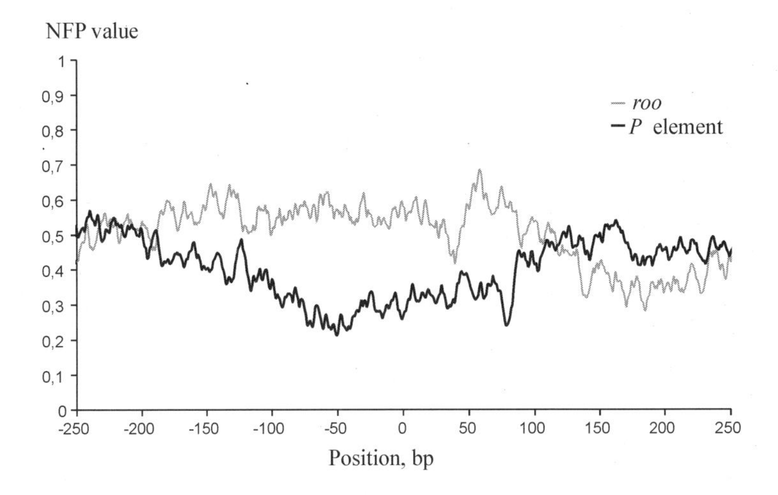

Figure 5. Nucleosome formation potential profiles of *Drosophila melanogaster* insertion sites for *P* element (DNA transposon) and *roo* (LTR retrotransposon): the horizontal axis, positions numbered with reference to the insertion point of mobile elements, assumed as 0.

In the case of *P* element insertion region, a decrease in the NFP values is evident in a range of ±100 bp compared with the fragments beyond this limit. This result suggests an "openness" of chromatin structure here, that is, the absence of a compact nucleosome packaging in this region (Becker, 2002), complying with the observation that *P* element inserts preferentially into gene promoters or 5'UTRs (Liao *et al.*, 2000), which, as a rule, are regions of "open" chromatin. However, the insertion regions of the LTR retrotransposon *roo,* rather inserting in an unselective manner (Kaminker *et al.*, 2002), display no decrease in NFP values.

4. CONCLUSION AND PROSPECTS

When developing the NPRD database, we are taking into account the world expertise in creating the databases on structure and functions of biological macromolecules, attesting that an adequate systematization of data is becoming an ever more topical problem under conditions of rapidly increasing volumes of information, resulting primarily from total genomic sequencing. Experimental studies, both *in vivo* and *in vitro,* generate very miscellaneous data on the patterns of nucleosome formation and basic level of chromatin organization. To make the primary information available for a large-scale computer research and theoretical generalization, we have developed the data representation format that reflects optimally the modern level of research into structure–function organization of chromatin within the curated databases we are elaborating. The entries of NPRD are supplemented with links not only to the commonly known gene indexes, but also to specialized databases, in particular, to the database Allgene.org on the structure of gene transcripts (Brunk *et al.*, 2002); TRRD, compiling the information on mutual arrangement and interactions of DNA functional sites and patterns of gene expression (Kolchanov *et al.*, 2002), etc. Moreover, NPRD is integrated with the analytical tools NucleoMeter and BDNA Sites Video, section Nucleosome Sites, to obtain prognostic estimates of both capabilities of arbitrary genomic sequences of positioning nucleosomes and the functional significance of such positioning basing on the information about characteristics of nucleosomal code generated using these tools. These estimates are of special importance when analyzing newly sequenced genomic regions yet lacking any experimental data on their nucleosomal organization and may contribute essentially to planning and optimization of experimental studies.

ACKNOWLEDGEMENTS

The work was supported by the Russian Foundation for Basic Research (grants Nos. 01-07-90376-в, 02-07-90355, 03-07-96833-p2003, 03-07-96833, 03-07-90181-в, 02-04-48802-a, 03-04-48829, 03-07-06078-мас, 03-04-48555-a); grant PCB RAS (No. 10.4); the Siberian Branch of the Russian Academy of Sciences (integration project No. 119); Russian Ministry of Industry, Science, and Technologies (grants Nos. 43.073.1.1.1501 and 43.106.11.0011, subcontract No. 28/2003); NATO (grants Nos. LST.CLG.979816 and PDD(CP)–(LST.CLG 979815)); the U.S. Civilian Research & Development Foundation for the Independent States of the Former Soviet Union (CRDF) No. NO-008-X1.

USING CHANGE IN LOCAL DNA SEQUENCE COMPLEXITY AS A POINTER TO THE MECHANISM OF MUTAGENESIS IN INHERITED DISEASE

N.A. CHUZHANOVA[1]*, D.N. COOPER[2]

[1]*Department of Computer Science, Cardiff University, PO Box 916, Cardiff, CF24 3XF, UK, e-mail: nadia.chuzhanova@cs.cardiff.ac.uk;* [2]*Institute of Medical Genetics, University of Wales College of Medicine, Cardiff, CF14 4XN, UK, e-mail: cooperdn@cardiff.ac.uk*

* *Corresponding author*

Abstract The vast majority of mutations causing human genetic disease are single base-pair substitutions and micro-deletions. However, a substantial proportion of the remainder involves the insertion of novel bases that serve to alter the reading frame leading to premature termination of translation. A relatively uncommon type of mutation is the *indel,* a complex lesion that appears to represent a combination of micro-deletion and micro-insertion. In the present study, we examine previously postulated mechanisms underlying micro-deletions and micro-insertions such as slipped mispairing and strand switching, secondary loop excision and quasi-palindrome correction, Moebius loop resolution and excision in terms of local DNA sequence complexity. Indels are regarded as being the result of a two-step deletion/insertion process. Data from the *Human Gene Mutation Database* (*HGMD* www.hgmd.org) were used to compare and contrast 3767 micro-deletions, 1960 micro-insertions, and 211 indels. The change in complexity was found to be indicative of the type of repeat sequences involved in mediating the deletion/insertion event, which in turn provided pointers as to the possible pathways through which the mutations could have arisen.

Key words: sequence complexity, micro-deletion, micro-insertion, indel, mutagenesis

1. INTRODUCTION

The most common types of mutation causing human genetic disease are single base-pair substitutions and micro-deletions (Antonarakis *et al.*, 2001).

The remainder comprises an assortment of larger deletions, insertions, inversions, repeat expansions and complex rearrangements. One relatively uncommon type of mutation is the indel, a combined micro-deletion/micro-insertion that results in the apparent replacement of one or more base pairs by others, not necessarily the same number. Although it is likely that the component micro-insertion and micro-deletion events occur contemporaneously (i.e. as part of the same complex mutational event), this need not necessarily be so. The study of mutational lesions causing human genetic disease has revealed that, irrespective of their type, the nature, frequency, and location of mutations are invariably non-random (Cooper and Krawczak, 1993; Antonarakis *et al.*, 2001) and strongly influenced by the complexity of the local DNA sequence environment (Ripley, 1982; DeBoer and Ripley, 1984; Glickman *et al.*, 1984; Krawczak *et al.*, 2000b). Complexity analysis was therefore used to examine the change in complexity and the subsequent involvement of different types of repeat in the micro-insertion and micro-deletion and in the deletion/insertion events leading to indel formation.

2. METHODS AND ALGORITHMS

Complexity analysis, as devised by Gusev *et al.* (1999) and described in (Chuzhanova *et al.*, 2003), was used to examine the potential contribution of local DNA sequence complexity to postulated mechanisms of deletion and insertion mutagenesis, and the two-step process of indel formation. Complexity analysis is based upon the assessment of the occurrences of four different types of repetitive element, namely direct, inverted repeats, symmetric elements, and inversions of inverted repeats. Specimen examples of these types of repetitive elements are given in Figure 1. One can conceive of the sequence as being decomposed into "words", where each word is the longest among all possible words for which a direct or inverted repeat, or an inversion thereof, occurs somewhere upstream of the current position. An overlap between two repeat copies is also permitted. The length of the first fragment is always 1. It is apparent that this decomposition, H(S), contains the minimum number of words. The number of words in this minimal decomposition, H(S), is called the complexity C of S. We shall denote respectively by C1(S), C3(S), C4(S) and C5(S) the complexities or number of words in decompositions H1(S) computed with respect to direct repeats, H3(S) with inverted repeats, H4(S) with symmetric elements, H5(S) with inversions of the inverted repeats. Let us consider as an illustrative example a DNA fragment, S, from the *APC* gene with reported micro-deletion, *HGMD* ID CD982432 (the precise location of the deleted nucleotides is indicated by lower case letters). Decompositions of S by means of different types of repeat words are shown in Figure 2.

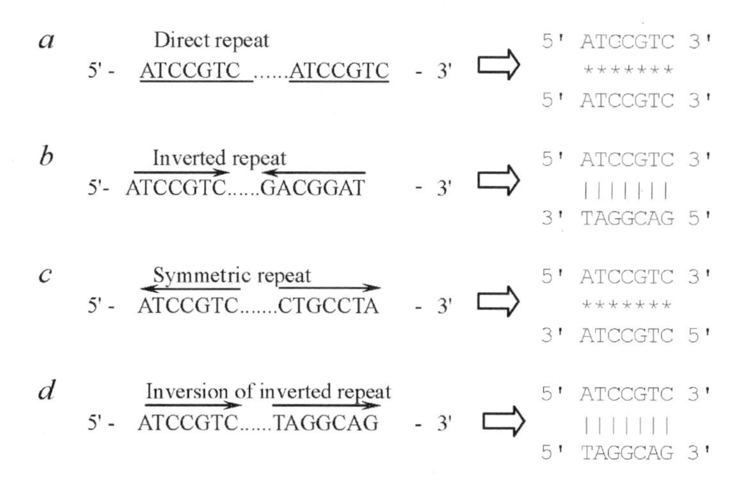

Figure 1. Examples of (*a*) direct repeat; (*b*) inverted repeat; (*c*) symmetric element and (*d*) inversion of inverted repeat. Repeated fragments are underlined whereas inverted repeats, symmetric elements and inversions of inverted repeats are marked by arrows in order to indicate their orientation. Identical and complementary base pairs are indicated by asteriks and bars respectively.

C-T-T-A-T- G-G-A-A-G-c-c-gg-gaag-GA-T-CT-G-TAT-C, C1(S)=20;

C-T-T-A-T-G-G-AAG-cc-gg-g-aag-G-ATC-T-G-T A-TC, C3(S)=18;

C-T-T-AT-G-G-AAG-ccg -gg-aagG-AT-CT-GTAT-C, C4(S)=14;

C-T-T-AT-G -GAA-G-cc-gg-gaa-gG-AT-CT-G-TA- TC, C5(S)=16.

Figure 2. Decomposition of a DNA fragment from the *APC* gene by reference to different types of repeat words in the vicinity of a micro-deletion. C1, C3, C4 and C5 are correspondingly the complexities with respect to direct repeats, inverted repeats, symmetric elements and inversion of inverted repeats.

H1 before deletion:
C-T-T-A-T-G-G-A-A-G-c-c-gg-gaag-GA-T-CT-G-TAT-C, C1=20;
H1 after deletion:
C-T-T-A-T-G-G-A-A-GGA-T-CT-G-TAT-C, C1=15.

Figure 3. Micro-deletion in the *APC* gene that may have involved a direct repeat.

Henceforth, repeat words are underlined. Different types of repeat are marked by arrows in order to indicate their orientation. Parameters C1–C5 represent suitable measures of complexity (regularity) of the sequence S, since any abundance in S of direct and inverted repeats and inversions thereof serves to reduce the corresponding complexity. As can be seen from the above example, the minimum complexity among C1, C3–C5 is achieved with measure C4. The second measure in ascending order is measure C5. Fragment S is thus comparatively rich in symmetric repeats and in inversions of the inverted repeats. To be able to compare the complexities of two or more sequences that differ in length, one can use the *complexity per base*, c = C/N, where N is the length of a given sequence in base pairs.

As shown by Chuzhanova *et al.* (2003), the type of repetitive element identified as making the maximum contribution to the change in local DNA sequence complexity is that which is most likely to be involved in mediating the micro-insertion or micro-deletion. In the example shown in Figure 2, the mechanism involving direct repeats, e.g. slipped mispairing, is the most likely mechanism underlying the micro-deletion in the *APC* gene (CD982432) because it yields the maximum change in complexity C1. The decompositions before and after deletion (Figure 3) indicate the disappearance of a direct repeat.

Each indel was regarded as having originated through a two-step deletion/insertion process; the first step transforms the wild-type sequence into an intermediate, the second step transforms the intermediate into the final mutated sequence (Chuzhanova *et al.*, 2003). There are three distinct paths for indel formation: the deletion occurs first generating a deleted intermediate, whereas the insertion occurs in a second step; the insertion occurs first with the inserted sequence being placed before the fragment about to be deleted, the deletion than occurs in a second step; the insertion occurs first with the inserted sequence being placed after the fragment about to be deleted, the deletion than occurs in a second step. It was assumed that the first step represents the initial mutational event whereas the second step constitutes the partially effective repair process that serves to fix the mutation (Chuzhanova *et al.*, 2003).

This means that the complexity of a fragment remains more or less the same before and after the deletion/insertion event, i.e. if during the first step, a decrease in complexity is observed, then the second step must reverse this process leading to an increase in complexity and *vice versa*. It was found that the most probable path and mechanism of mutation are those for which the sum of the changes in complexity in a two-step process is maximized. Let us now consider DNA fragment from the *RET* gene (CX942108) S = ACGAGCTGTGccGCACGGTGAT with reported indel delCCinsGG. The maximum change was observed in complexity C4 for the pathway "deletion first, insertion second". Decompositions (Figure 4) confirm the appearance of an extended symmetric element resulting from the excision of two bases from the middle of an imperfect symmetric element. Two bases at the end of the symmetric element become duplicated.

H4 before deletion: A-C-GAGC-T-GT-Gc-cG-CACG-GTG-A-T, C4=11;

H4 after deletion: A-C-GAGC-T-GT-GGCACGGTG-A-T, C4=8;

H4 after insertion: A-C-GAGC-T-GT-G-ggG-CACGG-TG-A-T, C4=11.

Figure 4. Indel in the *RET* gene that may have involved symmetric element correction and resolution.

3. RESULTS AND DISCUSSION

Overall, 3767 micro-deletions, 1960 micro-insertions, and 211 indels were obtained from the *Human Gene Mutation Database* (*HGMD*; www.hgmd.org; Krawczak *et al.*, 2000a). In almost all cases, 10 bp DNA sequence flanking the mutation on either side were analyzed. For two indels, however, only 5 bp flanking the indel were available. Direct repeats and symmetric elements were found to be involved in micro-deletions with approximately the same frequency (30 and 33%, respectively) whereas their involvement in mediating micro-insertions differs markedly (45 and 24%, respectively). One mechanism that is frequently implicated in the generation of micro-deletions and micro-insertions is slipped mispairing which involves the misalignment of short direct repeats. During DNA replication, the template strand can slip forward producing a single-stranded loop that can subsequently be excised and repaired thereby fixing a micro-deletion (Streisinger *et al.*, 1966; Efstratiadis *et al.*, 1980). Conversely, one of the nascent strands may slip backwards thereby templating a micro-insertion. These events serve to change the C1 complexity of a sequence. Thus, for example, a micro-deletion in the *ADRA2C* gene (CD002476) could have been mediated by a direct repeat (Figure 5). By contrast, the micro-insertion in the *MEN1* gene (CI983142) was found to be mediated by a direct repeat probably via slipped mispairing and strand switching (Figure 6).

A modified slipped mispairing model was proposed by Krawczak and Cooper (1991) to account for micro-deletions not readily explicable by the standard model. If the DNA sequence flanking the deleted bases also occurs as a contiguous sequence in the immediate vicinity, the intervening non-homologous bases may loop out thereby potentiating the formation of a second direct repeat copy. Transient misalignment of the two repeats may then allow the deletion of the intervening bases before strand alignment is restored. The juxtaposition of two repeat copies as a consequence of the deletion serves to decrease the complexity of the sequence. For example, a short deletion [ATTCTGTTCTcaGTTTTCCTGG] in the *CFTR* gene (CD920845) leads to the creation of a second copy of TTCTGTT with a concomitant decrease in complexity from 12 to 9 (complexity per base also decreases from 0.55 to 0.45).

H1 before deletion:
G-G-A-C-GG-G-C-A-Ggg-ggc-ggggc-cgGGG-GCGG-C-T-C, C1=16, c1=0.5;
H1 after deletion:
G-G-A-C-GG-G-C-A-GGG-GGC-GGC-T-C, C1=13, c1=0.65.

Figure 5. Micro-deletion in the *ADRA2C* gene that may have involved slipped mispairing.

H1 before insertion: A-G-C-CC-AGCCC-C-GCCCC-CG-A-C, C1=10, c1=0.5;
H1 after insertion: A-G-C-CC-AGCCCagccc-C-GCCCC-CG-A-C, C1=10, c1=0.4.

Figure 6. Micro-insertion in the *MEN1* gene that may have involved slipped mispairing.

Symmetric elements capable of forming a Moebius-loop structure (Cooper and Krawczak, 1993) have also been implicated in the generation of micro-deletions and micro-insertions. A symmetric element was found to be responsible for the micro-deletion in the *F9* gene (CD993401).

Mismatched bases looping out of a symmetric element TGGGGTGAAGagTGTGCAATGA can facilitate their own excision. The Moebius loop may partially resolve if one of the DNA strands disconnects and breaks. The repair of this region by DNA polymerase would generate the duplication of a sequence from the end of the symmetric element that was originally disconnected. This may have occurred in the case of a micro-insertion in the *ABCD1* (CI014470) gene CCACGCCTACcCGCCTCTACT. It appears likely that it was the first C rather than the second that was inserted. Mechanisms based upon either secondary loop excision or quasi-palindrome correction were found to be able to account for ~18% of micro-deletions and ~15% of micro-insertions. An inverted repeat allows hairpin loop formation and excision repair of such a loop may yield a micro-deletion. Inverted repeats may also promote slipped mispairing of the nascent strand and subsequent duplication of downstream sequence. In the case of a micro-deletion in the *LCAT* gene (CD951764), the deletion of "c" led to the appearance of a prominent inverted repeat (Figure 7). An inverted repeat could account for the micro-insertion in the *F9* gene (CI931084) that led to slipped mispairing and the duplication of downstream sequence (Figure 8). A knot structure formed by the inversion of inverted repeats may be implicated in the generation of both micro-deletions and micro-insertions (Chuzhanova *et al.*, 2003). The excision repair of such a loop may yield a micro-deletion whilst slipped mispairing of the nascent strand could lead to duplication of downstream sequence. It was estimated that the occurrence of ~11% of micro-deletions and ~8% of micro-insertions involved a knot structure. For example, in a micro-deletion in the *BRCA1* gene (CD961829), a mismatched base that looped out of the knot could have facilitated its own excision (Figure 9). The existence of this knot structure may help to resolve the ambiguity in deletion position that occurs as a consequence of the repetitive nature of the deletion-prone site. It would therefore appear likely that it was the third G of the triplet that was deleted.

H3 before deletion: A-G-G-A-G-A-T-GCA-c-GC-TGC-CT-AT-G, C3=14, c3=0.67;

H3 after deletion: A-G-G-A-G-A-T-GCA-GCTGC-CT-AT-G, C3=12, c3=0.6.

Figure 7. Micro-deletion in the *LCAT* gene potentially mediated by an inverted repeat.

H3 before insertion:T-ATA-C-C-A-A-GGTAT-CC-CGG-TAT-TG-CAA, C3=12, c3=0.5;
H3 after insertion:T-ATA-C-C-A-A-GGTAT-CC-C-a-a-ggta-cc-a-a-GGTAT-TG-T-CAA, C3= 20, c3=0.47.

Figure 8. Micro-insertion in the *F9* gene potentially mediated by hairpin loop formation.

H5 before deletion: AAAATATTTGgGAAAACCTAT, C5=12, c5=0.57;

H5 after deletion: AAAATATTTGGAAAACCTAT, C5=10, c5=0.50.

Figure 9. Micro-deletion in the *BRCA1* gene potentially mediated by a knot structure.

A knot structure could also account for the micro-insertion in the *ABCA4* gene (CI003643) which led to slipped mispairing of the nascent strand and the duplication of downstream sequence

TGTTTTCAAAcaaaGCCCCACCCC.

Repeats involved in indel formation and the possible path of mutation may also be explored by means of complexity analysis. Direct repeats, inverted repeats and symmetric elements were found to be involved in indel formation with approximately the same frequency (Chuzhanova *et al.*, 2003). A combination of mechanisms based upon strand switching and either secondary loop excision or slipped strand mispairing was found to be able to account for 32% of indels. Further, ~16% of indels were found to be compatible with mechanisms involving symmetric elements. The most frequent mechanism of indel formation is one that involves inverted repeats in both the deletion and the insertion event with the deletion occurring first (14 times). Mechanisms involved Moebius loop resolution and breakage or excision mediated by symmetric elements via paths "deletion first, insertion second" and "insertion occurs first, 5' to the fragment to be deleted" were assigned to 12 and 11 indels respectively. For example, reported indel delGTinsCC in the *CTNS* gene (CX994477) was found to be mediated by symmetric elements via path "insertion occurs first, 5' to the fragment to be deleted". A possible mechanism is shown in Figure 10. An inverted repeat-mediated hairpin loop formation and a direct repeat via pathway "insertion first, deletion second" where inserted base was positioned after the bases to be deleted, could account for the indel delGTAinsC that occurred in the *BRCA1* gene. The possible mechanism of indel formation shown in Figure 11.

H4 before insertion: CATGAAGTAAgtAACCAATCTT;

H4 after insertion: CATGAAGTAAccgtAACCAATCTT;

H4 after deletion: CATGAAGTAAccAACCAATCTT.

Figure 10. Possible mechanism of indel formation in the *CTNS* gene.

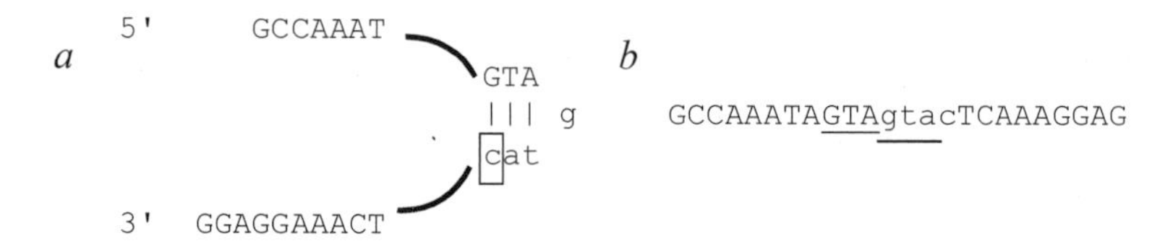

Figure 11. Possible mechanism of insertion/deletion events occurring to create an indel in the *BRCA1* gene: (*a*) palindrome correction due to insertion (inserted base "c" shown in a rectangle); (*b*) deletion mediated by a direct repeat GTA.

This analysis has demonstrated that changes in local DNA sequence complexity consequent to micro-insertion and micro-deletion mutagenesis or indel formation can be used to identify the repeat sequences involved in the mutational events and hence the probable underlying mechanism of mutagenesis; indeed, this approach appears capable of predicting both the number and identity of the bases deleted and/or inserted. It may in principle also be applied to the analysis of gross rearrangements, which have so far proved to be refractory to analysis.

ACKNOWLEDGEMENTS

We are most grateful to Emmanuel Anassis, Edward Ball and Peter Stenson for technical support and assistance.

PROPERTIES OF INSERTION REGIONS OF *DROSOPHILA* LTR RETROTRANSPOSONS

D.P. FURMAN[1,2]*, D.Yu. OSHCHEPKOV[1,2], M.A. POZDNYAKOV[1,2], A.V. KATOKHIN[1,2]

[1]*Institute of Cytology & Genetics, Siberian Branch of the Russian Academy of Sciences, prosp. Lavrentieva 10, Novosibirsk, 630090 Russia, e-mail: furman@bionet.nsc.ru;*
[2]*Novosibirsk State University, ul. Pirogova 2, Novosibirsk, 630090 Russia*
* *Corresponding author*

Abstract We applied *in silico* methods to analyze contextual and structural characteristics of *Drosophila* DNA in the regions of insertions of LTR retrotransposons belonging to 12 families, namely, *17.6*, *297*, *blood*, *copia*, *Dm412* (*mdg2*), *gypsy*, *HMSBeagle*, *mdg1*, *mdg3*, *roo* (*B104*), *tirant*, and *yoyo*, and to detect the properties that might determine why these regions are preferable for integration of a particular LTR retrotransposon. Samples of genomic sequences with a length of 60 bp containing 4–5-bp target sites *per se*, which are duplicated during integration of LTR retrotransposons, and the sequences flanking target site, were studied. The insertions regions fall into three groups with respect to variation pattern of mononucleotide context of the target sites and their flanking regions. Each of the three groups displays a characteristic distribution of the DNA local conformational and physicochemical properties in the regions of insertions, which might determine that this particular region is chosen as preferable for insertion. Preferential insertion of retrotransposons in the regions of host DNA displaying certain specific conformational and physicochemical properties may reflect the distinctions between the structures of DNA-recognizing domains of integrases of particular LTR retrotransposons.

Key words: *Drosophila melanogaster*, LTR retrotransposons, context analysis, conformational and physicochemical DNA properties

1. INTRODUCTION

Long Terminal Repeats (LTR) retrotransposons represent the most numerous group of mobile elements of invertebrates; in particular, these elements of *D. melanogaster* form 49 families with different copy numbers, constituting to 40% of the total number of mobile elements in this genome (Kaminker *et al.*, 2002, and references therein).

The structure and life cycle characteristics of LTR retrotransposons make them akin to retroviruses of vertebrates. Their transposition comprises formation of RNA intermediates, used as a template by reverse transcriptase to reproduce DNA copies of the elements that would then insert into a new site of the host genome. The LTRs, edging the retrotransposon's body, contain the element's transcription regulation sites, while the ORFs for genes *gag* and *pol*, providing for replication of the element and its insertion into the host genome, are located between the LTRs. Some LTR retrotransposons contain additionally the third ORF encoding *env* gene; this ORF is considered implicated in their infectious properties and the capacity for a horizontal (intra- and interspecific) transfer in addition to the vertical transfer from parents to descendants (Jordan *et al.*, 1999; Mejlumian *et al.*, 2002). These characteristics of LTR retrotransposons in combination with their ability to change actively the structure–function organization of genomes determine their role as one of the most important microevolution factors.

Integrase, encoded by LTR retrotransposon itself and composed of several domains, mediate transposition of these elements. The *N*-terminal domain of integrase binds specifically to the newly synthesized DNA copy of its own retrotransposon. The catalytic domain $DD_{35}E$, localized to the central region, cleaves the host DNA in a stepwise manner and inserts the DNA copy into it. The *C*-terminal domain displays a DNA-binding activity and has been implicated in interaction with the host DNA to a considerable degree in a sequence-independent manner (Malik and Eickbush, 1999). Integration is accompanied by duplication of a short region of the host DNA, now commonly known as the target site. Length of the target site corresponds to the length of the stepwise break and, as a rule, does not exceed 4–5 bp. Sequences of the target site edge the inserted LTR retrotransposon from both sides.

The specific DNA features of target sites of LTR retrotransposons belonging to different families and/or their flanking sequences that are capable of providing preferential insertion of the element into a particular genomic region are yet vague in the majority of cases. Preferential integration of the elements *297* and *17.6* into the target sites with the ATAT consensus sequence were reported (Whalen and Grigliatti, 1998; Bowen and McDonald, 2001). However, no strict consensus sequence of the target sites for other LTR retrotransposons has been discovered, despite certain selectivity of their localization (Dej *et al.*, 1998; Conte *et al.*, 2000; Glukhov *et al.*, 2000).

It has been demonstrated that integration of certain transposons and retrotransposons of yeast and nematode genomes as well as retroviruses of mammalian genomes correlates with particular physicochemical DNA properties at the sites of insertion (Muller and Varmus, 1994; Vigdal *et al.*, 2002), and the integration sites of transposons and retrotransposons differ in these properties. Similar data were obtained while studying several *Drosophila* mobile elements—transposons *P* and *pogo* and non-LTR retrotransposon *jockey* (Liao *et al.*, 2000; Kaminker *et al.*, 2002)—by experimental and *in silico* methods. The data on insertion sites of *Drosophila* LTR retrotransposons reported only for one element, *roo*, and only for one property, DNA denaturation temperature (Kaminker *et al.*, 2002).

Using *in silico* methods, we for the first time characterized contextual and conformational/physicochemical properties (CPCP) of insertion regions of *Drosophila* LTR retrotransposons belonging to 12 most representative families—*17.6*, *297*, *blood*, *copia*, *Dm412* (*mdg2*), *gypsy*, *HMSBeagle*, *mdg1*, *mdg3*, *roo* (*B104*), *tirant*, and *yoyo*. Further, we will call insertion regions those sequences of host DNA that contain target sites and the flanking sequences.

The LTR retrotransposons are classified into three groups according to the variation pattern of mononucleotide context and CPCP of the insertion regions. Each group displays its own profiles of local CPCP of dinucleotides located both in the target sites and outside. A set of these CPCP may determine choosing of the region as preferential for integration of a particular element. The preferential insertion may reflect distinctions in the mechanisms underlying integration events of LTR retrotransposons, including those resulting from different structures of DNA-recognizing domains of their integrases.

2. MATERIAL AND METHODS

2.1 Constructing samples

Reconstructed euchromatic sequence of the sequenced *D. melanogaster* genome Release 2 (Adams *et al.*, 2000) was used to construct the samples.

LTR retrotransposons of 12 families—*17.6*, *297*, *blood*, *copia*, *Dm412* (*mdg2*), *gypsy*, *HMSBeagle*, *mdg1*, *mdg3*, *roo* (*B104*), *tirant*, and *yoyo*— present in the genome in at least seven copies were chosen for the analysis. Their nucleotide sequences are available in the FlyBase (http://flybase. harvard.edu:7081/transposons/lk/melanogaster-transposon.html). With these sequences as a query, we localized the corresponding retrotransposons in the sequenced genome using BLASTn software (http://www.ncbi.nlm.nih.gov/ BLAST/ and http://www.fruitfly.org/blast/index.html) with the default parameters. For each of the 12 retrotransposon families, the coordinates of 5'-end of the left LTR and 3'-end of the right LTR in genomic sequence were determined. Then, the fragments adjacent to the left and right LTRs were extracted to be assembled in a contig. Duplicated target site was detected in the center of contig to delete one of the copies. Fragments with a length of 60 bp (± 30 bp relative to the center of target site), called further insertion regions, formed the sequence samples for analysis. Layout of construction of insertion regions is shown in Figure 1.

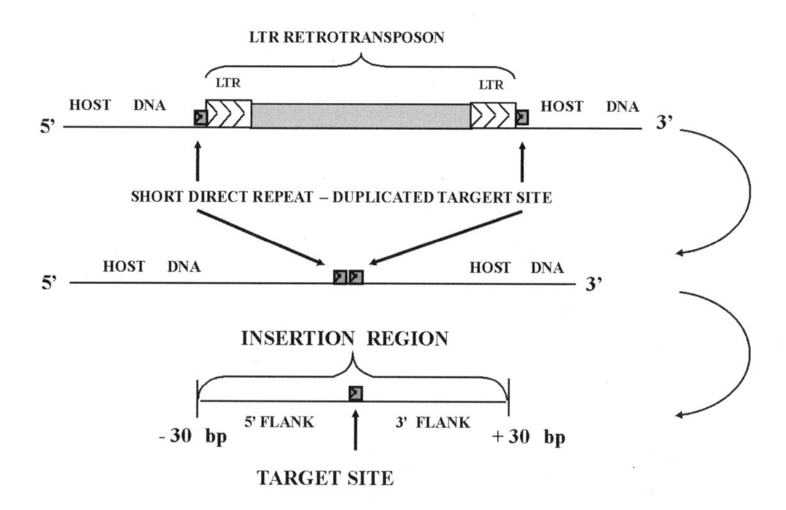

Figure 1. Layout of construction of insertion regions of LTR retrotransposons.

As the BLASTn search detected only one copy of *gypsy* in the genome sequenced, the sequences corresponding to its insertion regions were taken from the published data (Dzhumagaliev *et al.*, 1983; Freund and Meselson, 1984; Spana and Corces, 1990; Kuzin *et al.*, 1994; Udomkit *et al.*, 1995; Tsai *et al.*, 1997; Dej *et al.*, 1998).

2.2 Evaluating characteristics of mononucleotide context in insertion regions

It is assumed that nucleotide bases occur at each position of insertion regions independently of one another with frequencies calculated according to the Release 2 of *D. melanogaster* sequenced genome and amounting to 0.285 for A and T and 0.215 for C and G.

In this case, the set of nucleotide frequencies at each position of the 60-symbol sequence follows a multinomial distribution (Agresti, 1996). Sum of the probabilities to obtain the observed or larger deviations of nucleotide frequencies of each type from the expected values (W) according to random causes gives the estimate of context variation at each position of the insertion region sequence sample. For convenience of graphical representation, the reversed in sign value of natural logarithm of the probability to obtain the variation observed at each position of the sample according to random causes, $-\ln(W)$, was calculated. The value of $-\ln(0.05)$ was considered the threshold. If the function $-\ln(W)$ exceeded the threshold value, it was considered as an evidence of a significantly low context variation.

2.3 Characterizing CPCP of insertion regions

The sequence samples of insertion regions of retrotransposons of 12 families phased with respect to the target site were analyzed by the SITECON method (Oshchepkov *et al.*, 2002; Oshchepkov *et al.*, this issue), based on the data on 38 context-dependent CPCP of dinucleotides compiled in the database PROPERTY (http://wwwmgs.bionet.nsc.ru/mgs/gnw/bdna/; Ponomarenko *et al.*, 1997). In this database, a particular value of each of the 38 CPCP is associated with each dinucleotide. The SITECON method allows the positions of dinucleotides displaying low-variable CPCP values to be identified in short DNA fragments. The positions were considered with a step of one nucleotide.

Mean value of each of the 38 CPCP and its standard deviation were calculated for each dinucleotide position of the sample of aligned sequences. Statistical significance was assessed by chi-square test at a significance level of 0.05. Sequences of the same nucleotide composition generated by multiple random shuffling of nucleotides of the initial sequences were used as a control sample.

3. RESULTS

Described in this paper are the results obtained by *in silico* analysis of DNA contextual and structural characteristics of insertion regions of *Drosophila* LTR retrotransposons performed using the most representative data array with reference to both the number of retrotransposon families and sample volumes for insertion sites of each retrotransposon type. The numbers of sequences in target site samples of the 12 families are listed in Table 1.

Table 1. Numbers of analyzed sequences of insertion regions of LTR retrotransposons belonging to 12 families

Family	Sample size	Family	Sample size
17.6	7	*HMSBeagle*	7
297	20	*mdg1*	8
blood	13	*mdg3*	7
copia	14	*roo (B104)*	36
Dm412 (*mdg2*)	16	*tirant*	10
gypsy	11	*yoyo*	8

3.1 Contextual characteristics of insertion regions

Estimation of variation in mononucleotide context suggests classifying the insertion regions of LTR retrotransposons of 12 families into three groups. Characteristics of representatives of each group are shown in Figures 2*a–c*.

The first group is formed by insertion regions of the LTR retrotransposons *17.6* and *297.* The distinctive property of these insertion regions is the strict ATAT consensus sequence of the target sites. The plot of $-\ln(W)$ function for *297* insertion region is shown in Figure 2*a*. It is evident that the function reaches it maximal values at the positions corresponding to the target site, while in a wide range of positions 23–36, the function exceeds significantly the threshold value, indicating a considerable decrease in variation within this range. Outside this range, a low contextual variation is observed only at single positions (for example, position 17 in Figure 2*a*).

The insertion region of *17.6* element displayed similar profiles of stable contextual characteristics (data not shown).

The second group comprises the insertion regions of *gypsy, HMSBeagle, tirant,* and *yoyo* elements. Typical of insertion regions of the elements of this group is the absence of strict consensus sequence of the target site. In addition, a decreased variation of nucleotide composition is detected at two positions directly adjoining the target site. Moreover, some positions displaying low variation are also located at certain distances from the target site. The plot of $-\ln(W)$ function for *tirant* insertion region is shown in Figure 2*b*.

The third group consists of the insertion regions of *blood*, *copia*, *Dm412* (*mdg2*), *mdg1*, *mdg3*, and *roo* (*B104*) LTR retrotransposons. Characteristic of the representatives of this group is a high degree of variation both at positions of the target site itself and at the majority of positions in the flanking sequences (the values of $-\ln(W)$ are below the threshold). Profile of the contextual characteristics of this type of insertion regions illustrated by the example of *roo* (*B104*) is shown in Figure 2*c*.

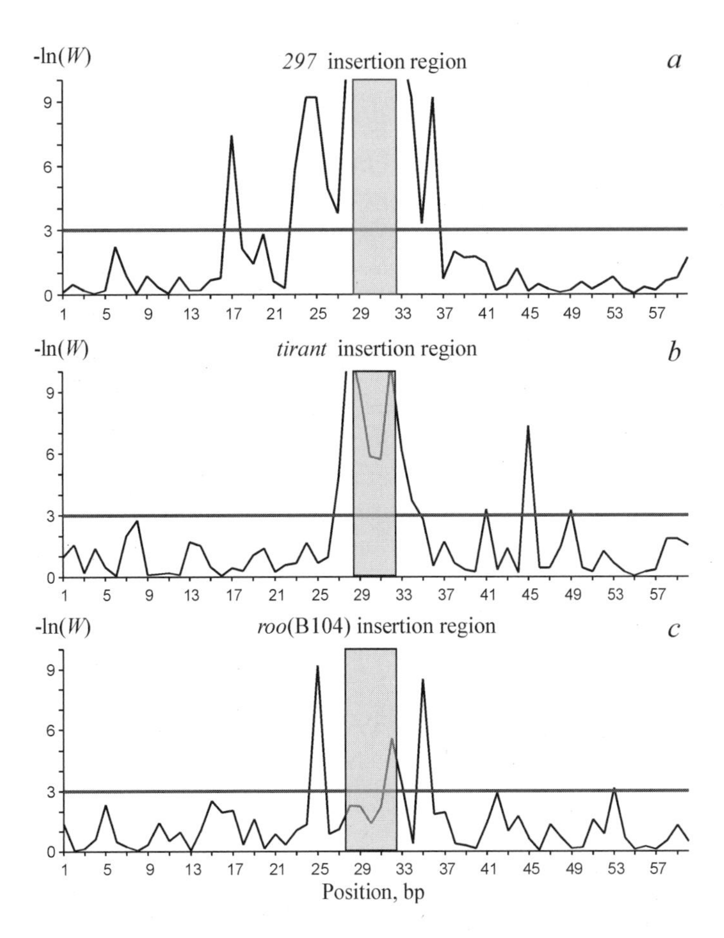

Figure 2. Plots of the function –ln(*W*), reflecting the degree of mononucleotide context variation, for the insertion regions of the LTR retrotransposons (*a*) *297*, (*b*) *tirant*, and (*c*) *roo*. Straight line indicates the threshold value of the function, –ln(0.05); rectangle, positions corresponding to the target site. The cut points determining the target sites for *297* and *tirant* are located between 28/29 and between 32/33 nucleotides. The length of these target sites is 4 bp. The cut points determining the target site for *roo* are located between 27/28 and between 32/33 nucleotides. The length of this target site is 5 bp.

3.2 Conformational and physicochemical properties of insertion regions

Further, the insertion regions of each type were characterized with reference to 38 context-dependent CPCP using the SITECON method. In this case, dinucleotides represented both the minimal context element and the unit of analysis.

Figures 3–5 illustrate certain patterns discovered while analyzing insertion regions of individual LTR retrotransposon groups. The first group – insertion regions of the elements *17.6* and *297*—displays an absolute invariance of all the CPCP in the target site. The profile of the property "DNA melting temperature" (No. 22 according to the nomenclature of the PROPERTY database) for the insertion region of *297* element is shown as an example (Figure 3). It is evident that this property displays a zero standard deviation within the ATAT target site due to the invariance in the nucleotide composition. A typical profile of the target site results from alternation of the dinucleotides AT–TA–AT.

Similar profiles reflecting distribution of invariant local CPCP were observed in insertion regions of element *17.6* (data not shown).

The second group—insertion regions of the elements *gypsy, HMSBeagle, tirant*, and *yoyo.* Distribution of CPCP along the insertion regions for the elements of this group is illustrated by the example of average values of the property No. 27 (tilt) of *tirant* insertion regions (Figure 4).

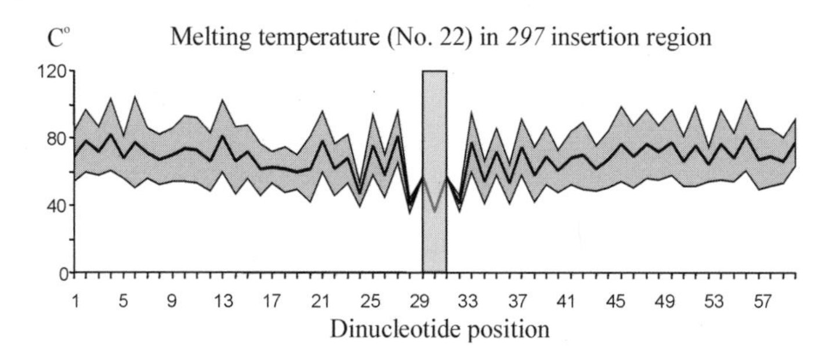

Figure 3. Profile of melting temperature (property No. 22 in the database PROPERTY) of the dinucleotides over the insertion region of LTR retrotransposon *297*. Firm line is the profile of average value; gray region around the firm line, range of variation. Rectangle indicates the target site positions. The cut points determining the target site are at dinucleotides 29 and 31.

It is evident that the property No. 27 at the positions adjoining the target site (27–28 and 32–33) displays a lower variation compared with the positions of the site *per se* and the majority of positions within the flanking sequences. These are just the positions of the stepwise break cleaved by integrase, which are characterized by a very low variation of the mononucleotide context (Figure 2*b*).

Insertion regions of the elements *gypsy, HMSBeagle*, and *yoyo* displayed similar distribution patterns of low-variable CPCP (data not shown).

The third group—insertion regions of the elements *blood, copia, Dm412* (*mdg2*), *mdg1*, *mdg3*, and *roo* (*B104*). A high variation in the context of insertion regions of these LTR retrotransposons determines the lack of distinct structural DNA characteristics at the majority of positions. Low variation of CPCP in virtually all the cases observed at positions located at a distance of two and more nucleotides from the target site. The plot of average values of the property No. 28 (roll) over the *roo* insertion region is an example (Figure 5). It is evident that the local decrease in variation is observed at positions 25 and 34.

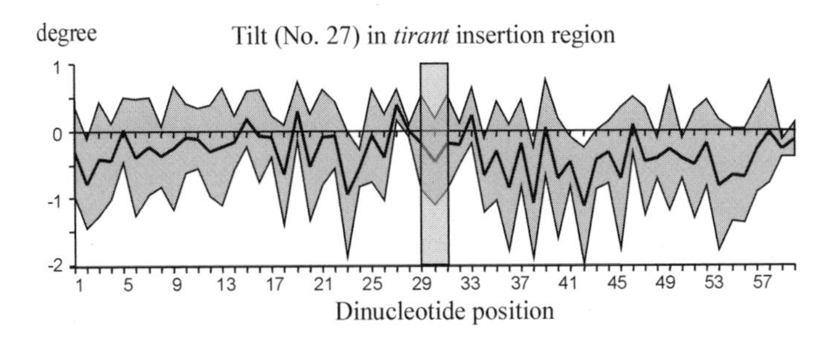

Figure 4. Profile of the property tilt (No. 27 in the database PROPERTY) of the dinucleotides over the insertion region of LTR retrotransposon *tirant*. For designations, see Figure 3. The cut points determining the target site are at dinucleotides 29 and 31.

Comparison of the plots for *roo* insertion region, shown in Figures 2*c* and 5, detected different variants of interrelation between the context and CPCP variations. For example, the peaks at positions 25 and 35 in Figure 2*c* (low variation in mononucleotide context) correspond to an essential decrease in variation of the property No. 28 at the same positions (Figure 5). However, the peak at position 32 in Figure 2*c* has none accompanying specific features at the same position in the plot shown in Figure 5.

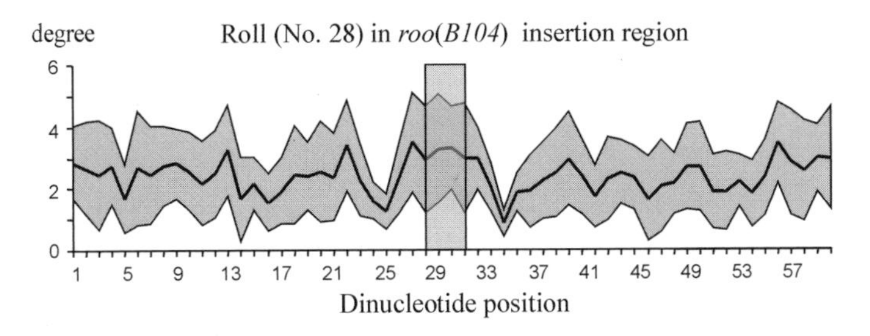

Figure 5. Profile of the property roll (No. 28 in the database PROPERTY) of the dinucleotides over the insertion region of LTR retrotransposon *roo*. For designations, see Figure 3. The cut points determining the target site are at dinucleotides 28 and 31.

The insertion regions of the elements *blood*, *copia*, *Dm412* (*mdg2*), *mdg1*, and *mdg3* displayed similar patterns in distributions of low-variation CPCP (data not shown).

4. DISCUSSION

Integration complex, containing along with certain other components, the element's DNA intermediate and integrase, encoded by *pol* gene (Labrador and Corces, 1997; 2001), is involved in the process of LTR retrotransposon integration. Presumably, integrase is the critical component for this stage, as it is this enzyme, displaying both DNA-binding and catalytic activities, that is responsible for interactions with the host DNA and, consequently, is capable of providing a preference, or even specificity, of insertion of the retrotransposon depending on characteristics of the host DNA, i.e. particular context and/or secondary structure characteristics in the insertion region that distinguish such regions from the others and that are recognized by the integration complex.

We performed a comparative contextual analysis of insertion regions and analysis of their CPCP to discover the properties that might determine why they are preferential for integration of a particular LTR retrotransposon.

The results of contextual analysis allow us to classify the insertion regions in question into three groups depending on context variability of the target site and its flanking regions. Two insertion regions—*17.6* and *297*, forming group I—have a strict ATAT consensus sequence of the target site, whereas the rest insertion regions either display certain stable contextual properties at the positions adjoining the target site (group II) or contain certain positions at various distances from the target site with a moderately pronounced context stability (group III). Groups I and II contain insertion regions only of LTR retrotransposons belonging to the *gypsy*-like clade, while group III collates insertion regions of the elements belonging to three different clades: *copia*-like (*copia*), *gypsy*-like (*mdg3*, *mdg1*, *Dm412* (*mdg2*), and *blood*), and *BEL*-like (*roo* (*B104*)) (Bowen and McDonald, 2001).

Certain local physicochemical characteristics of the DNA double helix within the target site and/or its neighborhood, which actually determine the possibility of a more or less specific interaction of the DNA region with proteins of the integration complex, might be important for an LTR retrotransposon in choosing the insertion region.

It is known that an efficient and specific binding of transcription factors to DNA demands certain properties of DNA secondary structure in the regions of interaction. As a rule, characteristic of these regions is the presence of positions in the dinucleotide context that display either invariant or low-variable CPCP (Starr *et al.*, 1995; Meierhans *et al.*, 1997). Therefore, detection of such positions in insertion regions would form the background for considering this fact as an evidence of possible specific DNA–protein interactions in these regions.

Comparing CPCP in the insertion regions of LTR retrotransposons, we found low-variable CPCP in insertion regions of various groups, which is an anticipated reflection of their context dependence. However, no univocal compliance between these characteristics was observed. For all the elements except for those from group I (*17.6* and *297*), the preference of insertion region correlates with low-variable CPCP of the dinucleotides located outside of the target site. In addition, length of the fragment containing such dinucleotides is different for LTR retrotransposons belonging to groups II and III: it is narrower in the former case and wider in the latter.

Thus, local CPCP represent a certain code determining interaction of the integration complex with genomic DNA in the insertion regions. Note that analysis of CPCP in insertion regions gives more information compared with contextual analysis, as it allows for detection of DNA secondary structure characteristics that are essential for interaction with the integration complex.

Preferential insertion of the elements into the regions displaying certain local CPCP may reflect the functional distinctions between integration events, in particular, resulting from different structures of DNA-recognizing domains of integrases of various LTR retrotransposons. A diversity of integrases eventuates from evolutionary changes in the integrase domain; and the integrase domain displays the highest evolutionary rate compared with the other domains of *pol* gene (Malik and Eickbush, 2001). Our classification of LTR retrotransposons belonging to 12 families with reference to genomic DNA characteristics in the regions of their insertion complies well with the phylogenetic analysis data on *pol* gene domains, which detected actually the same groups of elements (Whalen and Grigliatti, 1998; Malik and Eickbush, 1999; Bowen and McDonald, 2001; Malik and Eickbush, 2001). This fact may be considered as an evidence of the determining role of DNA-binding domains of LTR retrotransposon integrases in secondary-structure-dependent choosing of the genomic DNA regions for their insertion.

ACKNOWLEDGEMENTS

The work was supported by the Russian Foundation for Basic Research (grants Nos. 01-07-90376-в, 02-07-90355, 03-07-96833-р2003, 03-07-96833, 03-07-90181-в, 02-04-48802-а, 03-04-48829, 03-07-06078-мас, 03-04-48555-а); grant PCB RAS (No. 10.4); the Siberian Branch of the Russian Academy of Sciences (integration project No. 119); Russian Ministry of Industry, Science, and Technologies (grants Nos. 43.073.1.1.1501 and 43.106.11.0011, subcontract No. 28/2003); NATO (grants Nos. LST.CLG.979816 and PDD(CP)–(LST.CLG 979815)); the U.S. Civilian Research & Development Foundation for the Independent States of the Former Soviet Union (CRDF) No. NO-008-X1.

COMPOSITIONAL ASYMMETRIES AND PREDICTED ORIGINS OF REPLICATION OF THE *SACCHAROMYCES CEREVISIAE* GENOME

J.O. KORBEL[&]*, H.E. ASSMUS[&], S.M. KIELBASA, H. HERZEL
Institute for Theoretical Biology, Humboldt-University, Invalidenstr. 43, D-10115 Berlin, Germany, e-mail: j.korbel, h.assmus, s.kielbasa, or h.herzel@itb.biologie.hu-berlin.de
* *Corresponding author,* [&]*These authors contributed equally*

Abstract We analyze correlations of base skews and origins of replication in the chromosomes of the yeast *Saccharomyces cerevisiae*. Putative origins of replication are predicted by extracting all genomic matches of an 11-base-pair consensus sequence of yeast autonomously replicating sequences (ARS), `WTTTAYRTTTW`. Although distances between polarity switches of the GC-skew and predicted ARS are shorter than expected when assuming a uniform distribution, further analyses reveal that the observed tendency is mostly a result of a clustering of the determined base skew polarity switches. It is concluded that the GC-skew does not give additional information about the location of chromosomal origins of replication in *S. cerevisiae*.

Key words: compositional bias, base skew, origin of replication

1. INTRODUCTION

When both strands of a DNA molecule are analysed, equal frequencies of the bases C/G or A/T, respectively, are observed. However, asymmetries in the composition of long DNA single strand stretches have been observed in several prokaryotic organisms (see e.g. Lobry, 1996; Blattner *et al.*, 1997; Kunst *et al.*, 1997; Andersson *et al.*, 1998; Fraser *et al.*, 1998; Grigoriev, 1998; McLean *et al.*, 1998; Rocha and Danchin, 2001; Shioiri and Takahata, 2001). Possibly, different mutation rates on leading and lagging strand of the asymmetric replication fork, mutation rates during transcription, and a codon preference in connection to a favoured location of genes on the leading strand contribute to the so-called GC-skew (see Mrázek and Karlin, 1998; Frank and Lobry, 1999; Gautier, 2000 and references therein). Skewed base compositions have been used to localize origins of replication in eubacterial and large viral genomes (Lobry, 1996; Grigoriev, 1998; Mrázek and Karlin, 1998), and in the genomes of some archaea (Myllykallio *et al.*, 2000). Recent publications describing the completion of bacterial genome sequences commonly present predictions for origins of replication based on base skew analyses (see e.g. Read *et al.*, 2000; Kuroda *et al.*, 2001). In contrast to prokaryotic organisms that usually contain one single origin of replication, DNA synthesis in eukaryotes starts from a large number of replication origins, which are defined as specific sequence sites for initiation and regulation of DNA replication. Several elements capable of directing autonomous replication to plasmids (so-called autonomously replicating sequences or ARS) have been described in the yeast *S. cerevisiae* (Gierlik *et al.*, 2000; Wyrick *et al.*, 2001). Although only few have been mapped to chromosomal locations, it is estimated that the yeast genome contains several hundred origins of replication. Experimentally verified ARS normally contain a match to the well conserved ARS consensus sequence `WTTTAYRTTTW` (also referred to as 'ACS', the core of the so-called A domain or A element; the symbols W, Y, and R stand for: W = A or T; Y = C or T; R = A or G), together with usually only poorly conserved auxiliary flanking elements (Raychaudhuri *et al.*, 1997; Kohzaki *et al.*, 1999 and references therein). Grigoriev (1998) and Gierlik *et al.* (2000) studied correlations between base compositional asymmetries and experimentally verified origins of replication in the genome of *S. cerevisiae.* Both found no, or very weak, correlations of base skews and experimentally verified ARS elements. We present an analysis on correlations of asymmetric base compositions and yeast origins of replication predicted by extracting genomic matches to the 11-base-pair ARS consensus sequence. This study should help to elucidate whether there is any potential applicability of the base skew method for prediction of origins of replication in primitive eukaryotes.

2. METHODS

No bias conditions of long DNA stretches are characterised by equal frequencies of the bases A/T, and G/C, respectively. Lobry (1996) used non-overlapping moving windows of a size of 10 kilobases (kb) to analyse deviations from no-bias-conditions in three bacterial genomes. While small window sizes lead to strong fluctuations, too large window sizes might cause a loss of relevant zero crossings. Within a window, base frequencies are counted to obtain GC-content (equation (1)) and GC-skew (equation (2)):

$$\frac{G+C}{A+C+G+T}, \tag{1}$$

$$\frac{G-C}{C+G}. \tag{2}$$

When analysing the 4.6 mega base sequence of *Escherichia coli* K12 (Blattner *et al.*, 1997) from 5' to 3' as it is represented in the database (http://www.ebi.ac.uk/genomes), the origin of replication coincides with a polarity switch of the GC-skew from positive to negative values (Figure 1).

In order to measure the GC-skew in the genome of *S. cerevisiae* (Goffeau *et al.*, 1996; release 18/11/98), we tested different moving window sizes (i.e. 5 kb, 6 kb, 10 kb, and 15 kb), applying a step size of 1 kb.

A window size of 10 kb was selected, as it resulted in a final number of zero crossings close to the number of predicted ARS (see below). While the number of false positive as well as negative signals resulting from this strategy is unclear, we regard this window size as suitable for determining whether there exists a correlation between positions of predicted ARS consensus sequences and zero crossings of the GC-skew. ARS were predicted by extracting complete matches to ACS patterns. Although several are likely to be biologically inactive (Wyrick *et al.*, 2001), they should represent useful predictors for testing a putative correlation.

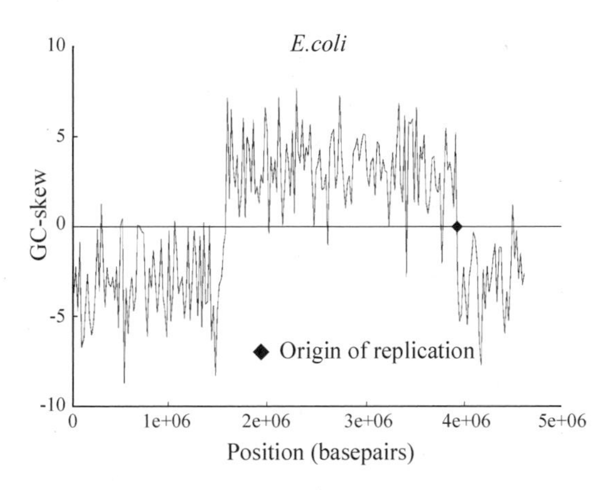

Figure 1. GC-skew of the *E. coli* genome. The location of the origin of replication as described in the literature (Grigoriev, 1998; here indicated by a black diamond) coincides with a GC-skew polarity switch from positive to negative values. A moving window of length 10 kilobases was applied for calculation of the GC-skew.

3. RESULTS AND DISCUSSION

An analysis of the GC-skew on all 16 yeast chromosomes indicates frequent local asymmetries in nucleotide composition (Figure 2). Consistent with previous observations (Gierlik *et al.*, 2000), we detected significant compositional bias at the ends of the linear chromosomes. Within the chromosomes, we found 620 transitions through the neutral axis from positive to negative values (Table 1). The prediction of ACS elements revealed 800 occurrences of the 11-base-pair core element. Figure 2 illustrates the positions of skew polarity switches and predicted ARS consensus sequences on yeast chromosome 14.

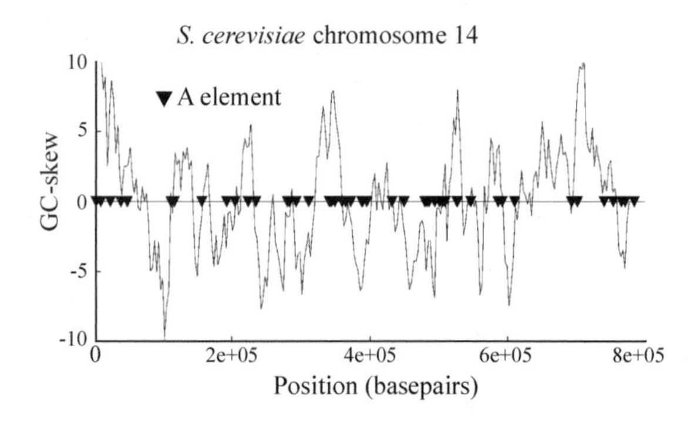

Figure 2. GC-skew for yeast chromosome 14 showing several polarity switches. Positions of predicted ARS (A elements) are indicated by triangles.

In order to ask the question whether there is a correlation between the observed local compositional bias and positions of predicted ARS, we measured distances between ACS elements and the corresponding nearest zero-crossings of the GC-skew. Observed distances on each chromosome were compared with a uniform distribution of distances.

Figure 3*a* indicates significant discrepancies between observed distances and a uniform distribution of distances between predicted ARS and zero-crossings of the GC-skew. Particularly large discrepancies were observed around 10–15 kb, a value close to the expected average distance of ARS and GC-skew polarity switches. In order to test whether these discrepancies are due to clustering of either the predicted ARS or of the GC-skew zero-crossings, we analysed distances between uniformly distributed ACS elements and fixed zero-crossings as extracted from the yeast chromosomes (Figure 3*b*) as well as distances between uniformly distributed GC-skew polarity switches and fixed ACS elements (Figure 3*c*). Figures 3BC show that the apparent clustering of ARS with GC-skew polarity switches remains when the chromosomal locations of the former are randomised, but disappears after randomising the latter. We conclude that the observed discrepancies (i.e. the trend observed in Figure 3*a*) are mainly due to clustering of the base skew polarity switches. We then studied the distributions of distances for ARS and zero crossings separately. While distances between predicted ARS resemble a uniform distribution, the GC-skew zero crossings are strongly clustered at distances around 10–20 kb (data not shown). We repeated the calculations applying larger and smaller window sizes (i.e. with polarity switch counts slightly above or below 800), with results supporting our conclusions.

Table 1. For each of the 16 chromosomes of *S. cerevisiae*, the sequence lengths in kilobases (kb), the number of predicted ACS elements (ARS), the sum of polarity switches from plus to minus (zero crossings), and the GC-content are given

Chromos.	Length (kb)	ACS	Zero-crossings	GC-content
1	230	15	9	39.27
2	813	56	46	38.34
3	315	26	14	38.55
4	1531	113	83	37.90
5	576	47	29	38.50
6	270	16	10	38.73
7	1090	64	56	38.06
8	562	28	29	38.49
9	439	37	25	38.90
10	745	52	33	38.36
11	666	40	28	38.06
12	1078	73	50	38.48
13	924	56	56	38.20
14	784	50	37	38.63
15	1091	63	61	38.16
16	948	64	54	38.06
Σ	12 067	800	620	38.42

The clustering of the GC-skew can be partly explained by the nature of the applied method. At sites of the yeast chromosomes where the skew is close to zero, polarity switches (i.e. transitions through the neutral axis from positive to negative values; see methods) are more likely to occur. Some clustered shifts of the skew may however be still informative and worth looking at in particular—they might have resulted from recent rearrangements within the genome, like Huang *et al.* (2003) proposed for a recently sequenced plasmid of *Streptomyces lividans*.

Grigoriev (1998) and Gierlik *et al.* (2000) both found only weak correlations of compositional bias and experimentally verified ARS elements in the *S. cerevisiae* genome. Our study shows significant discrepancies between observed distances of GC-skew polarity switches and predicted ARS, and a uniform distribution. Further analyses indicate, however, that the observed tendency is mainly due to clustering of the base skew polarity switches. We have thus presented further indication that—despite its usefulness for the prediction of replication origins in prokaryotes—the GC-skew does not provide relevant information about the location of origins of replication in eukaryotes. Eukaryotic genomes consist of several chromosomes, each containing a large number of origins of replication. These origins are present in excess—some are optionally used throughout replication cycles (Todorovic, 1999). Origin excess together with variation in origin usage blurs the measurable compositional bias, caused by similar mechanisms in eukaryotes as in bacteria—for instance by varying mutation rates on leading and lagging strand of the asymmetric replication fork.

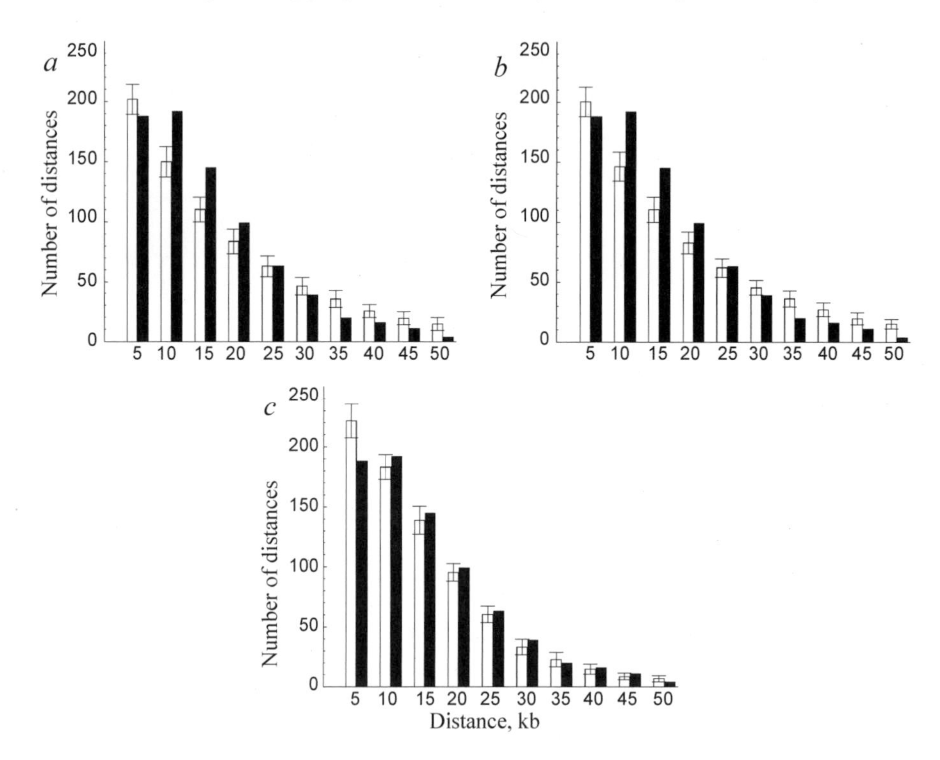

Figure 3. Analysis of the distribution of distances between predicted yeast ARS and the nearest polarity switches of the GC-skew: (*a*) distribution of distances between predicted yeast ACS elements and GC-skew polarity switches (dark bars; all 16 *S. cerevisiae* chromosomes were considered). A uniform distribution of distances is shown in white bars; (*b*) the distribution of distances is represented together with distances observed between uniformly distributed ACS elements and fixed GC-skew polarity switches (white bars); (*c*) here, GC-skew polarity switches were uniformly distributed, while the ACS remained fixed (white bars).

REVEALING AND FUNCTIONAL ANALYSIS OF tRNA-LIKE SEQUENCES IN VARIOUS GENOMES

F.E. FRENKEL*, M.B. CHALEY, E.V. KOROTKOV, K.G. SKRYABIN
Bioengineering Center, Russian Academy of Sciences, prosp. 60-letiya Oktyabrya 7/1, Moscow, 117312 Russia, e-mail: felix@biengi.ac.ru
* *Corresponding author*

Abstract tRNAs and their derivatives are known to be involved in various genetic processes from bacteria to primates. Revealing evolutionary ancestors and relatives of the modern tRNAs with further functional analysis could shed light on their role in expansion of interspersed repeats, their promoter activity, and so on. Although tRNA-like sequences were earlier found and described in many gene regions and most known SINE repeats contain tRNA part, there are still a lot of unknown entries. tRNA-like sequences could also help to find out new families of genetic repeats. Presence of tRNA-like sequences was detected in various genetic structures. Their relative distribution pattern over these sequences and preliminary functional analysis are presented.

Key words: tRNA-like sequence, interspersed repeats, functional analysis, computer analysis

1. INTRODUCTION

The goal of the work was to detect evolutionary ancestors and relatives of known tRNAs. In contrast to the majority of available tRNA search algorithms (like tRNA-Scan SE; Lowe, 1997) that are developed to detect functional tRNA genes by using information on their secondary structure and other tRNA-specific features, the method developed is intended only for showing up evolutionary relationships. It found most tRNA-derived sequences in genetic sequences available in the GenBank that extends available data on tRNA-like sequences without explicit structural and functional resemblance with tRNA genes (Mans, 1991).

2. METHODS AND ALGORITHMS

First, tRNAs available in Sprinzl tRNA database (Sprinzl, 1998) were collected. All of them where divided into 22 families by their isoacceptor specificity (20 amino acids plus methionine-initiator and selenocysteine). These families were converted into templates (positional frequency matrices). During this process, representation of each tRNA sequence was taken into account (Sibbald, 1990).

An original method of enlarged DNA nucleotide sequence similarity was used to detect highly divergent tRNA copies. It allowed us to detect sequences similar to the given template that could contain multiple insertions and deletions. Its core process lies in aligning the sequence sought for against the present template considering nucleotide distribution for each position in tRNA. Existing analogues of the method applied either use position invariant scores for symbol match/mismatch/deletion (Waterman, 1995) or proceed from template but require some strongly similar subsequence (PSI-BLAST).

The method applied (see detailed description of its ancestor in Chaley *et al.*, 2003) uses dynamic programming algorithm. The weight matrix for each position in the source tRNA was calculated as:

$$v(i,j) = f(i,j) \ln\{f(i,j)/p(i)\}, \tag{1}$$

where $f(i,j)$ is the frequency of the base i at the position j of the source tRNA template.

It indicates the likelihood that a given base is located at the given position. Dynamic programming algorithm implies finding the best alignment via filling the alignment weight matrix F as

$$F(i,j) = \max\{\max_{k=1,d_{\max}}\{F(i-k,j) - v_d(1+\log(k))\};$$
$$\max_{k=1,d_{\max}}\{F(i,j-k) - v_d(1+\log(k))\}; F(i-1,j-1) + v(S(i),j); 0.0\};$$
$$F(0,0) = 0.0;\ \ F(i,0) = F(0,0) - v_d(1+\log(i));$$
$$F(0,j) = F(0,0) - v_d(1+\log(j)),$$

where i is position in the template S(tRNA), j is position in the subsequence (window) of the analyzed DNA, $d_{\max}$ is the maximal number of allowed deletions/insertions, v_d is a deletion/insertion weight, and $v(i,j)$ is the previously defined (1) tRNA weight matrix.

After filling the matrix F, the optimal way from its maximum element to the first zero element is created. Weight of the found local alignment equals a difference between the corresponding last (maximal) and first (zero) element of F.

Statistical significance of the alignment was calculated by comparing the calculated weight of alignment assumed as a global one against the weights of tRNA global alignment along random sequences with the same nucleotide composition. Although global alignment differs mathematically from the local one just in the absence of zero member during filling alignment matrix (1), the reason to use global alignment lies deeper: local alignment does not allow its significance to be calculated because of non-linear process of building the best alignment without fixing its start position.

The significance score Z was found by Monte-Carlo simulations:

$$Z=\left(W_S - M\left(W_{rnd}\right)\right)/\sigma\left(W_{rnd}\right), \quad (3)$$

where W_S is the weight of found global alignment, W_{rnd} are weights of alignments along random sequences, and M and σ are their mean value and dispersion, respectively.

Simulations on random sequences with a length of $2 \cdot 10^6$ bp showed that for the threshold score Z = 7.0, the probability to reveal there a case of a higher similarity with the given template equals to $2 \cdot 10^{-7}$.

In order to describe functionally the found tRNA-like sequences, we have analyzed the GenBank's annotation data. For each tRNA-like sequence, the corresponding GenBank's feature was extracted if possible. In case of multiple functional descriptions felt on the tRNA-like sequence, the only one was taken into account by rejecting others as provided by predefined priorities. In that way, functional description was determined for each tRNA-like sequence.

Implementation of the method on personal computer showed that the computational time it requires is unsatisfactorily long. This problem becomes clear, if we remember its most time-consuming process—the Monte-Carlo simulations.

The way out was found in using computational cluster for implementation of this algorithm. To provide higher portability among computer platforms, the program complex was written in C language using MPI interface for interprocess communications.

3. RESULTS

All data concerning the tRNA-like sequences are presented in Tables 1–3. All the data listed in tables do not contain the known tRNA genes and tRNA-like sequences that were found in tRNA-derived repeats (like Alu, MIRs, B2, and so on). All these repeats where deleted using RepBase and GenBank annotations. Alu and MIR repeats were also excluded from findings using the method described above (Korotkov, 2000).

Table 1. The numbers of tRNA-like sequences detected in earlier characterized DNA regions of the GenBank's nine basic taxonomic divisions

Feature	Division								
	VRL	PHG	BCT	PLN	INV	VRT	ROD	MAM	PRI
3'UTR	–	–	–	1	2	3	22	4	287
5'UTR	–	–	–	4	–	2	6	2	78
CDS	363	14	398	358	103	45	107	87	9312
D-loop	–	–	–	–	1	299	–	11	12
Enhancer	–	–	–	–	–	–	–	–	3
exon	2	–	3	133	11	2	40	5	1512
gene	4	–	29	277	433	19	296	30	7851
iDNA	–	–	–	–	1	–	1	–	2
intron	–	–	6	64	26	12	311	112	1541
LTR	11	–	–	11	–	–	–	–	28
mat_peptide	2	–	1	–	–	–	–	–	9
misc_feature	21	11	490	348	29	60	153	18	6715
misc_RNA	21	–	281	9	2	5	–	–	8
misc_signal	–	–	1	–	–	–	–	–	9
mRNA	5	2	6	3	5	1	27	13	371
precursor_RNA	1	–	47	2	3	4	6	5	29
prim_transcript	–	–	4	6	–	4	11	2	289
primer_bind	–	–	–	5	1	–	–	2	1
promoter	–	1	–	8	–	1	80	4	206
protein_bind	–	–	–	–	–	–	4	–	17
repeat_region	1	–	8	31	54	12	421	111	29995
repeat_unit	–	–	–	16	20	104	42	72	29
rRNA	–	–	231	264	70	19	2	6	76
satellite	–	–	–	2	–	–	1	8	12
scRNA	–	–	1	–	1	–	14	–	1
sig_peptide	–	–	–	–	–	–	–	–	13
snRNA	–	–	–	–	–	–	4	–	–
stem_loop	–	–	2	1	2	–	–	–	–
STS	–	–	–	–	–	–	–	–	147
V_region	–	–	–	–	–	–	1	–	176
V_segment	–	–	–	–	–	–	–	–	4
unassigned	65	38	717	1214	2157	462	1900	263	104012
Total	496	66	2225	2758	2921	1054	3450	755	162749

The region definitions in the column "Feature" are presented as the GenBank's standard features. The "Division" columns correspond to the nine divisions of the GenBank: viral, phage, bacterial, plant, invertebrate, vertebrate, rodent, mammalian, and primate sequences. If a tRNA-like sequence overlapped several annotated regions or a region described by more than one feature, the only one feature was assigned to the tRNA-like sequence according to predefined priorities (see text for details).

Table 2. Percentage composition of the detected tRNA-like sequences according to their isoacceptor attribute over the GenBank's nine basic taxonomic divisions

Isoacceptor	Division								
	VRL %	PHG %	BCT %	PLN %	INV %	VRT %	ROD %	MAM %	PRI %
A	**15.3**	7.6	**20.4**	4.9	**13.0**	**8.3**	**39.1**	**15.0**	3.9
C	4.0	1.5	2.0	4.0	4.7	2.9	0.6	2.3	5.1
D	**13.3**	7.6	3.8	5.9	3.7	0.6	1.4	2.8	2.1
E	6.3	3.0	5.6	6.6	**8.2**	1.9	1.9	**13.2**	3.0
F	0.2	0.0	1.1	3.2	2.3	**9.0**	2.2	1.1	2.1
G	0.8	**18.2**	3.3	4.3	4.0	0.9	2.2	1.3	**8.6**
H	1.6	3.0	0.7	**11.5**	2.0	0.8	6.6	1.9	1.1
I	0.6	1.5	**18.4**	2.4	4.0	0.9	7.2	0.8	3.5
K	3.0	7.6	3.0	3.3	5.0	1.1	2.3	7.5	1.9
L	4.8	3.0	3.8	7.1	3.3	1.0	1.1	2.9	**8.5**
M	3.0	1.5	2.7	4.4	5.4	**10.2**	**12.2**	7.8	1.9
N	1.2	4.5	1.0	3.1	3.7	**10.0**	1.6	1.6	1.2
P	1.8	3.0	2.3	1.6	2.6	**14.3**	2.9	2.6	4.0
Q	7.1	3.0	3.5	**8.0**	4.8	0.7	0.9	0.7	2.1
R	3.2	**12.1**	3.5	4.8	**11.4**	0.3	3.3	**22.4**	2.8
S	**15.9**	3.0	4.6	6.9	4.8	0.6	3.0	1.3	6.9
T	2.4	16.7	3.9	5.0	**10.1**	**15.3**	7.4	**10.2**	1.6
V	4.8	0.0	3.5	3.8	2.8	2.3	0.9	1.1	1.1
W	7.7	1.5	7.9	4.2	1.4	**13.7**	0.6	1.9	3.0
X	1.8	1.5	1.3	1.1	0.7	0.0	0.3	0.3	**22.7**
Y	0.4	0.0	3.2	3.4	1.2	4.9	2.4	0.3	4.7
Z	0.6	0.0	0.6	0.5	0.9	0.3	0.1	1.2	**8.5**
Total	496	66	2225	2758	2921	1054	3450	755	162749

Numbers of the most representative groups of tRNA-like sequences are bold-faced. The line *Total* shows the total numbers of tRNA-like sequences detected over the GenBank's divisions. A common single letter alphabet for amino acids is used to indicate the isoacceptors' specificity of the findings.

Table 3. Percentage composition of the detected tRNA-like sequences according to their isoacceptor attribute in coding regions (CDS) over the GenBank's nine basic taxonomic divisions

Isoacceptor	Division								
	VRL %	PHG %	BCT %	PLN %	INV %	VRT %	ROD %	MAM %	PRI %
A	**14.9**	**21.4**	**11.1**	3.4	**9.7**	**8.9**	**9.3**	3.4	3.7
C	4.7	7.1	6.3	**9.5**	**17.5**	**40.0**	1.9	1.1	3.9
D	**15.2**	7.1	**9.3**	5.6	3.9	6.7	4.7	4.6	1.7
E	7.2	**14.3**	**10.1**	4.5	8.7	6.7	0.0	**57.5**	2.3
F	0.3	0.0	0.8	0.6	1.9	2.2	0.9	0.0	1.9
G	0.3	0.0	3.8	4.7	2.9	2.2	8.4	1.1	**9.8**
H	1.4	**14.3**	1.0	1.1	1.0	0.0	1.9	2.3	0.4
I	0.6	0.0	1.8	2.2	0.0	2.2	1.9	2.3	3.0
K	1.4	7.1	1.8	6.1	1.9	0.0	0.0	0.0	0.6

Isoacceptor	Division								
	VRL %	PHG %	BCT %	PLN %	INV %	VRT %	ROD %	MAM %	PRI %
L	5.5	0.0	6.0	2.5	1.9	2.2	0.9	0.0	7.1
M	1.9	0.0	4.0	4.7	6.8	0.0	2.8	0.0	1.4
N	0.6	0.0	1.8	7.8	0.0	0.0	0.9	1.1	1.0
P	1.7	0.0	2.0	1.4	0.0	2.2	**41.1**	2.3	2.1
Q	8.0	0.0	8.3	7.0	**9.7**	0.0	4.7	3.4	1.2
R	3.3	**14.3**	2.0	4.2	1.0	0.0	2.8	0.0	2.3
S	**20.4**	0.0	5.3	**20.1**	4.9	6.7	1.9	0.0	6.3
T	1.4	0.0	4.3	4.5	6.8	4.4	3.7	3.4	1.4
V	3.3	0.0	5.8	3.9	**9.7**	**8.9**	0.9	1.1	1.0
W	5.0	7.1	6.3	4.5	3.9	2.2	4.7	6.9	2.1
X	2.2	7.1	1.3	0.6	1.0	0.0	2.8	0.0	**26.9**
Y	0.3	0.0	5.0	0.6	2.9	0.0	1.9	0.0	5.4
Z	0.8	0.0	2.3	0.6	3.9	4.4	1.9	**9.2**	**14.3**
Total	363	14	398	358	103	45	107	87	9312

Numbers of the most representative groups of tRNA-like sequences are bold-faced. The line *Total* shows the total numbers of the tRNA-like sequences detected in coding regions (CDS) over the GenBank's divisions. A common single letter alphabet for amino acids is used to indicate the isoacceptors' specificity of the findings.

4. DISCUSSION

As we can see in Table 1 the tRNA-like sequences are non-uniformly distributed in functional regions of the genomes examined. The tRNA-like sequences were most numerous in the regions wherefrom proteins are translated (CDS). Regions signed as "gene" contain a somewhat fewer tRNA-like sequences and have not such a clear meaning as CDS regions do.

A lot of tRNA-like sequences in D-loop regions of vertebrates show that displacement loop in mitochondrial DNA has evident similarity to tRNA. A displacement loop region contains a site of mitochondrial heavy chain replication and promoters for transcription of heavy and light chains (Clayton, 1984). Although individual occurrence of structural similarity of a D-loop to tRNA was mentioned earlier, the similarity of D-loop primary sequences to each other and to tRNAs has not been noticed (Mans, 1991). The results listed in Table 1 suggest us a common evolutionary origin of D-loop regions from tRNA.

Detection of tRNA-like sequences in rRNAs corresponds to 16S and 18S ribosomal RNA. This similarity most likely appears to be rather a footprint of evolutionary rRNA formation than a significant functional feature.

Analysis of not so numerous tRNA-like sequences in promoter regions of rodents and primates showed that these promoters are specific of RNA polymerase II. We suppose that similarity of RNA polymerase II promoters to tRNA shows a transformation of RNA polymerase III promoters: almost 300 of such promoters could not be considered as a casual event. If these transformations are possible during evolutionary process, then we could assume the tRNA-like sequences in mRNA, 3'UTR, 5'UTR, or misc_RNA regions to correspond to some regulatory binding sites.

Surprisingly, numerous tRNA-like sequences have been found in repeat_region areas, especially in primates. The tRNA-like sequences have been detected in autonomous and non-autonomous transposons, endogenous retrovirus-derived repeats, LTRs, LINE repeats, and various satellites. The fact of occurrence of tRNA-like sequences in all these transposable elements brings us to assumption that distribution mechanism of tRNA-like sequences could be concerned with their transfer by certain repeats.

All newly revealed tRNA-like sequences have been analyzed in accordance with the 22 tRNA isoacceptor families they belong to. Table 2 refers to all the tRNA-like sequences whereas Table 3 only to those that were found in CDS regions (see Table 1) as far as these regions contain a large amount of tRNA-like sequences relatively to other functional regions. In Tables 2 and 3, we marked with bold those isoacceptor groups that have a significant share in the given GenBank taxonomic division. One could notice that although each division has its own group of predominant isotypes. alanine prevails in six of the nine divisions, including viruses, bacteria, invertebrates, vertebrates, rodents, and mammals.

We should also point to alanine group share in rodents that equals to significant 39% (see Table 2). Taking into account that recent publications report SINE repeats to have the maximum similarity with alanine tRNA in rodents, we could suppose that the leading position of the mentioned alanine group is caused by very ancient SINE copies. Concerning a significant share of alanine group in bacteria, we presume that their alanine tRNAs play special roles. For example, bacterial transmessenger RNAs (tmRNAs) have tRNA-like part of alanine isotype (Muto, 1998).

Compositions of the most significant specificity groups change in transition to CDS regions (Table 3), remaining the same only for viruses (alanine, aspartate, and serine) and for primates (glycine, leucine, initiator methionine, and selenocysteine).

Thus, our finding of ancient tRNA-derived sequences in genomes revealed their high content in gene coding regions, shown that diverged tRNA gene sequences could get a new behavior on regulatory sites (promoters, D-loops), and allowed us to suggest a possible mechanism for spreading of tRNA-like sequences over a genome within certain transposable elements (Waterston, 2002). Moreover, for each of the nine GenBank's taxonomic divisions (viruses, phages, bacteria, plants, invertebrates,

vertebrates, rodents, and primates) the sets of predominant isotypes of tRNA-like sequences were shown.

ACKNOWLEDGEMENTS

This work has been partially supported by the Project "Designing computer programs for analysis of structural and functional genome properties" of the Federal Program "Investigations and developments in priority directions of science and engineering".

TYPE-SPECIFIC FEATURES OF THE STRUCTURE OF THE tRNA GENE PROMOTERS

Yu.V. KONDRAKHIN[1,2], I.B. ROGOZIN[1,3], T.M. NAYKOVA[1], N.S. YUDIN[1], M.I. VOEVODA[1], A.G. ROMASHCHENKO[1,4]*

[1]*Institute of Cytology & Genetics, Siberian Branch of the Russian Academy of Sciences, prosp. Lavrentieva 10, Novosibirsk, 630090 Russia, e-mail: romasch@bionet.nsc.ru;* [2]*Ugra Research Institute of Information Technologies, ul. Mira 151, Khanty-Mansyisk, 628011 Russia;* [3]*National Center for Biotechnology Information, National Library of Medicine, National Institutes of Health, Bldg. 38A, Bethesda, MD 20894, USA;* [4]*Novosibirsk State University, ul. Pirogova 2, Novosibirsk, 630090 Russia*

* *Corresponding author*

Abstract A new method for the recognition of type 2 promoters in the eukaryotic tRNA genes, which takes into consideration not only nucleotide distribution at individual box positions, but also the multiple interactions between the different parts of the A and B boxes, is proposed. The recognition procedure is based on the module organization of the A and B boxes within the intragenic promoters of the tRNA genes. It was shown that each module of a box is represented by a limited number of the sequence variants. Their particular combinations are the characteristic features of every tRNA gene type, and they are conserved within the two kingdoms of multicellular organisms. The same module combinations were identified in the promoters of the tRNA genes of the other types in unicellular organisms. These results suggested that the sequence module variants might have recombined during the evolutionary transition from unicellular to multicellular organisms.

Key words: eukaryotic tRNA gene types, variability in the intragenic promoter structure, concerted substitutions

1. INTRODUCTION

Gene transcription in eukaryotes, in contrast to prokaryotes, is distributed among three RNA polymerases. The transcription of the genes of the large ribosome subunits is accomplished by RNA polymerase I (Bateman *et al.*, 1985). Almost all the genes encoding proteins and the small nuclear RNAs, except the *U6* gene, are transcribed by RNA polymerase II (Zawel and Reinberg, 1992).

The ribosomal 5S RNA (Pieler *et al.*, 1987), tRNA (Sharp *et al.*, 1981), *U6* RNA genes (Eschenlauer *et al.*, 1993), the *SINE* (Sutcliffe *et al.*, 1982) and LINE (Kurose *et al.*, 1995) repeats, as well as the adenovirus *VAI* genes (Fowlkes and Shenk, 1980), are transcribed by RNA polymerase III (Gabrielsen and Sentenac, 1991). Three promoter types in the genes transcribed by RNA polymerase III have been described. Of these, types 1 and 2 are intragenic. Type 1 promoter is characteristic of the *5S* RNA genes and type 2 of the tRNA genes, *SINE* and *VAI*. Type 3 promoter is located in the 5' flanking region of the *U6* genes in most of the studied species. Promoter elements were also found within the *U6* genes of yeast (Brow and Guthrie, 1990).

The tRNA gene promoters are composed of two spatially distinct elements, the A and B boxes (Geiduschek and Tocchini-Valentini, 1988). The A box occupies the region of the DNA sequence that corresponds to the D arm region of the spatial tRNA structure (Rich and Rajbhandary, 1976; Söll and Rajbhandary, 1995).

The location of this promoter element coincides with the tRNA region that encompasses the dinucleotide downstream of the acceptor stem, the strand of the D duplex and 3/4 of the D loop. The B box is in the DNA region corresponding to the tRNA T loop with dinucleotides of the tRNA T stem flanking it at both ends (Sharp *et al.*, 1981). The spacer between the A and B boxes contains the DNA region corresponding to the arm with the anticodon loop and the flanking tRNA nucleotide sequences up to the A and B boxes and possibly an intron. Only 10–20% of the tRNA genes have introns (Sharp *et al.*, 1985).

Figure 1 is a diagram showing the regions of the tRNA genes shared between the intragenic promoter elements (the A and B boxes) and the tRNA molecule loci they encode. In this way, in distinct DNA fragments of the tRNA genes, the information about two relatively independent genetic processes is encoded (Trifonov, 1997): the assembly of the protein initiation transcription complex, catalyzed by RNA polymerase III (Willis, 1993), and the structural features of the type-specific tRNA molecules, which supply the translation machinery by the particular amino acids.

Various consensuses have been used, so far, to recognize the A DNA box. The length of consensus sequences for the A box DNA differed by a single nucleotide and degeneracy degree at certain positions (Sharp *et al.*, 1985; Geiduschek and Tocchini-Valentini, 1988; Chalker and

Sandmeyer, 1993; Joazeiro *et al.*, 1996). For example, the generally accepted consensus for the A box DNA has the structure TRRYNNAGYGG (Sharp *et al.*, 1985). Our analysis of all the published consensuses revealed a high false negative rate in recognition without mismatches. Thus, only 40% of the actual boxes were found, using the consensus TRRYNNAGYGG. The remaining known consensuses were characterized by a poorer prediction accuracy.

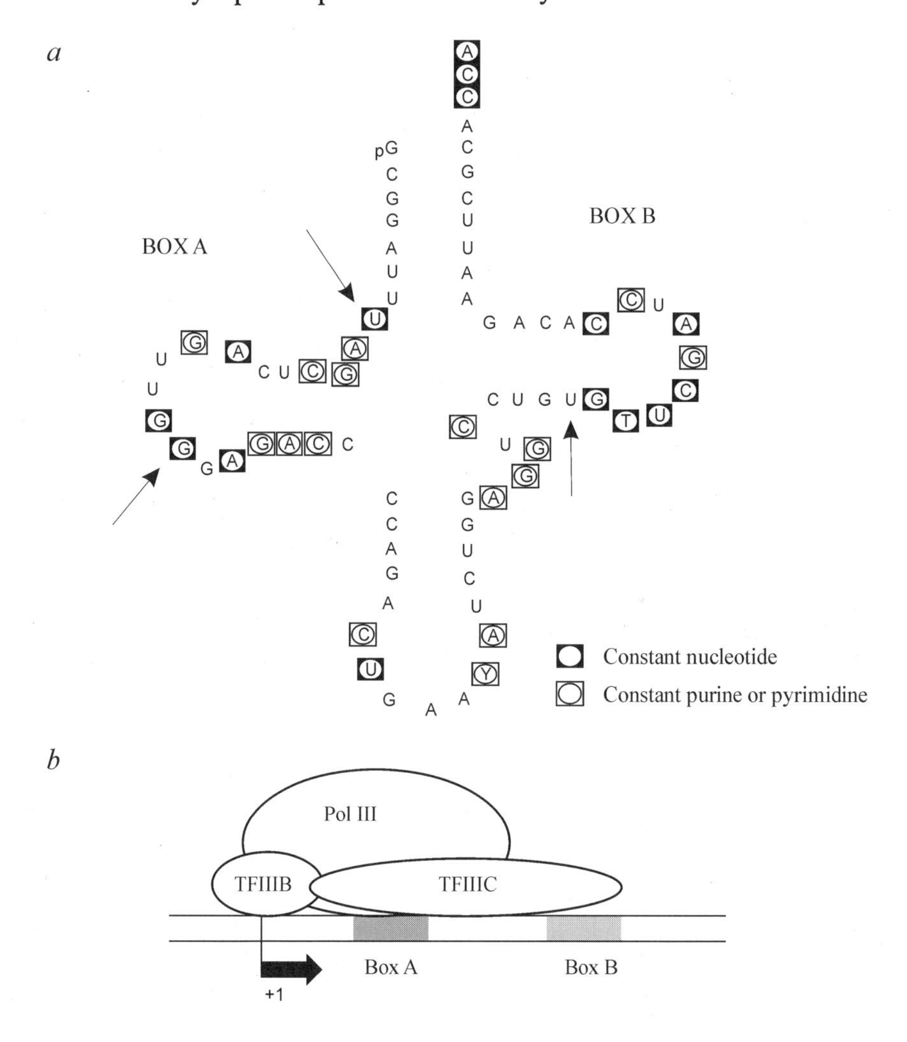

Figure 1. Structure of promoter of eukaryotic tRNA genes: (*a*) the regions of the tRNA molecule shared between the A and B box nucleotide sequences. Note: this figure presents the molecule of the yeast $tRNA^{Phe}$; (*b*) structure of the intragenic promoter of the eukaryotic tRNA genes (type 2 promoter of the RNA polymerase III genes).

To decrease the false negative rate to the acceptable 5–8%, we used a recognition procedure allowing 2–5 mismatches. However, in this case, the false positive rate for all the consensuses reached extremely high values (one false prediction for every 26–222 nucleotide, depending on the consensus). The prediction method we proposed is characterized by the following errors: a false negative rate of 7%, a false positive rate of 1 : 4100.

Obviously, the weight matrices comprising the diversity of the sequence variants of the site are more advantageous for describing the A and B box DNAs. The independence of a site position from each other underlies the consensus or matrix approaches, although even superficial analysis of the composite elements (Kel *et al.*, 1995) to which the boxes of the tRNA gene promoters are referred, indicates that there may possibly be relationships between the box positions, which are observable as interrelated substitutions in the common site structure.

As a result, the prediction accuracy yielded by standard methods may be unsatisfactory because a part of the information provided by the interacting individual site positions is bypassed. To describe more fully the promoter structure, we propose an advanced approach with a recognition algorithm additionally including statistically significant interactions revealed by analysis of the training samples.

2. METHODS AND DATA SOURCE

The tRNA gene sequences from the database available via the Internet at http://www.uni-bayreuth.de/departments/biochemie/trna/ were used to compile the training samples of the A and B box DNAs. The nucleotide sequence of the eukaryotic tRNA molecule contains all the information concerning the particular structural features of the A and B box promoter DNAs of the corresponding tRNA gene; the presence of tRNA in the cell cytoplasm is evidence that the promoter of this gene is functionally active. Each tRNA gene type is annotated with the anticodon sequence and amino acid symbol.

An interactive method was applied to align the 35 bp oligodeoxyribonucleotides encoding each half of the tRNA molecule from its ends (Naykova *et al.*, 2003). As a result, two sets of samples, each containing 544 nucleotide sequences of the eukaryotic A and B box DNAs were compiled. There were two sets of samples, each containing the 574 homologous prokaryotic nucleotide sequences.

3. RESULTS AND DISCUSSION

3.1 A new recognition method for the tRNA gene promoter DNA

The appearance of the promoter elements in the coding part of the tRNA genes in eukaryotes should be followed by an appropriate correction of the DNA structure; this is in contrast to the prokaryotic tRNA genes containing promoters outside the transcribed part of the genes. The aim was to reveal the differences in the DNA structure between the eukaryotic tRNA gene promoter regions and homologous counterparts in the prokaryotic tRNA genes. For this purpose, we compared the different features of the primary structure in the A and B box DNAs in eukaryotic species (Protozoa, unicellular fungi; Metazoa and Plantae) with the homologous counterparts of the tRNA genes of Archaebacteria and Eubacteria species, using the developed recognition procedure.

The procedure proposed for the box DNA recognition in an arbitrary nucleotide sequence is two-step.

Step 1. Based on the nucleotide frequency matrix, the arbitrary sequence was first analyzed by the standard weight matrix method. The weight matrix is $W = (w_{ij})$, i = A, C, G, T, $j = 1, \ldots, L$, where L denotes the length of the site in question. The element w_{ij} of the W matrix is the weight of the ith nucleotide at the jth box position. The matrix W was obtained by renormalization of the columns of the original nucleotide frequency matrix in such a way that the sum of the weights in each column was unity. The frequency matrix, in turn, was derived from the aligned sequences from the training sample. The site recognition procedure was centered on the estimation of the score $F(S)$ for each fragment $S = (s_1, \ldots, s_L)$ of the analyzed nucleotide sequence using the formula

$$F(S) = \sum_{j=1,\ldots,L} w_{s_j,j}.$$

The decision of whether the sequence fragment S is a box or not was made by comparing the calculated $F(S)$ score with the threshold. The threshold value was adjusted so that fragments least similar to the box represented by the W matrix were dismissed from further analysis and those contained by the training sample could be recognized. The rate of type I error was almost 0 and that of type II was high.

Step 2. It was applied to only the remaining candidate fragments in the boxes. Recognition was based on a 4-component additive function

$$F_0(S) = F_1 + F_2 + F_3 + F_4.$$

To calculate the first two components F_1 and F_2, the analyzed fragment S was first transformed from the nucleotide into the module form, $S \rightarrow S^* = (m_1, \ldots, m_k)$, where k denotes the number of site modules and m_j stands for the number of the variant of the jth module occurring in the S fragment. The first component F_1 expresses the degree to which each box module is similar to the corresponding region of the selected candidate fragment:

$$F_1 = \sum_{j = 1, \ldots, k} w^*_{mj, j},$$

where $w^*_{mj, j}$ is the weight of the m_jth variant of the jth module, $j = 1, \ldots, k$. The module variant weights $w^*_{mj, j}$ are pre-defined as the relative occurrence frequencies in the training sample. The sum of all the weights of the variants of a same module is 1. The value for the F_1 component is maximal for the S fragment that is composed of the most typical variants of each module.

The second component F_2 expresses the module interaction:

$$F_2 = \sum_{i, j = 1, \ldots, k} u_{mi, mj, i, j},$$

where $u_{mi, mj, i, j}$ is the weight of the interaction of the ith and jth modules when the ith module is represented by the m_ith variant and the jth module by its own m_j-variant. The maximum value for the F_2 component is assigned to the S fragments with the most typical module combinations preliminarily revealed by analysis of the box-training sample. It should be noted that 0 values were assigned to the $u_{mi, mj, i, j}$ weights of module interactions derived from analysis of the training sample, if module interaction was statistically insignificant. The significance of the interaction of two distinct module sequences was estimated by the chi-square test: the frequencies of the observed combinations of the module sequence variants were compared with those expected in the case of module independence.

The 3-rd and 4th components F_3 and F_4 take into account respectively the palindromic and mirror symmetrical structure DNA within a box. The F_3 component is used to recognize both the A and B box DNAs, while F_4 is utilized to recognize only the A_2 subclass of the A box DNA. Both components were determined by comparative analysis of the concentrations of 4 bp palindromes and mirror symmetrical oligonucleotides in the considered box DNAs and different regions of the pol II eukaryotic genes. The F_3 and F_4 components take maximum values for the tested S fragments in which the palindromes and mirror symmetrical oligonucleotides, respectively, are the most abundant at certain box DNA positions.

Recognition of the promoters was carried out on the basis of further analysis of individual box DNAs identified by the above two-step procedure. Certain stable combinations of the individual module sequences within the different type-specific box DNAs provided new information allowing for increasing the recognition ability. The additional component F_2* analogous to F_2 was used to take into account the box DNA interactions. F_2* operated with the interactions of module sequences belonging to either the A or B box DNA, not with the interaction of module sequences within the A box.

Thus, the box A nucleotide sequence in a particular type tRNA gene has a set of specific features, which include the presence of a particular subclass of the A box DNA (12 bp or 11 bp) consisting of three modules (Naykova *et al.*, 2003). Each module has a limited composition of sequence variants, dependent on the subclass of the A box DNA. The determined combinations of the different modules sequence variants are specific to each type of the tRNA gene. This information is used to further identify the type of the tRNA gene, whose promoter was predicted by the above recognition procedure. The computer program and the samples of the A and B box DNAs are available via the Internet at http://www.bionet.nsc.ru./ICIG/molgen/amgl/trna/ind001.html.

3.2 General features of the A and B box DNA structure

Cluster analysis, which took into account not only nucleotide composition at individual box positions, but also position interactions, identified the A_1 and A_2 subclasses of the A box. The subclasses were of different lengths because of the deletion/insertion of a nucleotide at position 9. They also differed by nucleotide distribution at certain box DNA positions. In eukaryotes, the size of the short (11 bp) A_2 subclass is much greater than that of the long (12 bp) A_1 subclass: the ratio being 61% : 39%. In the prokaryotic (Archaebacteria and Eubacteria) tRNA genes, the DNA sequences homologous to the A_1 and A_2 subclasses occurred at a 74% : 26% ratio, in favor of the long variants. The higher polymorphism level of these prokaryotic DNA sequences compared to their eukaryotic counterparts is noteworthy (Naykova *et al.*, 2003). Cluster analysis of the training sample of the B box DNAs failed to reveal its stable subclasses. Table 1 presents the weight matrices W for the B box and the two subclasses of the A box.

Table 1. The weight matrices W calculated for the eukaryotic B box and the A_1 and A_2 subclasses of the A box

Position	1	2	3	4	5	6	7	8	9	10	11	12
The A_1 subclass of the A box												
T	0.96	0.03	0.00	0.04	0.77	0.10	0.00	0.03	0.82	0.53	0.02	0.18
A	0.04	0.79	0.03	0.00	0.02	0.05	1.00	0.15	0.08	0.05	0.00	0.03
G	0.00	0.18	0.94	0.00	0.13	0.00	0.00	0.82	0.05	0.21	0.98	0.78
C	0.00	0.00	0.03	0.96	0.08	0.85	0.00	0.00	0.05	0.21	0.00	0.01
	T	R	G	C	K	Y	A	R	T	B	G	G
The A_2 subclass of the A box												
T	0.99	0.01	0.03	0.50	0.09	0.49	0.00	0.00	0.77	0.02	0.01	–
A	0.00	0.20	0.03	0.00	0.10	0.03	1.00	0.23	0.02	0.00	0.00	–
G	0.01	0.77	0.91	0.01	0.36	0.28	0.00	0.77	0.03	0.96	0.97	–
C	0.00	0.02	0.03	0.49	0.45	0.20	0.00	0.00	0.18	0.02	0.00	–
	T	R	G	Y	S	B	A	R	Y	G	G	
The B box												
T	0.04	0.00	0.92	1.00	0.01	0.00	0.00	0.35	0.62	0.01	0.17	–
A	0.16	0.01	0.08	0.00	0.00	0.23	1.00	0.44	0.09	0.00	0.04	–
G	0.78	0.99	0.00	0.00	0.00	0.77	0.00	0.14	0.01	0.00	0.01	–
C	0.02	0.00	0.00	0.00	0.99	0.00	0.00	0.07	0.28	0.99	0.78	–
	R	G	W	T	C	R	A	D	H	C	Y	

R = A or G; Y = T or C; W = A or T; S = G or C; K = G or T; M = A or C; B = T or C or G; H = A or T or C; V = A or C or G; and D = A or G or T.

3.3 Module organization of the A and B box DNAs

Statistical analysis of nucleotide distribution at positions within and between the boxes demonstrated that not individual positions, much rather groups of close positions (modules) served as information units in recognition of the promoter elements. In fact, multiple relations between the modules are characteristic of the tRNA gene of particular types in eukaryotes, in contrast with prokaryotes. Every A and B box DNA was subdivided into 3 modules. The subdivision took into account the pattern of relations between the individual box positions and locations of the relatively conserved nucleotides. The highly conserved adenosine at position 7 between modules 2 and 3 in the A box DNA was omitted. The most typical sequence variants of modules 1, 2 and 3 of the B box DNA occurring in combination with the polymorphic variants of modules 1, 2 and 3 of the A_1 and A_2 subclasses of the B box are given in Table 2.

The frequencies of the different module sequence variants and their combinations in the A and B box DNAs of the eukaryotic species were estimated using the SPSS 8.0 package of statistical programs. The comparative analysis revealed significant difference in the polymorphism level of the 5' half of the A box DNA between unicellular and multicellular organisms (Naykova *et al.*, 2003). The combinations of the sequence variants of the different modules in the A and B box DNAs in the promoters of the tRNA genes of different types in unicellular organisms were very different from those in multicellular organisms whose combinations were highly conserved (Table 2).

Table 2. The most typical sequence variants of the A and B box module DNA in the tRNA genes from multicellular organisms; the data for the other tRNA genes are not given because the nucleotide sequence samples were not available or too small

tRNA gene	The A box			The B box		
	Module 1: positions 1, 2, 3	Module 2: positions 4, 5, 6	Module 3: positions 8, 9, 10, 11, 12	Module 1: positions 1, 2, 3, 4, 5	Module 2: positions 6, 7, 8	Module 3: positions 9, 10, 11
1. tRNA^{Lys}(UUU)	-TAG-	-CTC-	-GTCGG-	-GGTTC-	-AAG-	-TCC-
2. tRNA^{Lys}(CUU)	-TAG-	-CTC-	-GTCGG-	-GGTTC-	-GAG-	-CCC-
3. tRNA^{Phe}(GAA)	-TAG-	-CTC-	-GTTGG-	-GGTTC-	-GAT-	-CCC-
					-AAT-	
4. tRNA^{Tyr}(GUA)	-TAG-	-CTC-	-GCTGG-	-GGTTC-	-GAT-	-TCC-
					-AAA-	
					-AAT-	
5. tRNA^{Ile}(AAU)	-TAG-	-CTC-	-GTTGG-	-GGTTC-	-GAT-	-CCC-
6. tRNA^{Ala}(AGC)	-TAG-	-CTC-	-GATGG-	-GGATC-	-GAT-	-ACC-
			-AATGG-			-GCC-
7. tRNA^{Thr}(AGU)	-TGG-	-CTT-	-GCTGG-	-AGTTC-	-GAA-	-TCC-
			-GTTGG-	-GGTTC-	-GAT-	
8. tRNA^{Asn}(GUU)	-TGG-	-CGC-	-ATTGG-	-GGTTC-	-GAG-	-CCC-
			-ATCGG-			
9. tRNA^{Met}(CAU)	-TGG-	-CGC-	-GCGG-	-GGATC-	-GAA-	-ACC-
			-GTGG-			
	-TAG-		-GTAGG-	-AGTTC-	-GAT-	-CCT-
10. tRNA^{Ala}(UGC)	-TAG-	-CTC-	-GTGG-	-GGTTC-	-GAT-	-CCC-
					-AAT-	
11. tRNA^{Thr}(UGU)	-TAG-	-CTC-	-GGGG-	-AGTTC-	-AAT-	-TCT-
12. tRNA^{Cys}(GCA)	-TAG-	-CTC-	-GGGG-	-GGTTC-	-AAA-	-TCC-
			-GTGG-	-TGTTC-	-GAA-	-TTC-
						-CCC-
13. tRNA^{Val}(AAC)	-TAG-	-TGT-	-GTGG-	-GGTTC-	-GAA-	-CCC-
14. tRNA^{Val}(CAC)	-TAG-	-TGT-	-GTGG-	-GGTTC-	-GAA-	-CCC-
			-GCGG-			-ACC-
15. tRNA^{Val}(UAC)	-TAG-	-TGT-	-GTGG-	-GGTTC-	-GAT-	-CCC-
			-GCGG-		-GAA-	
16. tRNA^{Asp}(GUC)	-TAG-	-TAT-	-GTGG-	-GGTTC-	-GAT-	-TCC-
					-AAT-	
17. tRNA^{Glu}(UUC)	-TGG-	-TCT-	-GTGG-	-GGTTC-	-GAT-	-TCC-
18. tRNA^{Glu}(CUC)	-TGG-	-TCT-	-GTGG-	-GGTTC-	-GAT-	-TCC-
19. tRNA^{Trp}(CCA)	-TGG-	-CGC-	-ACGG-	-TGTTC-	-GAA-	-TCA-
			-ATGG-	-CGTTC-	-GAT-	-CCG-
20. tRNA^{Arg}(ACG)	-TGG-	-CGC-	-ATGG-	-GGTTC-	-GAC-	-TCC-
					-GAA-	
21. tRNA^{Arg}(UCG)	-TGG-	-CCT-	-ATGG-	-GGTTC-	-GAA-	-TCC-

tRNA gene	The A box			The B box		
22. $tRNA^{Arg}$(UCU)	-TGG-	-CCT-	-ATGG-	-GGTTC-	-GAC-	-CCC-
		-CGC-			-GAG-	-TTC-
23. $tRNA^{Leu}$(CAG)	-TGG-	-CCG-	-GCGG-	-GGTTC-	-GAA-	-TCC-
			-GTGG-			
24. $tRNA^{Leu}$(AAG)	-TGG-	-CCG-	-GCGG-	-GGTTC-	-GAA-	-TCC-
25. $tRNA^{Leu}$(CAA)	-TGG-	-CCG-	-GTTG-	-GGTTC-	-GAA-	-TCC-
			-GCGG-			
26. $tRNA^{Gln}$(UUG)	-TGG-	-TGT-	-ATGG-	-AGTTC-	-AAA-	-TCT-
27. $tRNA^{Gln}$(CUG)	-TGG-	-TGT-	-ATGG-	-AGTTC-	-AAA-	-TCT-
28. $tRNA^{Ser}$(AGA)	-TGG-	-CCG-	-GTGG-	-GGTTC-	-GAA-	-TCC-
29. $tRNA^{Ser}$(UGA)	-TGG-	-CCG-	-GTGG-	-GGTTC-	-GAA-	-TCC-
30. $tRNA^{Ser}$(CGA)	-TGG-	-CCG-	-GTGG-	-GGTTC-	-GAA-	-TCC-
31. $tRNA^{Ser}$(GCU)	-TGG-	-CCG-	-GTGG-	-GGTTC-	-GAA-	-TCC-
32. $tRNA^{Pro}$(AGG)	-TGG-	-TCT-	-GGGG-	-GGTTC-	-AAA-	-TCC-
33. $tRNA^{Pro}$(UGG)	-TGG-	-TCT-	-GGGG-	-GGTTC-	-AAA-	-TCC-
			-GTGG-		-AAT-	-CCC-
34. $tRNA^{Pro}$(CGG)	-TGG-	-TCT-	-GGGG-	-GGTTC-	-AAA-	-TCC-
35. $tRNA^{Gly}$(UCC)	-TGG-	-TAT-	-GTGG-	-GGTTC-	-GAT-	-TCC-
36. $tRNA^{Gly}$(CCC)	-TGG-	-TGT-	-GTGG-	-GGTTC-	-GAT-	-TCC-
			-ATGG-			
37. $tRNA^{Gly}$(GCC)	-TGG-	-TTC-	-GTGG-	-GGTTC-	-GAT-	-TCC-
38. $tRNA^{His}$(GUG)	-TCG-	-TAT-	-GTGG-	-GGTTC-	-GAA-	-TCC-

Table 2 shows that the particular long variants (the A_1 subclass) occur only in certain tRNA types to which the $tRNA^{Lys}$(UUU; CUU), $tRNA^{Phe}$(GAA), $tRNA^{Tyr}$(GUA), $tRNA^{Asn}$(GUU), $tRNA^{Thr}$(AGU), $tRNA^{Ala}$(AGC) and $tRNA^{Ile}$(AAU) genes belong. Remarkably, most tRNA gene types of the A_1 subclass differ by the structure of module 3 of the A box DNA at positions 8, 9 and 10. The differences are characteristic of a particular type of the tRNA genes. The module 2 sequence variant is also different (see the $tRNA^{Thr}$(AGU) and $tRNA^{Asn}$(GUU) genes) in a part of the tRNA genes of this subclass with -TAG- substituted by -TGG- in module 1.

In the $tRNA^{Met}$ genes from different multicellular species, both long and short variants of the A box were identified. In the case of the short variants of the A box DNA in the $tRNA^{Met}$ genes from different species , the structure of all the modules of the two boxes were different as compared with the long, except for the sequence of module 2. The promoters of the $tRNA^{Lys}$(UUU) and $tRNA^{Lys}$(CUU) genes have an identical A box, but differ by combinations of modules 2 and 3 sequence variants of the B box DNA.

As seen in Table 2, the short variants of the A box DNA are the prevailing for most tRNA gene types. The A_2 subclass with the -TAG- variant of module 1 falls into two groups. One group combines with the -CTC- variant of module 2, like in the long sequences of the A box (for example, the $tRNA^{Cys}$(GCA), $tRNA^{Ala}$(UGC) and $tRNA^{Thr}$(UGU) genes); the other group combines with the variants -TGT- (the tRNA^{Val}(AAC;UAC;CAC) genes)) and -TAT- (the tRNA^{Asp}(GUC) gene). Most short sequence variants of the A box DNA have the -TGG- variant of module 1, which combines with module 2 variants (-CGC-, -CCG-, -TGT-, -TCT-, -TAT- and -TTC-), the very rarely occurring in the DNA sequences of the A1 subclass and only in the combination with the -TGG- variant. Exceptional are the tRNA^{Met} genes in which the variant -CGC- is present in both isoforms of the A box DNA. The tRNA^{His}(GUG) gene is outstanding among those listed in Table 2 in that its A box DNA contains the sequence variant of module 1 -TCG-. This variant is conserved in the tRNA^{His}(GUG) genes isolated from the genomes of *Drosophila*, mouse, sheep, and human. The -TCG- variant combines with the -TCT- variant of module 2 in *Drosophila* and with the -TAT- variant in the other species.

The sequence variants of module 3 in the short A box DNA, in contrast to those of the A_1 subclass, occur, as a rule, in the tRNA genes of several types. Thus, the -GGGG- variant was found in the tRNA^{Thr}(UGU), tRNA^{Pro}(AGG) and tRNA^{Cys}(GCA) genes. The -ATGG- variant was found in the tRNA^{Arg}(ACG;UCG;UCU), tRNA^{Trp}(CCA), tRNA^{Gln}(UUG;CUG) genes and so on.

Summarizing the data in Table 2, it is evident that particular types of the tRNA genes differ from each other by the structure of at least one module. The tRNA^{Leu} and tRNA^{Ser} genes are exceptions.

3.4 Relations between combinations of the module sequence variants and secondary structure features of the A and B box DNAs

Elements of the secondary structure at the junction of 2 modules were most prominent in the A box DNA for a number of identified combinations of the sequence variants (Table 2). For example, when combined with the variants -CGC-, -CCT- and -CCG- of module 2, the variant -TGG- of module 1 forms palindromes -GCGC-or -GGCC- respectively; however, when combined with the module 2 variants -TGT-, -TAT- and -TCT-, it forms the repeat with mirror symmetry -TGGT-. The frequently occurring variant of module 1 -TAG- and the variant of module 2 -CTC- form the palindrome -AGCT-. In the case of the tRNA^{Val} gene, the combination of the -TAG- (module 1) and -TGT- (module 2) variants forms the tetraoligonucleotide -TAGT- repeated after one position in the next part of the A box DNA. The -TAT- variant of module 2, in combination with adenine at position 7 of the A box, form the palindrome -TATA- in the tRNA^{His}(GUG), tRNA^{Gly}(UCC) and tRNA^{Asp}(GUC) genes. In the tRNA^{Arg}

genes, when combined with adenine at position 7, and with the -ATGG- variant of module 3, the -CCT- variant of module 2 forms a repeat with mirror symmetry -TAAT-. The same patterns were true for the structure of the B box DNA. All the module 1 variants of the B box DNA form palindromes, when combined with variants of module 2 having guanine at position 6. The palindrome -TCGA- forms in the case of the variants -GAG- and -GAT-, and the hexanucleotide -TTCGAA- forms in the case of the variant -GAA-.

Table 3 lists the observed (p_1) and expected (p_2) occurrence probabilities of palindromes at the listed positions in the samples of the eukaryotic and prokaryotic tRNA genes in the case of the A_2 subclass of the A box DNA. From comparative analysis, it follows that palindrome occurrence frequencies increase from prokaryotes to unicellular eukaryotes, reaching maximum values in multicellular organisms. The same pattern was observed for the occurrence frequencies of repeats with mirror symmetry (Table 4). It may be concluded that DNA sequences in the region of the tRNA genes promoters possessing palindromes and repeats with mirror symmetry as the characteristic features of their secondary structure were preferentially selected during evolution. These features of the secondary structure are inherent in cores of many known *cis*-elements to which the specific trans-acting protein bind (Wingender *et al.*, 1996). It cannot be ruled out that the abundance of such elements in the coding part of the tRNA genes is associated with the formation of an intragenic promoter with whose transcriptional factors from the RNA polymerase III machinery (Kassavetis *et al.*, 1989). The concentration of palindromes and repeats with mirror symmetry on a relatively short DNA stretch of the tRNA genes is highly probable because more stringent requirements are imposed on nucleotide sequences harboring both the promoter features and elements providing the functioning of the tRNA molecule itself.

Table 3. The observed (p_1) and expected (p_2) probabilities of palindromes at different positions of the A_2 subclass of the A box

Kingdom	Probabilities	Box positions							
		1–4	2–5	3–6	4–7	5–8	6–9	7–10	8–11
Eukaryotes	p_1	–	0.258	0.156	0.084	0.174	–	–	–
	p_2	–	0.170	0.049	0.126	0.177	–	–	–
	p_1/p_2	–	1.52	3.20	0.67	0.98	–	–	–
Eukaryotes without single cell	p_1	–	0.248	0.214	0.092	0.168	–	–	–
	p_2	–	0.185	0.055	0.130	0.176	–	–	–
	p_1/p_2	–	1.34	3.89	0.71	0.96	–	–	–
Prokaryotes	p_1	–	0.078	0.01	0.04	0.127	–	–	–
	p_2	–	0.122	0.025	0.098	0.141	–	–	–
	p_1/p_2	–	0.64	0.044	0.418	0.90	–	–	–

Table 4. The observed (p_1) and expected (p_2) probabilities of the repeats with mirror symmetry at different positions of the A_2 subclass of the A box

Kingdom	Probabilities	Box positions							
		1–4	2–5	3–6	4–7	5–8	6–9	7–10	8–11
	p_1	0.309	0.033	0.197	–	–	0.141	–	0.046
Eukaryotes	p_2	0.356	0.012	0.073	–	–	0.099	–	0.035
	p_1/p_2	0.87	2.71	2.69	–	–	1.41	–	1.32
Eukaryotes	p_1	0.344	0.042	0.206	–	–	0.143	–	0.046
without	p_2	0.390	0.014	0.072	–	–	0.091	–	0.035
single cell	p_1/p_2	0.88	2.91	2.86	–	–	1.58	–	1.31
	p_1	0.123	0.052	0.071	–	0.090	0.071	–	0.041
Prokaryotes	p_2	0.150	0.031	0.0079	–	0.075	0.048	–	0.032
	p_1/p_2	0.82	1.71	0.90	–	1.20	1.46	–	1.28

The eukaryotic tRNA genes are, as a rule, reiterated many times and scattered throughout many chromosomes as individual copies or clusters of different sizes (Sharp *et al.*, 1985). The identification of the acting tRNA genes among the multiplicity of homologous nucleotide sequences (pseudogenes) in the genomes requires a more efficient method for recognizing the functional promoters. Many of the RNA polymerase II transcribed genes contain interspersed repeats (SINEs) with nucleotide sequences homologous to the A and B box DNAs of the tRNA genes (Geiduschek and Tocchini-Valentini, 1988).

Some of the repeats function to provide the synthesis of RNA that possibly ensures the epigenetic control in the genome (Wolffe and Matzke, 1999). Recognition of the function of the type 2 promoters of the RNA polymerase III genes in these occasionally highly diverged repeats would provide a better understanding of its role within the RNA polymerase II genes. In the current study, we thoroughly analyzed the structure of the intragenic promoters of the tRNA genes of different types, as well as the prokaryotic tRNA genes whose promoters are outside the genes. The observed differences between eukaryotes and prokaryotes are interpretable rather in terms of the promoter than translational functions of tRNA.

Thus, it appears that the structural diversity of the A and B box DNAs in the eukaryotic tRNA genes have mainly resulted from combinations of a limited set of sequence variants of each module. The striking observation was that the same types of the tRNA genes in unicellular and multicellular organisms differed from each other by combinations of sequence variants of particular modules. It seems reasonable to assume that during the transition from unicellular to multicellular eukaryotes there occurred a grand reshuffling of the sequence variants of the modules in the tRNA genes of different types.

As noted above, particular combinations of the module variants in each tRNA gene type were conserved in remote species of multicellular organisms.

4. CONCLUSION

The module organization of the promoter tRNA genes in eukaryotes is advantageous in that it provides an approach to the specific features of the function of each tRNA gene type alone, as well as of the jointly functioning gene types, which is essential for the translation of the mRNA pool in a cell of the multicellular organism.

ACKNOWLEDGMENTS

This work was supported by the Russian Foundation for Basic Research (grants Nos. 01-07-90376-в, 02-04-48508, 02-07-90359, 03-07-96833-p2003, 03-07-96833, 03-04-48506-a, 03-07-06077-мас, 03-01-00328); grant PCB RAS (No. 10.4); Russian Ministry of Industry, Science, and Technologies (grants Nos. 43.073.1.1.1501; 43.106.11.0011 and Sc.Sh.-2275.2003.4); the Siberian Branch of the Russian Academy of Sciences (integration project No. 119); NATO (grant No. LST.CLG.979816).

MATHEMATICAL TOOLS FOR REGULATORY SIGNALS EXTRACTION

M. RÉGNIER

INRIA 78153 Le Chesnay-FRANCE, e-mail: Mireille.Regnier@inria.fr

Abstract Statistical techniques provide an efficient way to analyze "in silico" the huge amount of data from large-scale sequencing. The key idea is to search for regulatory signals among exceptional words, e.g. words that are either underrepresented or overrepresented. We provide a few mathematical results to assess the significance of an exceptional word.

Key words: regulatory signals, statistics, overrepresentation, protein binding sites, Markov models

1. INTRODUCTION

Statistics for the number of occurrences of a set of words has numerous applications and a great interest arose recently (van Helden *et al.*, 1998; Apostolico *et al.*, 2000; Regnier, 2000; Reinert *et al.*, 2000; Hampson *et al.*, 2002). The underlying assumption is as follows: a biological function that is enhanced or avoided is associated to a word, or a set of words that is overrepresented or underrepresented. This regulatory signal can be for example, a *protein binding site*. We provide below a brief formalization of the main issues addressed in biological applications, which we classify into two main classes.

One studies a long text—typically, a genome—and assumes a probability model on it. The goal is the extraction of exceptional words—either overrepresented or underrepresented. To achieve this goal, one needs an algorithm that searches for candidate motifs and mathematical tools to assess statistical significance of these candidates. A recent survey on algorithms can be found in Lonardi (2001) and Marsan (2002).

Set of independent identically distributed sequences. A set of sequences is given. These sequences are relatively short, generated independently according to a common distribution, but may have different lengths (van Helden *et al.*, 1998; Buhler *et al.*, 2001). One looks for exceptional signals, a word H, or a set of words $\mathbb{H}$. One counts the number of sequences where H, or $\mathbb{H}$, is actually observed and compares this number with its expected value. A typical example is the characterization of polyadenylation signals in human genes (Beaudoing *et al.*, 2000). This scheme underlies the software RSA-oligonucleotides of van Helden (van Helden *et al.*, 1998). An important related problem is the alignment of two (or more) sequences.

In both cases, one needs a probability model on the input texts. Different background probability models are discussed in (Hampson *et al.*, 2002). In this paper, we assume that the model is either Markovian, or Bernoulli. We derive new *tractable* formulae and provide efficient algorithms that actually compute them. We also discuss widely used approximations. We state their validity domains and occasionally extend them. We discuss the critical domains—or phase transition phenomena—that are observed (van Helden *et al.*, 1998; Buhler *et al.*, 2001; Robin *et al.*, 2001) and suggest a few solutions.

2. LARGE SEQUENCES

In this section, we concentrate on the statistical significance of word avoidance or overrepresentation in a large sequence. We assume that this sequence is randomly generated according to a Bernoulli or Markov model.

For a given word H, one denotes $P(H)$ its probability under the model. The probability of a set of words $\mathbb{H}$ is denoted by $P(\mathbb{H}) = \Sigma P(H)$ with H in $\mathbb{H}$. The expectation $E(\mathbb{H})$ and the variance $\mathrm{Var}(\mathbb{H})$ of a given set of words $\mathbb{H}$ have been extensively studied by various authors (Pevzner *et al.*, 1989; Bender *et al.*, 1993; Regnier *et al.*, 1997; Regnier, 2000; Reinert *et al.*, 2000) and closed formulae have been derived. This allows for an efficient computation of the Z-score:

$$Z = \frac{O(\mathbb{H}) - E(\mathbb{H})}{\mathrm{Var}(\mathbb{H})},$$

where $O(\mathbb{H})$ is the observed number of occurrences from $\mathbb{H}$ in the sequence under study. A highly positive (respectively negative) Z-score shows that a word is exceptional. Still, this does not provide valuable information on the probability that this exceptional event occurs, the so-called P-value. All results rely on the overlapping structure of the words:

Definition 2.1 (Guibas *et al.*, 1981). *Given two words H and F, the correlation set $A_{H,F}$ is the set of proper suffixes w of F such that F is a suffix of $H \cdot w$. The correlation polynomial is, in the Bernoulli case:*

$$A_{H,F}(z) = \sum_{w \in A_{H,F}} P(w) z^{|w|},$$

where $|w|$ is the length of w. When $H = F$, one denotes $A_H(z) = A_{H,F}(z)$.

Example: Let H be the *Chi* motif GCTGGTGG. The correlation set is $A_{H,H} = \{\text{CTGGTGG}\}$ and the correlation polynomial is $A_{H,H}(z) = 1 + P(\text{CTGGTGG})z^7$. Let $\mathbb{H}$ be the generalized *Chi* motif GNTGGTGG. Then $A_{\text{GN}_1\text{TGGTGG,GGTGGTGG}} =$ $= \{\text{TGGTGG,GTGGTGG}\}$, $A_{\text{GN}_1\text{TGGTGG,GN}_2\text{TGGTGG}} =$ $\{\text{N}_1\text{TGGTGG}\}$ if $\text{N}_2 = \{A, C, T\}$. The associated correlation polynomials are $1 + P(\text{TGGTGG})z^6 + P(\text{GTGGTGG})z^7$ and $1 + P(\text{GTGGTGG})z^7$.

Now, our recent combinatorial results on the distribution allow for proving (Denise *et al.*, 2001; Regnier *et al.*, 2003):

Theorem 2.1. *Let H be a given pattern. Let **a** be a real positive number such that $0 < a < 1$ and $a = P(H)$. Let z_a be the largest real positive root satisfying $0 < z_a < 1$ of*

$$D(z)^2 - \big(1+(a-1)z\big)D(z) - az(1-z)D'(z) = 0, \quad (1)$$

if $a > P(H)$, or the smallest real positive solution that satisfies $1 < z_a$, if $a < P(H)$, when

$$D(z) = (1 - z)A(z) + P(H)z^m. \quad (2)$$

Then:

$$\text{Prob}(N_H \geq na) \approx \frac{1}{\sqrt{n}\sigma_a} e^{-nI(a)+\delta_a}, \quad (3)$$

where

$$I(a) = a \log\left(\frac{D(z_a)}{D(z_a) + z_a - 1}\right) + \log z_a \quad (4)$$

and δ_a, σ_a *and* α *are rational functions of a and* z_a.

Extending the reasoning in (Regnier, 2001), we observe here that Theorem 2.1 steadily extends to some sets of words: e.g. when the correlation sets A_{HF} do not depend on H. This is the case for the generalized *Chi* motif.

Example:

1. The *Chi* motif is overrepresented in many bacterial genomes. The motif GCTGGTGG occurs 499 times in *E. coli*, where the Bernoulli distribution of the nucleic acids is $p_A = 0.2461913$, $p_C = 0.2542308$, $p_G = 0.2536579$, $p_T = 0.2459200$. The P-value computed by Theorem 2.1 is: $1.417e^{-227}$. A computation by induction by the software GDon (Nuel, 2001) yields a P-value: $1.403e^{-227}$.
2. The generalized *Chi* motif GNTGGTGG occurs 159 times in *H. influenzae*. The P-value is $.8597547482e^{-36}$ under the uniform model.

Approximation quality. It turns out (Denise *et al.*, 2001) that expression (3) is very closed to the exact expression of this probability provided in (Regnier *et al.*, 1997), computed by the software *Excep* (Klaerr-Blanchard *et al.*, 2000). Still, as the computation reduces to the numerical solution of a polynomial equation, this computation is much faster and numerically very stable. Moreover, the possible range for the length n of the text and the probability $P(H)$ is much larger. For example, the computation of (3) for the *Chi* motif in *E. coli*, where the size of the genome is $n = 4639221$, exceeds the computational limits of *Excep*. Applications with avoided words in restriction–modification systems for bacteria can be found in (Vandenbogaert and Makeev, 2002).

This can be compared to some common approximations for the p-values, e.g. the p-values computed for the normal law with mean $P(H)$ and variance $V(H)$ or for the (compound) Poisson distribution with parameter $P(H)$. In these cases, the rate function is well known (Dembo *et al.*, 1992), e.g.:

(i) $\frac{(a-P(H))^2}{2\mathrm{Var}(H)}$ for the normal law, (ii) $\frac{(a-P(H))^2}{2\mathrm{Var}(H)}$ for the Poisson law.

When $a–P(H)$ is not too large, a local development shows that:

$$z_a \approx 1 - \frac{a-P(H)}{2A_H(1)-1-A-1mp}.$$

This yields the following approximation for the rate function:

$$I(a) = a\log\frac{a}{p} + (p-a) - [(A_H(1)-1)]\frac{(a-p)^2}{p}. \quad (5)$$

This proves theoretically that the Poisson approximation (or the compound Poisson approximation) is very tight, especially when the motif is not self-overlapping (e.g. $A_H(1) - 1 = 0$). This fact was experimentally observed for some ranges of n and p in (Robin *et al.*, 2001; Nicodeme, 2001). Equation (5) also shows that the normal approximation is tight only for word occurrences na that satisfy $a - p << p$. This equation implies that a is close to p. Hence, such words are neither overrepresented nor underrepresented: they are not exceptional! For exceptional words (with a high Z-score), the ratio between the exact probability and the normal approximation is very small, e.g. the relative error is very high. Nevertheless, the difference, used for instance in (Robin *et al.*, 2001) is upper bounded by the maximal value. Hence, we point out the paradox: the smaller the difference, the worst the approximation... As a conclusion, we claim that the normal approximation is always poor.

Restriction enzymes. For a given genome, the palindromic sites associated to the restriction enzymes of the genome are avoided (Vandenbogaert, 2003). In Table 1, these avoided words are sorted by decreasing Z-score. Their Z-rank, in column 4, is significantly different from their P-rank (column 6) derived from a sort by P-values. Column 2 yields the number of occurrences in *E. coli.*

Table 1. Avoided palindromic hexanucleotides in *E. coli*

Hexamer	Occ	Z-score	Z-rank	P-val (M)	P-rank
GGCGCC	94	–41.9	1	.3385654443e-3	1
GCATGC	588	–39.9	2	.2514096486e-3	2
GCCGGC	293	–37.4	3	.2358169663e-3	3
CGGCCG	284	–35.1	4	.2050106769e-3	4
CACGTG	143	–32.3	5	.1849237452e-3	5
GGGCCC	68	–30.7	6	.1763897841e-3	6
TTGCAA	1024	–30.2	7	.1280869614e-3	7
ATGCAT	839	–28.0	8	.1095910354e-3	9
CCGCGG	657	–26.2	9	.980276091e-4	10
GCTAGC	157	–26.1	10	.1133125447e-3	8
CAATTG	986	–25.7	11	.904588694e-4	12
GAGCTC	152	–23.9	12	.9436368542e-4	11
GTGCAC	575	–23.1	13	.753139868e-4	14
TGGCCA	630	–23.0	14	.738769289e-4	15
GAATTC	645	–21.9	15	.664817246e-4	19
AAGCTT	556	–21.9	16	.671111153e-4	18
CTCGAG	177	–21.3	17	.7121997639e-4	17
CCTAGG	16	–21.2	18	.8539762880e-4	13

Hexamer	Occ	Z-score	Z-rank	P-val (M)	P-rank
ACATGT	477	−20.4	19	.5812779786e-4	20
TCTAGA	39	−19.8	20	.7189624736e-4	16
TTCGAA	800	−19.4	21	.498331920e-4	22
CCATGG	612	−19.4	22	.5086498391e-4	21
CGCGCG	2127	−118.7	23	.3641666775e-4	23

The first two binding sites that exchange their ranks are GCTAGC and ATGCAT for enzymes *NheI* and *Ava*III: (8, 10) changes to (9, 8). Similarly, CCTAGG (enzyme *Avr*II) goes from rank 18 up to rank 13, and TCTAGA (enzyme *XbaI*) goes from rank 20 up to rank 16. For similar Z-scores, the P-value, in column 5, may reverse such order when the maximal Z-order is different. This is the case, for example, the three binding sites TCTAGA, TTCGAA and CCATGG for enzymes *XbaI, Asu*III and *Nco*I. It is interesting to notice, that the last hexamer, CGCGCG, keeps its rank, with a sensible difference in P-value (a 0.13e-4 difference in the logarithms represents a 10^{-26} factor for the P-value), but is not a binding site.

3. SMALL SEQUENCES

In this context, one searches for a signal that occurs in a set of L small sequences more often than expected. Typically, these sequences are upstream regions of (possibly co-regulated) genes. The signal may be a single word H, such as the G-box CACGTG, or a consensus word, such as the cleavage site CCWGG for the restriction enzyme *Eco*RII. It can also be a structured motif (Marsan *et al.*, 2000) or a dyad (Eskin *et al.*, 2002): for example, the motif TCGGCGGCTAAAT N_{20}GATTCGGAAGTAAA, where N_{20} denotes a sequence of 20 unspecified nucleotides, in gene *YDR285W* in *S. cerevisiae*. One proceeds in two steps:

1. Compute for a given signal, say a word H, the probability p that it occurs at least once in the text;
2. Compute $P_L(k)$, the probability that k out of L sequences contain that motif at least once.

When the P-value $P_L(k)$ is very small, one concludes that the extracted signal is relevant.

Step (i): the combinatorial results on words (Guibas *et al.*, 1981; Regnier, 2000) allow for an exact computation of p. With an indexation by the size n of the sequence, one proves (Guibas *et al.*, 1981) that the probabilities p_n satisfy:

$$\sum_{n\geq 0}(1-p_n)z^n = \frac{P(H)z^n}{D(z)}.$$

This extends to the Markov model (Regnier *et al.*, 1997). A formal development of the right hand size yields $1 - p_n$ for any n. Practically, when n is small, this is easily computed by a symbolic computation system. But, the larger n is, the trickier the computation, and, even worse, numerical instability may occur. The software *Excep* (Klaerr-Blanchard *et al.*, 2000) relies on a careful implementation of the Fast Fourier Transform, and allows a computation for middle n, Still, it is stucked for large n by the numerical problems and the computational costs. The computation by induction in (Robin *et al.*, 1999), suffers, even more deeply, of the same drawbacks.

A very common approximation (Beaudoing *et al.*, 2000; Hart *et al.*, 2000; Buhler *et al.*, 2001; Vinogradov *et al.*, 2002) for p_n is:

$$P_n = 1 - (1 - P(H))^{n-m+1},$$

where n is the size of the sequence. It has been observed by many people (van Helden *et al.*, 1998) that this approximation is bad when H is a self-overlapping word. Incidentally, the exact results show that, anyhow, the Poisson approximation is rough for small n. When n is larger, we extend this formula for a set of possibly overlapping words, $\mathbb{H}$. The case of a unique self-overlapping word is addressed in (Vandenbogaert, 2003), and a formal proof, including the Markov case, can be found in (Clement *et al.*, 2003).

Theorem 3.1. *Let $\mathbb{H}$ be a set of patterns and $P(\mathbb{H})$ be the probability of the words in $\mathbb{H}$: the probability p_n to find at least one word from $\mathbb{H}$ in a sequence of size n, satisfies, in the Bernoulli case, when $nP(\mathbb{H})$ is small:*

$$P \sim 1 - e^{-nC(\mathbb{H})}, \tag{6}$$

where $C(H) = \sum_{H\in H} P(H) \sum_{F\in H} \left(2A_{H,F}(1) - 1\right)$. *For a single pattern H, a tighter approximation holds:*

$$C(H) + \frac{P(H)}{A_H(1)}.$$

Hint for the proof. Now classical combinatorial analytics (Flajolet *et al.*, 1995) provide a good asymptotic expression for p_n. For a single word, $p_n \; \rho_1 - e^{-n} \log \rho$, where ρ is the smallest real positive root of $D(z) = 0$. It is easily checked that. The approximation is very tight. For a set of words, one defines the matrix $A(z)$ of the autocorrelation polynomials, and one computes approximately $Trace(HA(1))^{-1}$, where H is a matrix where all rows are identical to the vector $(P(H_1), \ldots, P(H_i), \ldots)$.

Step (ii): our main observation here is that $P_L(k)$ is the tail distribution of a Bernoulli process. Hence, it is known (Waterman, 1995; Dembo *et al.*, 2002) that $\frac{\log P_L(k)}{n} \approx -I(a)$ with

$$I(a) = a \log \frac{a}{P(H)} + (1-a) \log \frac{1-a}{1-P(H)}. \tag{7}$$

This approximation is very tight. It is easier and faster to compute than the binomial formula for $P_L(k)$. Moreover, the numerical computation of these small numbers is very stable. It was recently implemented in *RSA-tools*.

4. CONDITIONAL EXPECTATION

The overrepresentation—or under representation—of a signal modifies the statistical properties of the sequence. It is valuable to eliminate artefacts of a strong signal in order to extract a weaker signal. It is also of interest to cluster similar motifs in a single degenerate signal.

Theorem 4.1 (Regnier *et al.*, 2003). *Let H and F be two patterns. Assume that k occurrences of H are found in the text. Then, the expected number of occurrences of F, knowing that H occurs k times is:*

$E(FN_H = k) \sim n\nu(a)$, *where*

$$\nu(a) = \frac{z_a\left[(1-z_a)A_{H,F}(z_a) + P(F)z_a^m\right]\cdot\left[(1-z_a)A_{H,F}(z_a) + P(H)z_a^m\right]}{D(z_a)\left(D(z_a) + z_a^{-1}\right)}$$

and $P(H)$ and $P(F)$ are their probabilities of occurrence, $A_{H,F}(z)$ and $A_{F,H}(z)$ are their correlation polynomials.

We briefly present below two applications. Similar problems are studied in (Blanchette *et al.*, 2001).

Polyadenylation. In (Beaudoing *et al.*, 2000), short (around 50 bp) upstream regions of EST of human genes are searched for polyadenylation signals. The largest Z-score is assigned to AAUAAA, which actually is the dominating signal. The computation of a Z-score using E(FAAUAAA) as the new expected value for each pattern *F* drastically reduces the Z-scores of artefacts AUAAAN and NAAUAA and assigns the 2nd rank in Z-scores to AUUAAA, that actually is the second (weak) signal (Denise *et al.*, 2001).

Arabidopsis thaliana. RSA-tools were "blindly" used on upstream regions of *Arabidopsis thaliana*, by M. Lescot. In one test, the two highest Z-scores were assigned to H = ACGTGG and F = CACGTG. The number of occurrences were $N_H = 32$ and $N_F = 52$. The conditional expectations are: $E(HF) = 24$ and $E(FH) = 20$. Our conclusion is that H is better explained by F than F is explained by H. This leads to chose F = CACGTG as the significant signal, although its Z-score is smaller. As a matter of fact, the biological knowledge confirms that CACGTG is the regulatory signal, the so-called G box.

5. CONCLUSION

It is an interesting challenge to derive the rate function for a set of consensus words, or degenerated words (IUPAC code). The special case of double strand counting is addressed in (Lescot *et al.*, 2003). One can also extend the approximation (5). Another interesting issue is the derivation of the conditional expectation when a set of words is overrepresented. One expects a simplification of the formulae for consensus words. Finally, it is worth extending (7) for sequences of different lengths. These formulae are implemented in the QuickScore library (http://algo.inria.fr/regnier/QuickScore/).

ACKNOWLEDGEMENTS

I am grateful to Magali Lescot (Marseille) for fruitful discussions and for providing data on *Arabidopsis thaliana*. This research was partially supported by French–Russian Liapunov Institute and INTAS grant No. 99-1476.

ARGO_VIEWER: A PACKAGE FOR RECOGNITION AND ANALYSIS OF REGULATORY ELEMENTS IN EUKARYOTIC GENES

O.V. VISHNEVSKY[1,2]*, E.A. ANANKO[1,2], E.V. IGNATIEVA[1,2], O.A. PODKOLODNAYA[1,2], I.L. STEPANENKO[1,2]

[1]*Institute of Cytology & Genetics, Siberian Branch of the Russian Academy of Sciences, prosp. Lavrentieva 10, Novosibirsk, 630090 Russia;* [2]*Novosibirsk State University, ul. Pirogova 2, Novosibirsk, 630090 Russia*

* *Corresponding author: e-mail: oleg@bionet.nsc.ru*

Abstract The task of reliable recognition of promoters in eukaryotic genomes has not been yet solved completely. This is largely due to poor understanding of the features of the structure–function organization of eukaryotic promoters essential for their function and recognition. Five groups of promoters of tissue-specific genes were analyzed and vocabularies of region-specific oligonucleotide motifs were constructed. We suggest a procedure to recognize tissue-specific gene promoters basing on the presence and distribution of certain oligonucleotide motifs.

Key words: structure–function organization of eukaryotic regulatory regions, recognition of eukaryotic promoters, oligonucleotide vocabularies

1. INTRODUCTION

Promoters of eukaryotic genes transcribed by RNA polymerase II consist of hierarchically arranged domains. Assemblage of the basal transcription complex and the tissue- and stage-specific features of eukaryotic gene transcription depend on the context and structural organization of the promoter core and the presence of transcription factor binding sites (TFBS) in the 5' regulatory region of the gene (Ignatieva *et al.*, 1997). Most approaches to promoter recognition are based on detection of potential

TFBSs by using weight matrices, consensuses, and other methods taking into account both their densities (Prestrige, 1995) and their distribution over promoters (Kondrakhin *et al.*, 1995). The efficiency of these approaches depends highly on the completeness of TFBS databases, reliable recognition of putative TFBSs, etc. Therefore, approaches to promoter recognition not involving data on TFBSs have been developed (Hutchinson, 1996; Solovyev and Salamov, 1997; Knudsen, 1999; Scherf *et al.*, 2000). They are based on analysis of the oligonucleotide composition of promoters and consideration of oligonucleotide distribution over promoters. It has been shown that the consideration of oligonucleotide distribution over promoters increases the accuracy of promoter recognition (Zhang, 1998). Alternative approaches are based on information on promoter location in orthologous genes (Solovyev and Shahmuradov, 2003) and on combination of independent methods for promoter recognition (Liu and States, 2002). However, in spite of the diversity of the approaches, the task of reliable recognition of promoters in eukaryotic genomes has not been yet solved completely (Fickett and Hatzigeorgiou, 1997). Recent data point to a certain similarity in promoter organization in genes with similar expression patterns (e.g. promoters of genes expressing in particular tissues). This manifests itself as the presence of similar TFBS sets in promoters of such genes (Ananko, 1997; Ignatieva *et al.*, 1997; Podkolodnaya and Stepanenko, 1997). Elaboration of recognition methods for certain promoter groups responsible for transcription of genes with similar expression patterns and, therefore, showing similar features in their contextual organization appears to be promising for increasing the accuracy of recognition of these regulatory regions.

2. METHODS AND ALGORITHMS

2.1 Promoter sequences analyzed

We investigated five samples of tissue-specific promoters from the TRRD database of regulatory sequences (http://wwwmgs.bionet.nsc.ru/mgs/dbases/trrd4) in the region from –300 to +100 bp relative to the transcription start (Table 1). Five sets of 1000 400-bp-long random sequences with mononucleotide compositions typical of the corresponding promoter groups were used as negative training samples.

The total samples of promoter sequences were divided into training and control subsamples. The control samples for each group included 20% of randomly chosen tissue-specific promoters. The training samples included the remaining 80% of promoters. Type I errors were estimated in the control samples, and type II errors, in extended randomly generated sequences.

Table 1. Analysis and recognition of tissue-specific promoters in five gene samples

Promoters of tissue-specific genes	Number of sequences	Number of motifs found	Recognition errors	
			False negative	False positive
Heat shock-induced genes	34	45	0.09	$\sim 10^{-4}$
Interferon-inducible genes	41	131	0.07	$<10^{-5}$
Erythroid-specific genes	26	78	0.08	$\sim 10^{-5}$
Genes of lipid	50	281	0.04	$<10^{-5}$
Genes of endocrine system	78	814	0.05	$<10^{-5}$

2.2 Recognition of degenerate oligonucleotide motifs

Promoters were analyzed with the ARGO software (http://wwwmgs.bionet.nsc.ru/mgs/programs/argo; Vishnevsky and Vityaev, 2001). It allows oligonucleotide motifs with the following properties to be recognized: (1) degeneracy, i.e. use of the extended IUPAC code (A, T, G, C, R = G/A, Y = T/C, M = A/C, K = G/T, W = A/T, S = G/C, B = T/G/C, V = A/G/C, H = A/T/C, D = A/T/G, N = A/T/G/C); (2) region-specificity, i.e. preferred occurrence in a certain region of a promoter; (3) quasi-invariance, i.e. occurrence only in certain promoter subgroups; and (4) contrast, i.e. much more frequent occurrence in promoters than in random sequences.

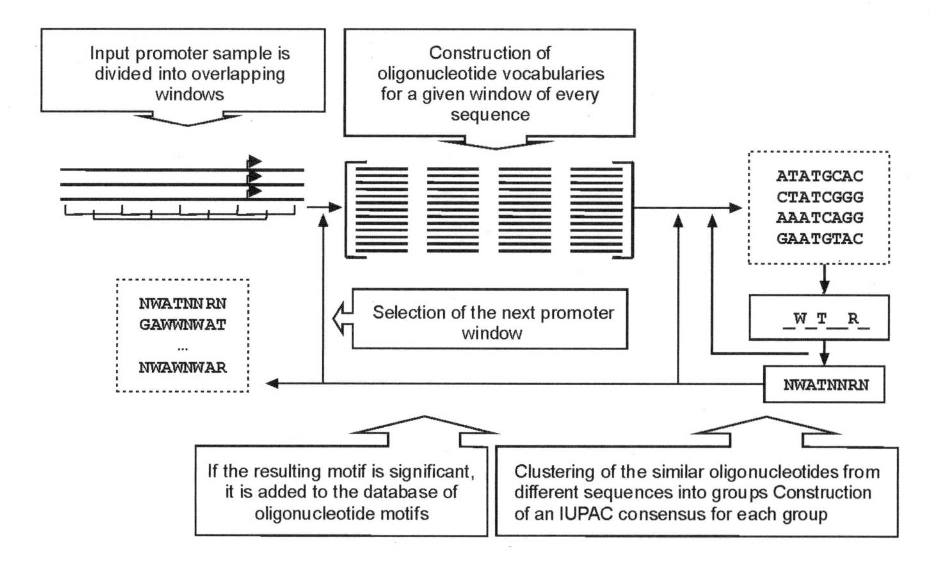

Figure 1. Layout of the algorithm for recognition of degenerate oligonucleotide motifs in a promoter sample.

The region-specific significant oligonucleotide motifs in samples of promoter sequences are sought according to Figure 1.

1. Source data: (i) promoter sequences of length L_{prom}, phased with reference to the transcription start (positive sample) and (ii) random sequences with the same nucleotide frequencies as in promoters (negative sample).
2. Specification of the starting position of the scanning window. The window length is L nucleotides. At the start, its first nucleotide matches the first nucleotide at the 5'-end of the promoter.
3. Construction of initial oligonucleotide vocabularies. An initial vocabulary of perfect oligonucleotides of length l is constructed for each sequence within the window considered.
4. Clustering of oligonucleotides. Similar oligonucleotides are selected from vocabularies corresponding to different sequences and clustered. A group comprises oligonucleotides differing in not more than R positions. ($R < r_0$, where r_0 is a similarity threshold).
5. Construction of consensuses for groups of similar oligonucleotides. Oligonucleotides of each group obtained at step 4 are phased. Consensuses are constructed for them in the extended IUPAC code by an iteration method. To find significant letters assigned to consensus positions, the probability $W(k, K, x_i)$ of accidental occurrence of each of the 15 letters x_i at least in k of K oligonucleotides of the group under consideration $W(k, K, x_i)$ is evaluated by the binomial test:

$$W(n, N, x_i) = \sum_{i=k}^{K} C_K^i P^i(x_i)(1 - P(x_i))^{K-i}. \qquad (1)$$

Here, $P(x_i)$ is the frequency of occurrence of the letter x_i in the 15 single letter-based code in the promoter sequences calculated as:

$$\begin{aligned}
&P(x_1) = P(A);\ P(x_2) = P(T);\ P(x_3) = P(G);\ P(x_4) = P(C);\ P(x_5) = P(A) + (G);\\
&P(x_6) = P(T) + P(C);\ P(x_7) = P(A) + P(C);\ P(x_8) = P(T) + P(G);\\
&P(x_9) = P(A) + P(T);\ P(x_{10}) = P(G) + P(C);\ P(x_{11}) = P(T) + P(G) + P(C);\\
&P(x_{12}) = P(A) + P(G) + P(C);\ P(x_{13}) = P(A) + P(T) + P(C);\\
&P(x_{14}) = P(A) + P(T) + P(G);\ P(x_{15}) = P(A) + P(T) + P(G) + P(C).
\end{aligned} \qquad (2)$$

By exhaustive search, we find the letter and position with the minimal $W(k, K) < W_{crit}$ (threshold probability value). This most reliable letter is placed to the corresponding position of the consensus. All the oligonucleotides not fitting the consensus under construction are removed from the group. Then, letters and positions with minimal $W(k, K, x_i)$ are sought for the remaining $l - 1$ positions of the consensus. The iteration goes on until all the reliable ($W(k, K, x_i) < W_{crit}$) letters in the 15 single letter–based code are found at positions of the consensus. After discovery of all the reliable letters, the remaining positions are filled with Ns.

6. Evaluation of the significance of consensus constructed. An oligonucleotide motif obtained at step 5 is considered significant, if it meets the following criteria:

$$\left.\begin{array}{ll} \text{a)} & F > f_0 \\ \text{b)} & P(n,N) < p_0 \\ \text{c)} & Q < q_0 \end{array}\right\} . \qquad (3)$$

Here, F, the proportion of promoters containing the motif; f_0, threshold level of the motif occurrence in the promoter sample; $P(n, N)$, probability of accidental occurrence of the promoter in a certain window in $\geq n$ sequences of N; p_0, threshold probability level (see the estimation method below); Q, proportion of sequences from the negative sample containing the motif; and q_0, threshold level of the motif occurrence in the negative sample.

The probability $P(n, N)$ is calculated as follows. Let us consider the oligonucleotide motif $M = m_1, m_2, ..., m_l$ of the length l in the extended 15 single letter-based code. The probability of occurrence of the motif at a certain position of a sequence of length L is evaluated as:

$$P(M) = \prod_{i=1}^{l} P_i ,$$

where P_i is the frequency of the letter m_i calculated from the mononucleotide composition of promoters (equations (2)).

The binomial probability of the occurrence of the motif M in $\geq n$ sequences of N, $P(n, N)$, equals

$$P(n,N) = \sum_{i=n}^{N} C_N^i P^i (1-P)^{N-i} \text{, where } P = 1 - e^{-(L-l+1)P(M)} .$$

The calculated probability $P(n, N)$ is used for estimating the significance of the motif according to equation (3b).

7. Construction of current oligonucleotide vocabularies. Oligonucleotides present in the newly found motif are removed from oligonucleotide vocabularies of all the sequences. The vocabularies constructed in this

way are clustered, and a new motif is constructed according to step 4. The clustering and construction of new motifs are repeated while new motifs meeting conditions (3) can be constructed on the basis of constantly reducing current glossaries.

8. Change of the position of the scanning window. The scanning window is shifted by t nucleotides. Steps 3–8 are repeated in the new window position. Hence, all the positions of the scanning windows of length L with a step t are considered along a promoter of the length L_{prom} ($t < L < L_{prom}$).

3. RESULTS AND DISCUSSION

3.1 Search for signals

Degenerate oligonucleotide motifs in tissue-specific promoters were sought four in 14 overlapping windows of a length of 50 bp with a step of 25 bp. The motifs of length $l = 8$ meeting the following parameters of conditions (3) were regarded as significant: $P(n, N) < 10^{-8}$; $f_0 = 20\%$; and $q_0 = 10\%$.

Examples of oligonucleotide motifs found in the sample of erythroid-specific gene promoters are shown in Table 2.

Table 2. Features of degenerate oligonucleotide motifs in promoters of erythroid-specific genes

Motif	Promoter region where the motif was found	Occurrence in promoters	Occurrence in random sequences	Probability of accidental occurrence of the motif in promoters
RRCCAATN	–100 : –50	0.38	0.009	$10^{-14.4}$
YTTATCWN	–75 : –25	0.27	0.007	$10^{-11.9}$
WTAWAWRN	–50 : +1	0.46	0.017	$10^{-15.3}$
CWSHCANW	–25 : +25	0.54	0.06	$10^{-10.5}$
DVNBCWGG	+1 : +50	0.73	0.14	$10^{-9.5}$
YHRSTBCW	+1 : +50	0.54	0.09	$10^{-8.3}$
YSMWGSTG	+25 : +75	0.50	0.02	$10^{-14.4}$
TRNYNCWG	+25 : +75	0.58	0.09	$10^{-10.5}$
CTKCYSWN	+50 : +100	0.54	0.04	$10^{-12.5}$
SWDSNCTG	+50 : +100	0.46	0.03	$10^{-10.5}$

As an example, let us consider the first oligonucleotide listed in Table 2: RRCCAATN = (G/A)(G/A)(C)(C)(A)(A)(T)(A/T/G/C), found in the region [–100: –50] with reference to the transcription start. This motif was found in 10 promoters of 26 (38%), exceeding the threshold (20%) approximately twofold. The probability of accidental occurrence of this motif in 10 or more of 26 promoters equals $10^{-14.4}$. In the negative sample, this motif occurred in the region under consideration only in 9 random sequences of 1000 (2%). Hence, this motif is significant according to criteria (3).

To understand the nature of the oligonucleotide motifs found, they were compared with the known transcription factor binding sites from the Transfac database (Heinemeyer, 1998). Many oligonucleotide motifs proved to be similar to known TFBSs (Table 3).

Table 3. Examples of transcription factor binding sites with context similarity to region-specific motifs

Promoter region	Transcription factor binding sites
–100 : –50	CP1, NFE-6, GATA-1, CDP
–75 : –25	CTF, NF-1, AP-1, Sp1, GATA-1, NFE-6, CP1, USF
–50 : +1	TBP, TFIID, GATA-1
–25 : +20	GATA-1, CP1, CTF, NF-1, USF

3.2 Promoter recognition

Promoters of the types investigated can be recognized in extended genomic sequences using the set R of region-specific motifs found by the method described above and the ARGO_VIEWER software (Vishnevsky and Vityaev, 2001; http://www.mgs.bionet.nsc.ru/mgs/programs/argo/argo_viewer.html). The recognition is performed in a 400-bp-long scanning window moving with a specified step along the sequence under investigation. At each position of the window, it is divided into overlapping subwindows, similarly to promoter division. The corresponding region-specific motifs from the set R are sought for in these subwindows (Figure 2). Then the score F (similarity between the distribution of the motifs found in the window and the distributions of the motifs in promoter groups considered) is calculated, and the promoter with the greatest score is found. If this score exceeds the threshold F_b value, we assume that a promoter of the group considered occurs in the examined window. The score of similarity between the motif distributions within the examined window in the sequence under study and in promoter groups considered was calculated as follows.

Let us consider the examined sequence in a scanning window of $L = 400$ bp, equal to the length of promoters in the training sample. Let us specify S regions whose positions match the positions of regions in promoters of the training sample. A set of region-specific motifs $\{r_{i*}\}$ ($r_{i*} \subset R$, $i = 1, ..., S$) is found for each region of the sequence.

Then, let us consider each promoter of the training sample. Compare the motifs r_{ij} ($r_{ij} \subset R$) present in the ith region of the jth ($j = 1, ..., N$) promoter with the motifs r_{i*} present in the ith region of the sequence examined. We choose the motifs present in both r_{i*} and r_{ij}. Let us construct the set $r_{ij*} = r_{ij} \cap r_{i*}$. Thus, the distinctive feature of each element of the set r_{ij*} is that it is present in both the ith region of the examined sequence and in the ith region of the jth promoter.

Let us calculate the measure of similarity between the sequence under examination and the jth promoter on the base of the motif set r_{ij*}. For this purpose:

1. For each motif of the set r_{ij*} ($i = 1, \ldots, S$), determine its location in the sequence under examination. This would yield a set of overlapping motifs. A typical pattern is shown in Figure 2.
2. For the kth ($k = 1, \ldots, L$) position of the sequence, determine the set of allowed nucleotides (Figure 2). A nucleotide is considered allowed if its presence in the kth position does not contradict to the letters present in all the motifs involving the kth position. For example, for position (*) in Figure 2, the nucleotide set {A,T,G,C} is allowed. For position (^) in Figure 2, the set of allowed nucleotides assumes the form {C}, because this position falls into three motifs containing the following set of letters: C, Y = (T/C), and S = (G/C). Apparently, the nucleotide C does not contradict to any letter of the set.
3. The probability of accidental presence of the set of the allowed nucleotides in a specified position p_k is calculated with regard to the mononucleotide composition of the promoter regions according to equations (2). Obviously, if the position falls to no motif, $p_k = 1$.
4. The similarity between the jth promoter and the sequence under examination is estimated by means of the value $P_j = -\sum_{k=1}^{L} \log p_k / L$, where L is the length of the sequence. The greater is P_j, the lower is the probability of accidental occurrence of the motif set characteristic of the jth promoter in the sequence.

Of all the promoters analyzed in the training sample of tissue-specific promoters, we choose the promoter j_+ with the greatest value of P_j: $P_{j+} = \max_j \{P_j\}$. The same procedure, including steps 1–4, is performed for all sequences of the negative sample. Among them, the sequence with the greatest $P_{r-} = \max_r \{P_r\}$, where r is the number of a sequence from the negative sample, is found. Then $P = P_{j+} - P_{r-}$ is calculated.

The value of the recognition function F for a promoter is determined as:

$$F = \begin{cases} P \text{ for } P > 0 \\ 0 \text{ for } P \leq 0. \end{cases}$$

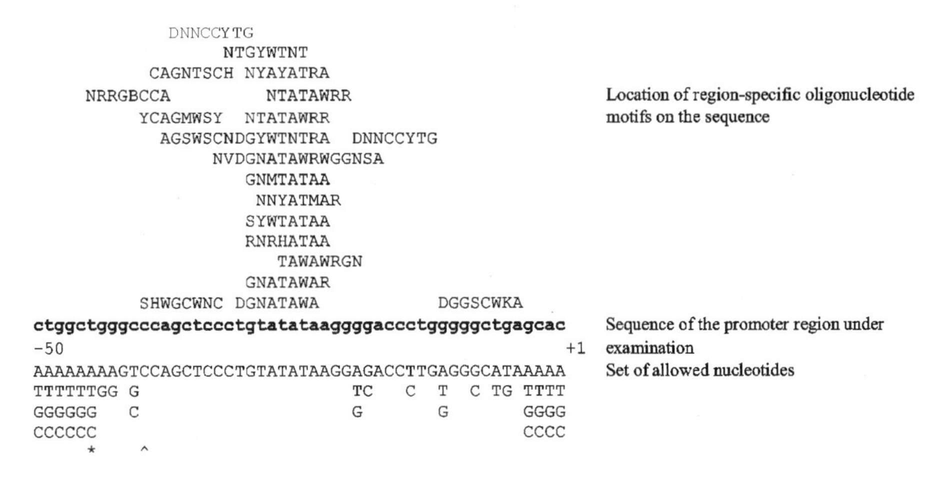

Figure 2. Example of determination of the set of allowed nucleotides for each position of the region [–50; +1] of an erythroid-specific promoter.

The value of the recognition function F is calculated in this way for each position of the scanning window of length L in an extended sequence. If $F > F_b$ for a certain position of the window, where F_b is the threshold value determined using control samples, the sequence entering the window can be assigned to corresponding promoter groups. Otherwise, it is regarded as a non-promoter.

Table 1, showing results of the analysis and recognition of several groups of tissue-specific promoters performed by the method described, demonstrates that the level of overprediction errors for the four of five promoter groups is 1 per 100 000 bp, with the level of type I errors amounting to 4–8%. Generally, the number of motifs recognized increases with the size of the training sample. An explanation for this fact is recognition of weaker functional signals with larger samples, whereas with smaller samples, their significance fails to exceed the threshold value. This rule is violated only in the case of heat shock-induced genes. This can be explained by a lower similarity of these promoters or by the fact that the site of a heat-shock element responsible for the specific operation of promoters of this type can contain several (one to nine) identical subunits occurring in both the direct and complementary strands. Such signals are difficult to describe in terms of degenerate motifs of a fixed length, which increases the frequency of types I and II errors.

ACKNOWLEDGMENTS

We are grateful to Prof. N.A. Kolchanov and Dr. E.E. Vityaev for fruitful discussions. The study was supported in part by the Russian Foundation for Basic Research (grants Nos. 01-07-90376-в, 02-07-90355, 03-07-96833-p2003, 03-07-96833, 03-07-90181-в, 02-04-48802-a, 03-04-48829, 03-07-06078-мас, 03-04-48555-a, 03-07-06082-мас); grant PCB RAS (No. 10.4); the Siberian Branch of the Russian Academy of Sciences (integration project No. 119); Russian Ministry of Industry, Science, and Technologies (grants Nos. 43.073.1.1.1501 and 43.106.11.0011, subcontract No. 28/2003); NATO (grants Nos. LST.CLG.979816 and PDD(CP)–(LST.CLG 979815)); the U.S. Civilian Research & Development Foundation for the Independent States of the Former Soviet Union (CRDF) No. NO-008-X1.

TRANSCRIPTION REGULATORY REGIONS DATABASE (TRRD): DESCRIPTION OF TRANSCRIPTION REGULATION AND THE MAIN CAPABILITIES OF THE DATABASE

E.V. IGNATIEVA[1,2], E.A. ANANKO[1,2], O.A. PODKOLODNAYA[1,2], I.L. STEPANENKO[1,2], T.M. KHLEBODAROVA[1,2], T.I. MERKULOVA[1,2], M.A. POZDNYAKOV[1,2], A.L. PROSCURA[1,2], D.A. GRIGOROVICH[1,2], N.L. PODKOLODNY[1,2], A.N. NAUMOCHKIN[1], A.G. ROMASHCHENKO[1,2], N.A. KOLCHANOV[1,2]*

[1]Institute of Cytology & Genetics, Siberian Branch of the Russian Academy of Sciences, prosp. Lavrentieva 10, Novosibirsk, 630090 Russia, e-mail: kol@bionet.nsc.ru; [2]Novosibirsk State University, ul. Pirogova 2, Novosibirsk, 630090 Russia

** Corresponding author*

Abstract The main goal of TRRD (http://www.bionet.nsc.ru/trrd/) construction is a complete and precise description of the structure–function organization of the regions of eukaryotic genes involved in transcription regulation. The current release of TRRD contains information on 2186 genes, 3254 regulatory units, 9480 transcription factor binding sites, 14 locus control regions, and 13 173 expression patterns. The data inputted into TRRD have been extracted from 7301 scientific papers. They are stored in seven SRS tables: TRRDGENES, TRRDUNITS, TRRDEXP, TRRDSITES, TRRDFACTORS, TRRDLCR, and TRRDBIB. The main tool for search of and navigation through TRRD is the Sequence Retrieval System (SRS). A hierarchical structure of dictionaries and thesauri has been developed to provide additional options for data retrieval.

Key words: database, transcription regulation

1. INTRODUCTION

The regulatory regions of eukaryotic genes are intricately organized. They are located mainly in noncoding regions of the genes. The core promoter, a short region (~ 50 bp), plays a key role in transcription initiation by RNA polymerase II. It is located immediately before the transcription start. The core promoter is the site of assemblage of the preinitiation complex, which includes, in addition to RNA polymerase II, common transcription factors: TFIIA, TFIIB, TFIID, TFIIE, TFIIF, and TFIIH (Reese, 2003). The transcription of a gene is controlled by regulatory units (promoter region, enhancers, and silencers) (Merika and Thanos, 2001). Each regulatory unit contains binding sites for transcription factors (TFBSs) of various types. The transcription factors affect the rate of transcription by interaction with proteins of the basal transcription complex, either direct or mediated by other proteins (coactivators and corepressors) (Featherstone, 2002). Coordinated expression of several genes within a single locus is determined by locus control regions (LCRs) (Festenstein and Kioussis, 2000). An LCR may contain enhancers, positive, or negative regulatory regions, and TFBSs (Podkolodnaia and Stepanenko, 1997). The rate of gene transcription also depends, to a marked extent, on the density of DNA nucleosome packing (Luger, 2003), degree of acetylation of histone proteins (Yang and Seto, 2003), and degree of DNA methylation (Finnegan, 2000). The rapid accumulation of experimental data on the transcriptional level of gene expression regulation generates a need for development of databases for their arrangement, storage, and utilization. There are several databases on the regulation of eukaryotic gene transcription. The Eukaryotic Promoter Database (EPD) includes sequences of eukaryotic promoters of genes transcribed by POL II, for which the transcription start has been experimentally determined (Praz *et al.*, 2002). The Mammalian Promoter Database (MPromDb; http://bioinformatics.med.ohio-state.edu/MPromDb/) includes data on human, murine, and rat promoter sequences and on location of TFBSs. TRANSFAC database family contain data on both prokaryotic and eukaryotic transcription factors and on the nucleotide sequences of their binding sites (Wingender *et al.*, 2001). COMPEL stores data on composite regulatory elements: pairs of closely spaced binding sites of various transcription factors (Kel-Margoulis *et al.*, 2002). Each of the mentioned databases includes data on a special aspect of the problem without providing a general view of mechanisms governing gene transcription. The Transcription Regulatory Regions Database (TRRD) is a unique informational resource designed for combined description of regulation of eukaryotic gene transcription mediated by RNA POL II. TRRD contains only experimentally verified data, which have been inputted by experts in biology through annotating published scientific papers.

2. TRRD FORMAT

The format of TRRD (Kolchanov *et al.*, 1999; 2000; 2002) provides extensive potentials for describing the structure–function organization of regulatory regions of eukaryotic genes. TRRD accumulates both structural and functional data on the following hierarchical levels of transcription regulation: (i) transcription factor binding sites (TFBSs); (ii) regulatory units (promoters, enhancers, and silencers), which may occur in various regions of the gene (5'-flanking or 3'-flanking regions, exons, or introns); and (iii) LCRs.

An entry of TRRD represents a gene. Information on transcription regulation is stored in seven linked tables: TRRDSITES, TRRDUNITS, TRRDLCR, TRRDFACTORS, TRRDEXP, TRRDGENES, and TRRDBIB.

TRRDSITES includes data on the first-level regulatory elements, TFBSs.

TRRDUNITS describes the next organization level: structural units (promoters, enhancers, and silencers), which include functionally related site sets. Regulatory units are located in various gene regions: 5'-flanking and 3'-flanking regions, exons, or introns.

TRRDLCR describes the structure–function features of the next regulatory level, locus control regions (LCRs):

1. TRRDFACTORS includes data on transcription factors interacting with the sites.
2. TRRDEXP stores the data on qualitative features of gene expression depending on developmental stage, cell cycle stage, cell type and differentiation level, external actions, etc., which are accumulated as expression patterns.
3. TRRDGENES stores general information on individual genes and a hierarchical representation of the regulatory elements of all levels.
4. TRRDBIB contains references to the papers annotated.
5. The types of experiments in which the data were obtained are entered into special table fields as numerical codes.

The seven SRS tables of TRRD are interlinked and integrated with other GeneExpress-2 informational and program modules (Kolchanov *et al.*, 1999) with the use of SRS v. 6.

3. EXAMPLES OF DESCRIPTION OF THE FEATURES OF GENE TRANSCRIPTION REGULATION IN TRRD

On the average now, there are two regulatory units and five annotated sites for each gene stored in TRRD. However, many genes have much more intricate regulatory regions. Consider the structure of the regulatory regions of human apoB gene (TRRD accession No. A00149), coding for apolipoprotein B. According to TRRDEXP, this gene is expressed in the liver, intestine, placenta, and macrophages. TRRDGENES contains data on eight regulatory units controlling apoB expression (Figure 1). Five regulatory units are located in the 5'-flanking region: promoter [–128; –1], two negative regulatory elements [–3678; –1802] and [–261; –129], a regulatory region [–898; –262], and a 300 bp long intestine-specific enhancer in the remote 5'-region of the gene [–57 000 bp]. Three regulatory units are located downstream of the transcription start: a regulatory unit in the first exon [+1; +128], liver-specific enhancer in the first intron [+346; +521], and liver- and intestine-specific enhancer in the second intron [+621; +1064]. The regulatory units of the apoB gene contain 21 TFBSs. The site positions in this gene are counted from two reference points: the transcription start of the gene (STexp) and the start of the remote (–57 000 bp) enhancer (SSexp). The TRRD format includes the field "DNABankLink", which contains links to entries from EMBL or GenBank, storing the nucleotide sequences of the regulatory units of the gene. This field also indicates the positions of the reference points in the EMBL or GenBank sequences used for describing the positions of regulatory units and related binding sites. For apoB, this field contains the positions of STexp and SSexp (Figure 1).

It is known that the bulk of apolipoprotein B is produced in the liver and intestine. The high tissue-specific level of its expression there is ensured by the liver- and intestine-specific enhancers. The enhancers contain binding sites for the proteins C/EBP, HNF1, HNF3, and HNF4, whose levels are very high in the liver and intestine. This ensures a high tissue-specific expression of apoB.

TRRD accumulates data on the functional features of LCRs, regulatory units, and TFBSs. Consider the structure–function organization of the locus control region of the human growth hormone cluster (TRRDLCR accession No. C0008). The cluster (Figure 2) comprises five genes: growth hormone (GH-1), chorionic somatomammotropins (CS-5 and CS-1), placental lactogen A (CS-1), growth hormone variant I (GH-V), and placental lactogen B (CS-2). The LCR comprises five DNase I hypersensitive sites (I to V), located from –32 000 to –15 000 bp with reference to the transcription start of the first gene of the cluster, GH-1. This LCR controls the tissue-specific activation of the genes of the cluster. In pituitary cells LCR is marked by the hypersensitivity sites I, II, III, and V. In this case, only GH-1 is intensely expressed (Figure 2*a*). In placenta cells, LCR is marked by hypersensitivity sites III, IV, and V. There, CS-1 and CS-2 are intensely expressed (Figure 2*b*). The expression of two other genes, CS-5 and GH-V, is very low. The GH-1 gene is entirely repressed.

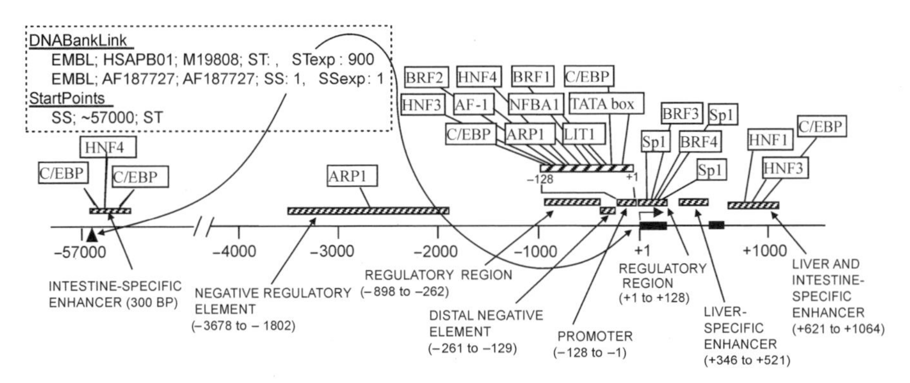

Figure 1. Organization of the regulatory regions controlling transcription of the gene for human apolipoprotein B. The transcription start (ST) is indicated with curved arrows; start of remote (–57 000 bp) enhancer (SS), with a black triangle; black bars, exons; and diagonally hatched bars, regulatory units. The box bounded by the dotted line contains some information fields from TRRDGENES records corresponding to apoB gene.

TRRDUNITS and TRRDSITES have fields that contain data on tissue specificity and inducibility of regulatory units and on the effect of transcription factors on expression (increase or decrease). Expression patterns (TRRDEXP), demonstrating features of gene expression (tissue specificity, inducibility, and expression increase/decrease at certain developmental stages) include links to regulatory units and TFBSs responsible for a given expression feature. Consider the description of transcription regulation of the rat hydratase–dehydrogenase (*HD*) gene (TRRD accession No. A00254) under the effects of various inducers.

The rat *HD* gene has an enhancer [–3178; –2281, TRRDUNITS accession No. P00566], which contains binding sites for PPAR/RXR, CAR/RXR, RevErb, and COUP-TF (Figure 3*a*). Peroxisome proliferators (clofibrate, ciprofibrate, and WY-14.643) activate the heterodimeric transcription factor PPAR/RXR, which interacts with corresponding binding site and activates *HD* transcription. This case is described as pattern A00254.002 (Figure 3*b*), which contains a link to the PPRE binding site (TRRDSITES accession No. S968). Another substance, 5 alpha-androstan-3 alpha-ol, activates the transcription factor CAR (constitutive androstane receptor), which, as part of the heterodimer CAR/RXR, interacts with the corresponding binding site (S9266) from the same enhancer. The gene expression pattern A00254.004, describing this case, is shown in Figure 3*c*.

In this example, the binding sites for PPAR/RXR and CAR/RXR overlap, and this allows for a quick control of the expression by displacement of the less active transcription factor from the complex with DNA.

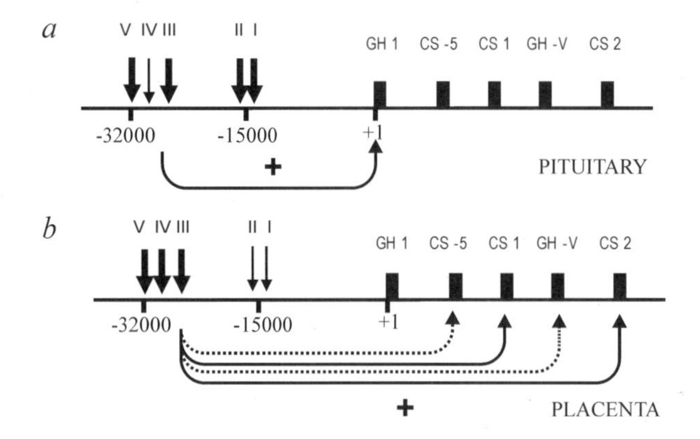

Figure 2. The gene cluster of human growth hormone regulated by locus control region: (*a*) LCR functioning in pituitary; (*b*) LCR functioning in placenta. Arrows numbered from I to V indicate hypersensitive sites marking the LCR; solid and dotted curved arrows, transcription activation.

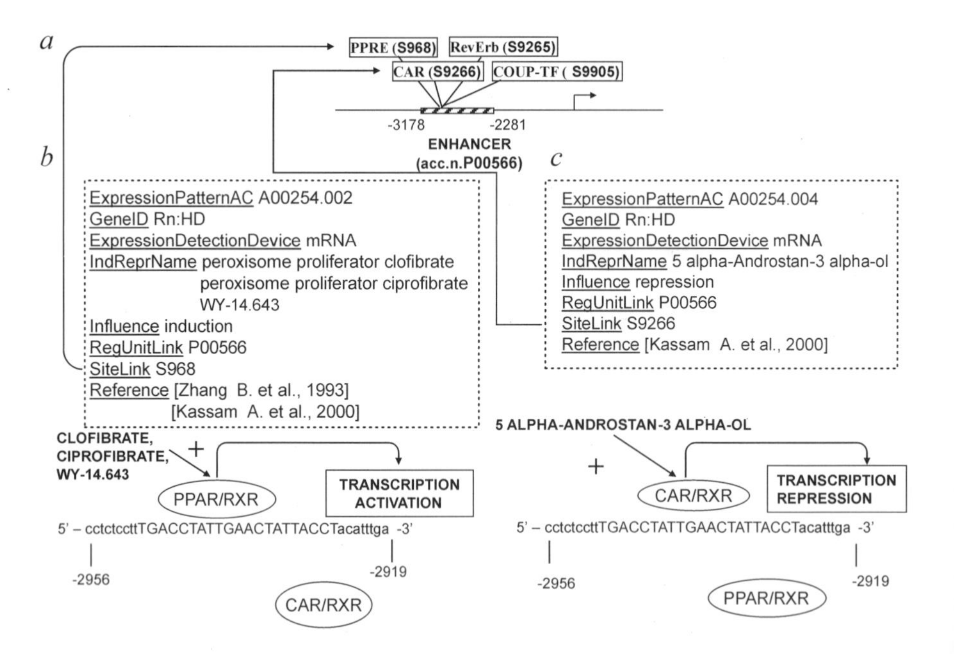

Figure 3. Regulation of rat *HD* gene by various inducers: (*a*) structure of the regulatory regions of the gene; (*b*) activation by peroxisome proliferators; and (*c*) repression by 5 alpha-androstan-3 alpha-ol. Expression patterns of *HD* gene from TRRDEXP are shown in boxes bounded with dotted lines.

4. THE BODY OF INFORMATION STORED IN TRRD

TRRD is the largest informational resource comprising data on structural and functional organization of transcription regulatory regions of eukaryotic genes. New information is inputted into TRRD monthly. The numbers of entries in the TRRD release 4.2.5 (Kolchanov *et al.*, 2000) and in the current release are shown in Table 1.

Thematic sections designed within TRRD include genes grouped according to their function: Heat Shock-Induced Genes (HS-TRRD), Interferon-Inducible Genes (IIG-TRRD), Erythroid-Specific Regulated Genes (ESRG-TRRD), Genes of Lipid Metabolism (LM-TRRD), Endocrine System Transcription Regulatory Regions Database (ES-TRRD), Glucocorticoid-Regulated Genes (GR-TRRD), Plant Genes (PLANT-TRRD), Cell Cycle-Dependent Genes (CYCLE-TRRD), Redox-Sensitive Genes (ROS-TRRD), Genes Expressed in Endocrine Pancreas (EP-TRRD), Macrophage-Expressed Genes MG-TRRD, Genes Controlling Blood Coagulation and Fibrinolysis (BCF-TRRD), Apoptosis Genes (Apoptosis-TRRD), Genes Controlling the Circadian Rhythm and Genes with Circadian Expression (CLOCK-TRRD), and Genes Encoding Proteins Involved in *Fe* Metabolism (FM-TRRD).

Table 1. Information stored in TRRD

Table name	Number of entries in release 4.2.5*	Number of entries in the current release **	Number of indexed fields in the current release
TRRDGENES	760	2186	24
TRRDUNITS	–	3254	11
TRRDEXP	3403	13173	17
TRRDSITES	3604	9480	16
TRRDFACTORS	2862	8218	14
TRRDLCR	–	14	40
TRRDBIB	2537	7301	9

* Information as of January 01, 2000; ** information as of August 22, 2003.

5. VISUALIZATION TOOL

Information on the structural organization of regulatory regions is represented in TRRD-Viewer as maps of regulatory regions of a gene. An example of a map is shown in Figure 4*a*. The Viewer interface includes three windows: (1) navigation window, (2) text and designations, and (3) maps of regulatory regions. The left and right borders of the region to be shown in the lower window are specified in the navigation window with movable brackets. The lower window of Figure 4*a* shows the map of the regulatory regions of human CYP7 gene, including an enhancer in the 5'-region and promoter. These regions contain four and two TFBSs, respectively (short stretches below the scale). The images of regulatory units and sites are links to their complete textual descriptions in the TRRDUNITS and TRRDSITES tables (Figures 4*b*, *c*), available by a double click of the left mouse button.

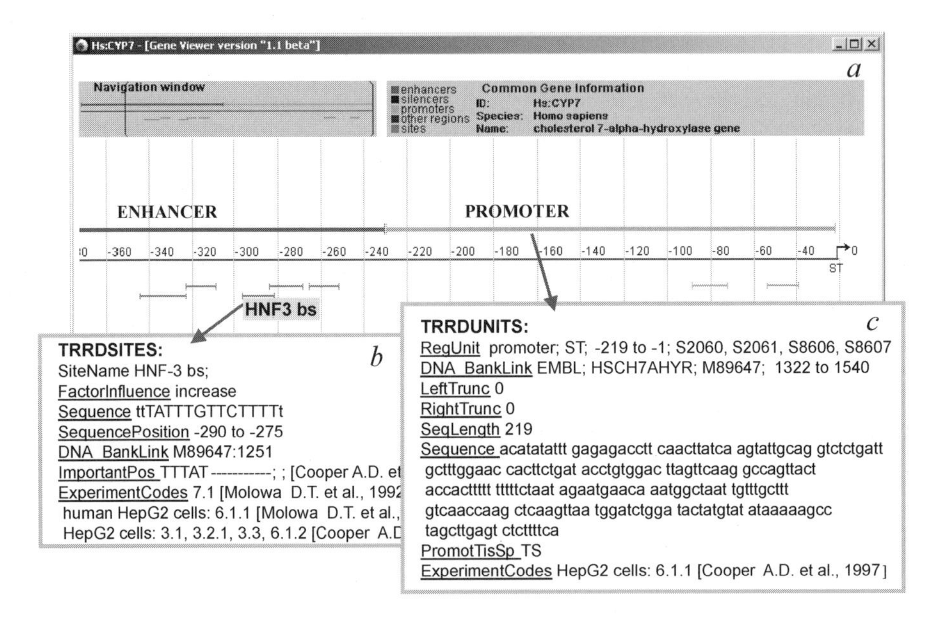

Figure 4. TRRD-Viewer Interface: (*a*) map of the regulatory regions of a gene (ST, the transcription start; box designated as HNF3 bs, tooltip with brief textual description of the binding site); (*b*) textual description of HNF3 site in TRRDSITES; and (*c*) textual description of the promoter in TRRDUNITS.

6. COMMON TOOLS FOR DATA SEARCH IN TRRD

The main tool for search by keywords in TRRD is the Sequence Retrieval System. Search can be performed over 131 indexed fields (Table 1).

Genes can also be sought for with browsers by gene and species names. Quick access to genes from the functionally significant groups listed above can be performed through thematic sections of TRRD. Regulatory regions similar to DNA sequence of interest can be sought for in the sequence database with the use of the program BLAST (http://wwwmgs.bionet.nsc.ru/mgs/systems/fastprot/units_blast.html).

7. HIERARCHICALLY ARRANGED DICTIONARIES AND THESAURI AND DATA RETRIEVAL FROM TRRD

A technique has been developed for creation and support of controlled dictionaries in TRRD (Ananko *et al.*, 1998). Hierarchically arranged dictionaries have been elaborated for tissues, cells, organs, developmental stages, and external effects as well as dictionaries of synonymic terms. In addition, there are thesauri on mammal tissues and organs, which contain additional information on cell composition, location, and origin of tissues as well as on functions of various organs and their parts. The thesauri are cross-linked html pages (http://wwwmgs.bionet.nsc.ru/mgs/gnw/trrd/thesaurus/). Figure 5 shows a sample html page of the thesaurus containing information on the entry "Kidney".

KIDNEY

AUTOMATED QUERY TO THE TRRD

Query to the TRRD database:
genes expressing in KIDNEY
Select species:
All Human Murine
Do Query

Location: bean-shaped organs on either side of the spinal column in the retroperitoneal tissue of the posterior abdominal cavity

Function:

- Conservation of water, essential electrolytes, and metabolites, and removal of certain waste products of metabolism from the body
- Synthesis and secretion of erythropoietin, renin, and hydroxylation of Vitamin D

Anatomic features

- **Cortex**, the outer part, consists of renal corpuscles, along with the convoluted (distal and proximal) and straight (distal and proximal) tubules of the **nephron**, the collecting tubules,

Figure 5. The thesaurus page describing the entry "Kidney".

We developed thesaurus-based search of TRRD and constructed a special-purpose search system, in which the relations "general–particular", "part–whole", "synonym", etc., are supported. The retrieval system links the thesauri with TRRD and immediately provides the list of genes expressed in a tissue or an organ specified (Figure 5).

The system allows simultaneous queries to the SRS version of TRRD for an entered word and all the associated words in a corresponding dictionary as well as for all synonyms. In this process, two SRS tables of TRRD,

TRRDEXP4 and TRRDGENES4, are linked automatically (Kolchanov *et al.*, 2002). The user's work is facilitated by the fact that the system provides the list of genes with all synonyms and organism species rather than expression patterns as a result of such a query (Figure 6), unlike the conventional SRS system. The same method was used for developing systems for search for genes either inducible or repressible by certain stimuli and genes expressed at certain developmental stages.

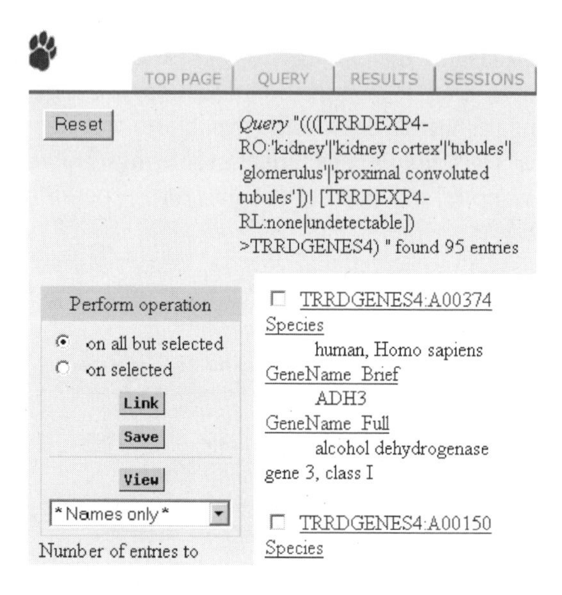

Figure 6. The result of search by the word "kidney".

8. ANALYSIS OF DNA SEQUENCES

DNA sequences are analyzed with the BinomSite program. It looks for DNA sequences similar to TFBSs present in TRRD (http://wwwmgs.bionet.nsc.ru/mgs/programs/mmsite/).

9. RELATIONAL VERSION OF TRRD

A relational version of TRRD has been developed in the medium Oracle8i. An XML presentation is used as an exchange format. Data are loaded from a flat file with a specially designed loader, which converts the data to the XML format. By now, the relational TRRD version contains 102 tables (52 data tables and 50 link tables). The pattern of its data is available at http://www.bionet.nsc.ru/trrd/RelScheme/.

10. CONCLUSION

TRRD offers ample possibilities for theoretical and experimental studies. Each TRRD entry, corresponding to one gene, simultaneously presents information on long regulatory regions, TFBSs, and features of expression of various eukaryotic genes. The TRRD information system has more than 130 indexed fields (Table 1) distributed over seven SRS tables. The collection of annotated regulatory units of eukaryotic genes stored in TRRD is the richest in the world (3254), as well as the collection of TFBSs (9480). For the first time the availability of data on gene expression patterns and regulatory elements responsible for their realization in one informational resource allows the molecular genetic systems of organisms to be analyzed at the level of gene networks.

ACKNOWLEDGMENTS

The work was supported in part by the Russian Foundation for Basic Research (grants Nos. 01-07-90376-в, 02-07-90355, 02-07-90359, 03-07-90181c, 03-07-96833-p2003, 03-07-96833, 03-07-90181-в, 02-04-48802-a, 03-04-48829, 03-07-06078-мас, 03-04-48555-a, 03-04-48506a, 03-04-48469-a); grant PCB RAS (No. 10.4); the Siberian Branch of the Russian Academy of Sciences (integration projects Nos. 119, 142 and 145); Russian Ministry of Industry, Science, and Technologies (grants Nos. 43.073.1.1.1501 and 43.106.11.0011); Russian Federal Research Development Program Research and Development in Priority Directions of Science and Technology (contract No. 28/2002); NATO (grants Nos. LST.CLG.979816 and PDD(CP)–(LST.CLG 979815)); the U.S. Civilian Research & Development Foundation for the Independent States of the Former Soviet Union (CRDF) No. NO-008-X1.

SITECON—A TOOL FOR ANALYSIS OF DNA PHYSICOCHEMICAL AND CONFORMATIONAL PROPERTIES: E2F/DP TRANSCRIPTION FACTOR BINDING SITE ANALYSIS AND RECOGNITION

D.Yu. OSHCHEPKOV[1,3]*, I.I. TURNAEV[1], M.A. POZDNYAKOV[1,3], L. MILANESI[2], E.E. VITYAEV[1,3], N.A. KOLCHANOV[1,3]

[1]*Institute of Cytology & Genetics, Siberian Branch of the Russian Academy of Sciences, prosp. Lavrentieva 10, Novosibirsk, 630090 Russia, e-mail: diman@bionet.nsc.ru;* [2]*CNR Institute of Biomedical Technology (CNR-ITB), via Fratelli Cervi, 93, 20090 Segrate (Milan), Italy;* [3]*Novosibirsk State University, ul. Pirogova 2, Novosibirsk, 630090 Russia*

* *Corresponding author*

Abstract Research into molecular mechanisms underlying DNA–protein interactions using statistical analysis of nucleotide sequences of binding sites is most important for understanding the principles of gene expression regulation. The local conformation of transcription factor binding sites determined by their context is a factor responsible for specificity of the DNA–protein interactions. Analysis of the local conformations of a set of functional DNA sequences allows the conservative context-dependent conformational and physicochemical properties (CDCPP) reflecting molecular mechanisms of interactions to be determined. A set of conservative CDCPP specific of sites for binding a particular transcriptional factor can be effectively used for their recognition. The methods for determining the conservative CDCPP in short regions of aligned functional DNA sequences and recognizing potential transcription factor binding sites basing on a set of these conservative CDCPP determined are proposed in this work. To demonstrate implementation of the method, the binding sites of the heterodimeric complex E2F/DP were analyzed as an example. The discovered specific CDCPP for a set of these binding sites reflect the molecular mechanism of their interaction.

Key words: site recognition, conformational and physicochemical DNA properties, DNA–protein interactions, transcription factor

1. INTRODUCTION

As a rule, each transcription factor is capable of interacting with regulatory DNA sequences differing in their context. Specific features of these interactions impose limitations on the sequences of binding sites, manifesting themselves in a partial conservation of the nucleotide sequence. These suggestions formed the background of numerous methods for detecting and predicting regulatory regions, such as consensus method (Mulligan *et al.*, 1984), weight matrices (Stormo *et al.*, 1986), and the method of weight matrices, based on statistical physics approach to DNA–protein interactions (Berg and von Hippel, 1988). Presently, the weight matrix-based approach dominates among the methods for transcription factor binding site recognition. Among the best known tools for recognition of transcription factor binding sites, ooTFD (Ghosh, 2000), TESS (Stoeckert *et al.*, 1999), TRANSFAC (Heinemeyer *et al.*, 1999), RegulonDB (Salgado *et al.*, 2001), and GeneExpress (Kolchanov *et al.*, 1999) are widely used, most of them manipulating successfully hundreds of weight matrices. Nevertheless, recent evaluation of the computer tools for transcription factor binding site recognition within genomic DNA (Roulet *et al.*, 1998) unexpectedly demonstrated that the matrix scores correlate better with each other rather than with DNA–protein affinity magnitudes.

On other hand, an increasing volume of experimental data suggests that the function of transcription factor binding sites is determined to a considerable degree by the conformational and physicochemical DNA properties (Starr *et al.*, 1995; Meierhans *et al.*, 1997). Moreover, it was been previously shown that CDCPP might be significant for the site function (Ponomarenko M.P. *et al.*, 1997). The developed computer system ACTIVITY (Ponomarenko M.P. *et al.*, 1999), based on utility theory for decision making (Fishburn, 1970), demonstrated a successful application of CDCPP for site activity prediction from sequence context (Ponomarenko J.V. *et al.*, 1999). That is why we believe that CDCPP may be an alternative approach to detection of specific features of transcription factor binding sites and their successful recognition.

Dickerson and Drew (1981) were first to discover the dependence between the DNA conformation and its context using X-ray analysis of DNA dodecamers. An ever-increasing data of structural analyses demonstrate both heterogeneity of conformational and physicochemical properties and their dependence on the nucleotide sequence (Frank *et al.*, 1997; Suzuki *et al.*, 1997). Thus, one can suppose that the local conformation of DNA molecules determined by the context is a factor affecting the specificity of DNA–protein recognition. This suggests that certain conformational and physicochemical properties of the variants of genomic sequences interacting with a certain regulatory protein should be preserved.

Two approaches are possible here. First, the mean value of a property over a sequence region of 10–30 nucleotides long that either interacts with a protein or is located in the vicinity of the protein bound is analyzed and compared with random sequences set. As it was previously mentioned, this approach appeared very advantageous and allowed several methods for recognition and activity determination to be designed. This approach was realized in the system B-DNA-Video (Ponomarenko J.V. *et al.*, 1999). Second, the sequence of a site may be analyzed in more detail, considering conservation of the properties over shorter regions of the functional sequence or at individual positions within the site (Oschepcov *et al.*, 2002). In this case, analysis of a more intricate pattern of conformational properties over the site sequence may elucidate the molecular mechanisms involved in the interactions of the functional sequence analyzed with the corresponding protein. Thus, such analysis allows us to determine the allowable conformations of the DNA regions whose similarity suggests most efficient binding, i.e. to determine finally those regions that are potential binding sites for the corresponding regulatory element.

A more detailed analysis of binding site sequences and study of specific DNA helix properties over shorter fragments of functional sequences can aid the recognition accuracy. For example, analyzing the binding site for transcription factor MetJ, Liu *et al.* (2001) demonstrated that the discriminatory power might be increased when DNA helix conformational properties were taken into account in addition to conventional methods based on contextual analysis.

In this work, a method for recognizing regulatory DNA regions is proposed. This method is based on comparison of context-dependent conformational and physicochemical properties of short fragments within functional DNA regions with the properties conserved in a set of transcription factor binding site sequences. The efficiency of this method is tested by an example of the binding site for heterodimer E2F/DP (Zheng *et al.*, 1999).

2. METHODS AND ALGORITHMS

For the analysis, we selected binding sites for the heterodimer E2F/DP; the structure of E2F4/DP2–DNA complex was clarified by X-ray structure analysis (Zheng *et al.*, 1999). E2F controls transcription of a group of genes whose expression is essential for the normal course of cell cycle, i.e. is actually a key regulator of the cell cycle. E2F forms heterodimers with proteins of a related DP family. The structure of this DNA-binding domain corresponds, to a winged-helix type (Jordan *et al.*, 1994; Zheng *et al.*, 1999); amino acid residues within the structurally similar domains belonging to this type bind invariantly to DNA.

This suggests that the winged-helix type domains of other E2F/DP combinations display similar specificities of DNA binding. A set of 40 sites for binding E2F/DP heterodimer with a length of 49 bp was analyzed. The sequences with experimentally confirmed binding were extracted from TRRD (http://wwwmgs.bionet.nsc.ru/mgs/gnw/trrd/; Kolchanov *et al.*, 2002) and phased with respect to the consensus TTTCGCGCG without gaps. While analyzing sets of aligned DNA sequences of the sites, 38 conformational and physicochemical DNA properties compiled in the database Property (http://wwwmgs.bionet.nsc.ru/mgs/gnw/bdna/) were used (Figure 1, Table 1).

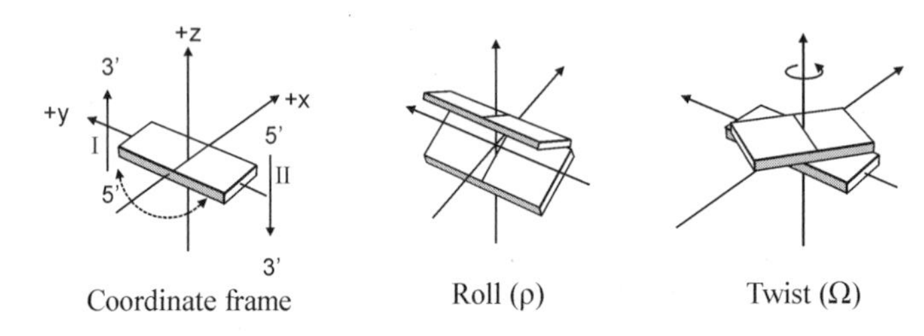

Figure 1. An examples of conformational properties presented in PROPERTY Database.

Table 1. An examples of conformational properties values presented in PROPERTY database

Dinucleotide	Roll value, degree	Twist value, degree
AA	0.3	35.3
AT	–0.8	31.2
AG	4.5	31.2
AC	0.5	32.6
TA	2.8	40.5
TT	0.3	35.3
TG	0.5	32.6
TC	–1.3	40.3
GA	–1.3	40.3
GT	0.5	32.6
GG	6.0	33.3
GC	–6.2	37.3
CA	0.5	39.2
CT	4.5	31.2
CG	–6.2	36.6
CC	6.0	33.3

The essence of our approach was as follows. A set of N aligned (phased) DNA sequences with the length L (without gaps) is considered. A value of a certain physicochemical or conformational property F_i is ascribed to each dinucleotide. Consequently, the matrix with a size $N\times(L-1)$ is formed. An element of this matrix F_{ikl} corresponds to the value of this particular property

F_i of dinucleotide at the kth position in the sequence l. The mean value of the property i at position l amounts to

$$\overline{F_{il}} = \frac{1}{N}\sum_{k=1}^{N} F_{ikl}\ . \qquad (1)$$

Variance is used as a measure of conservation of the ith property for each position l:

$$\sigma^2_{F_{il}} = \frac{1}{N-1}\sum_{k=1}^{N}(F_{ikl} - \overline{F_{il}})^2\ . \qquad (2)$$

It is assumed that if a particular property F_{il} at particular location l within the nucleotide sequence is important for the function of the binding site, the value of this property is conserved for all the sequences of the set, providing a low value of the variance compared with a set of random sequences. Thus, a low variance of a particular property indicates its conservation at a particular position l. The significance of $\upsilon_{F_{il}}$ is estimated using χ^2 test (Anderson, 1958). Then, we are assuming that the probability P_{il} of the ith property at position l of the sequence analyzed to take the value F_{il} required for the function at the value F_{il} follows a Gaussian distribution:

$$P_{il} = \frac{1}{\sqrt{2\pi}\sigma_{F_{il}}}\exp\left((\overline{F_{il}} - F_{il})/\sigma_{F_{il}}\right)^2. \qquad (3)$$

Let us select a sum P_{il} of all the significantly conservative properties normalized to the number of these properties as a measure of similarity between the sequences of the set and the sequence analyzed.

$$P_{\Sigma} = \frac{\sum_{i=0}^{I}\sum_{l=0}^{L}\delta_{il}P_{il}}{\sum_{i=0}^{I}\sum_{l=0}^{L}\delta_{il}}, \qquad (4)$$

where $\delta_{il} = 1$, if $\upsilon_{F_{il}}$ is significantly low, otherwise $\delta_{il} = 0$. The value P_{Σ} corresponds to the probability of the properties of the DNA sequence analyzed to be close to the detected conservative properties of the sequences forming the initial set. Let us designate this value as the level of required conformational similarity or, in other words, this value is considered to be a "score" value and is compared with the particular "threshold" value to decide whether this sequence could be a "site" or "not site". For testing the recognition quality, we compared the method proposed with weight matrix method. The following equation was used to construct the weight matrix for the set of aligned E2F binding sites (Hertz et al., 1999):

$$w_{ij} = \ln\left(\frac{(n_{ij} + p_i)/(N+1)}{p_i}\right), \tag{5}$$

where n_{ij} is the nucleotide frequency i (i = A,T,C,G) at the position j for aligned set of sites; p_i, prospective frequency of nucleotide i in the human genome and N, number of site sequences in the set.

3. RESULTS AND DISCUSSION

3.1 Analysis of conservative features of E2F/DP binding site

Analysis of the set of E2F/DP binding sites for conservation of all the 38 CDCPP allowed us to detect a number of specific DNA conformational properties, four of them being most interesting. The properties major/minor groove width (Ponomarenko *et al.*, 1997) and bendability towards the major/minor groove (Gartenberg *et al.*, 1988) appeared conserved at all the positions within the site (8 bp at a 99% significance level according to chi-square test). Interestingly, conserved values of the major groove width (Figure 2) for the region TTTCGCGCG and the minor groove width for the region TTTCGCGCG display the values exceeding considerably the mean values of these properties for a random sample. This suggests that the properties in question are crucial for the DNA–protein interaction. However, the bendability profile is insufficiently coordinated, suggesting that the conservation discovered for this property at some positions is likely to result from the interdependence between bendability and the major groove width values (correlation coefficients of these properties are 0.70 and 0.85, respectively).

The observations on the sizes of major and minor grooves comply with the X-ray structure analysis data. As was noted above, the transcription factor E2F belongs to a winged-helix type (Jordan *et al.*, 1994). While binding, two recognition α helices, one from each of the heterodimer E2F/DP constituents, are localized to the major groove (Figure 3).

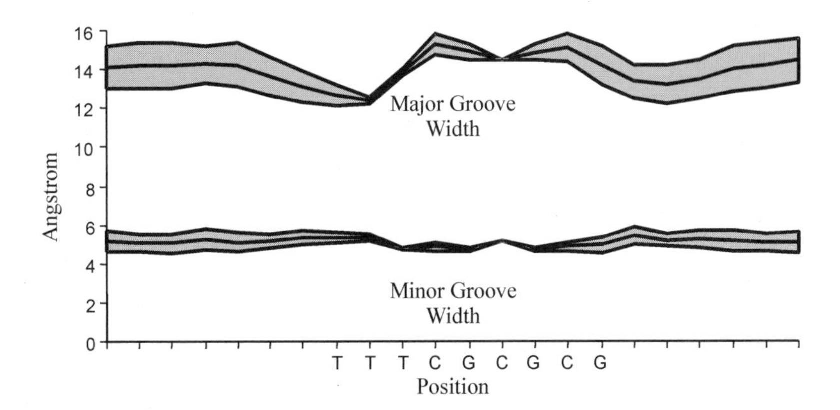

Figure 2. The plots of major (upper) and minor groove (lower) width at positions of the aligned E2F/DP biding sites, and its dispersion.

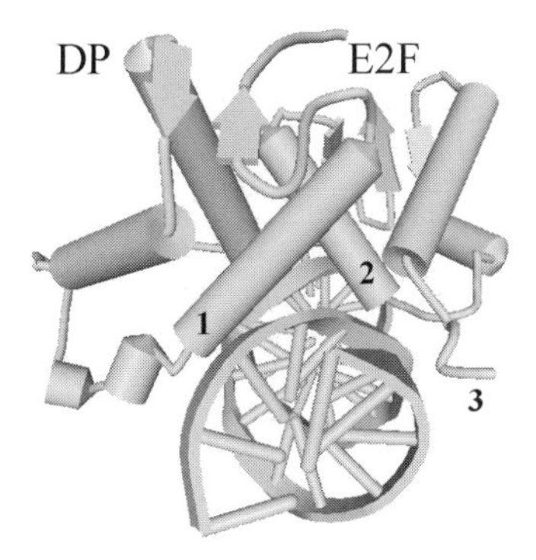

Figure 3. E2F/DP complex with its target DNA sequence: (1 and 2) are recognition α helices from DP and E2F and (3) *N*-terminal of E2F recognition domain.

In this process, the α helix of E2F DNA-binding domain (Figure 3, *2*) lies in the region TTT<u>CGC</u>GCG, while the α helix of DP DNA-binding domain (Figure 3, *1*) in the region TTTCGC<u>GCG</u>. Thus, a widened major groove (Figure 2) in the region TTT<u>CGCGCG</u> is likely to be a necessary condition for the heterodimer E2F/DP to recognize the binding site. Note that the dependence of this type is indeed the feature specific of E2F/DP binding sites, since the recognition α helices of both constituent proteins are spatially located one after the other and the bases contacting with these helices are adjacent too.This complies with the data of Zheng *et al.* (1999) who demonstrated that these proteins bind to DNA already as heterodimer. It was demonstrated that the T tract in <u>TTT</u>CGCGCG is critical for the binding (Jordan *et al.*, 1994) and necessary for insertion of the *N*-terminus of E2F recognition domain (Figure 3, *3*; Zheng *et al.*, 1999), complying with the obtained data on a widened minor groove in the region TTTCGCGCG (Figure 2).

3.2 Recognition of E2F/DP binding sites

To test the recognition method proposed, we used the same set of E2F/DP binding sites. The overall distributions of the values of the required conformational similarity for the YES (positive) and NO (negative) sets are shown in Figure 4. As is evident (Figure 4), overlapping of the two sets is minimal, suggesting that the quality of distinguishing between the positive and negative sets is high. Sequences for the negative set were generated by random shuffling of nucleotides in the initial site sequences; thus, the nucleotide compositions of both the positive and negative samples were identical.

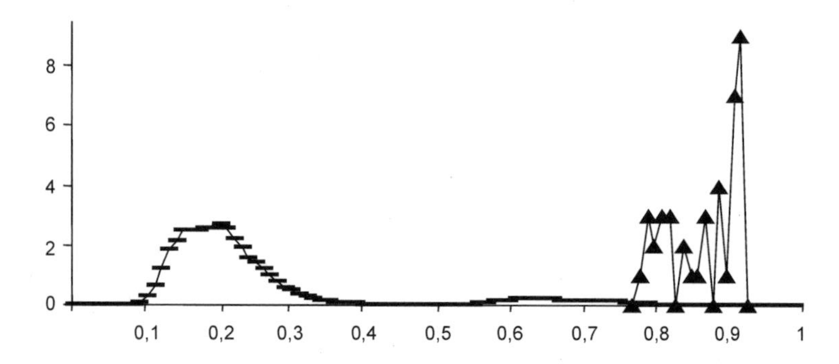

Figure 4. Score distributions for positive and negative sites. Triangles correspond to positive sites score value; lines to negative, respectively.

We have compared the recognition quality of the method proposed and the weight matrix method described by Hertz and co-authors (1999). The error determination technique for both methods, described below, was the same. The recognition quality for type I errors (true negatives) was checked by the jack-knife method with exclusion of one of the sequences from the training sample in series. For type II errors (false positives), the control recognition of binding sites in a randomly generated sequence with a length of 500 000 bp was performed. The recognition was carried out in both directions; the sequence was generated by random shuffling of nucleotides in the initial site sequences. The resulting plots for type I and type II errors are shown in Figure 5. For low values of type I errors (true negatives), SITECON demonstrates up to severalfold better recognition quality. For higher (up to 30%) values of type I errors, the methods demonstrate comparable quality.

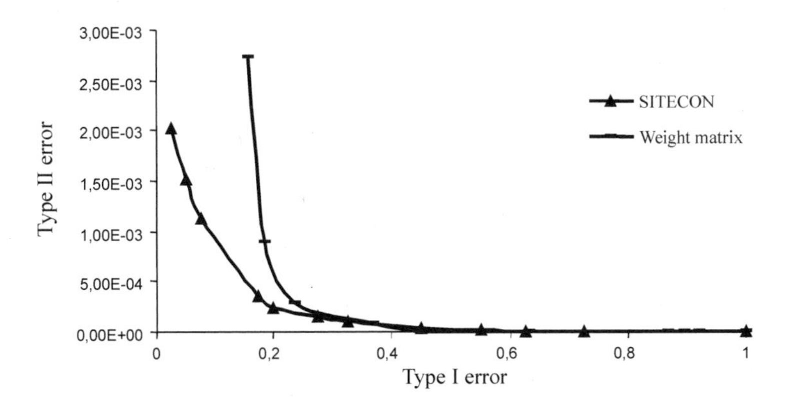

Figure 5. Type I and type II errors scatter plot for SITECON method and weight matrix method. Errors for both methods were estimated as it is described in text.

4. DISCUSSION

Several mechanisms that should be taken into account while analyzing the DNA sequences of functional sites and regulatory regions can provide preservation of specific CDCPP. (1) Conservation of the nucleotide context, manifesting itself as a fixed pattern of nucleotides at certain positions in different site variants. The methods involving contextual analysis are suitable for studying this type of functional sites. (2) Conservation of certain conformational and physicochemical properties of particular regions within the site provided by the substitution pattern preserving these properties. Such regions of functional sites may be detected using the methods that take into account the CDCPP. Thus, conservation of nucleotide context may actually result from the limitations imposed on the values of particular properties at particular regions of the sequence. Hence, we hypothesize that analysis of local properties instead of the corresponding context may provide essentially more information on the structure of the site. As a result, it could provide better recognition quality and give an opportunity to analyze the underlying molecular mechanisms of DNA–protein interaction. The method described in this work may be one of the possible approaches, satisfying these terms and allowing for recognition quality increase in comparison with the weight matrices (Figure 5). The method proposed was successfully applied to recognition of some other transcription factor binding sites: IRFs, ISGF3, STATs, NF-κB (Ananko *et al.*, 2002), COUP-TF, PPRE (Proscura *et al.*, 2002), and HSF (Furman *et al.*, 2002).

5. CONCLUSION

A method for detection and analysis of significant conformational DNA properties was developed. E2F heterodimer binding site was analyzed. The significant properties—sizes of major and minor grooves of two regions within the site—comply with the data on the mechanism of DNA–heterodimer binding. A method for recognizing potential transcription factor binding sites basing on a set of conservative context-dependent conformational and physicochemical properties determined for short fragments of aligned functional DNA sequences is proposed and the effectiveness of the method is demonstrated.

ACKNOWLEDGMENTS

The authors are grateful to V.G. Levitsky for fruitful discussion of the study and G.B. Chirikova for assistance in translation. The work was supported by the Russian Foundation for Basic Research (grants Nos. 01-07-90376-в, 02-07-90355, 03-07-96833-p2003, 03-07-96833, 03-04-48506-a, 03-07-90181-в, 03-04-48469-a, 03-04-48555-a, 03-04-48829); grant PCB RAS (No. 10.4); the Siberian Branch of the Russian Academy of Sciences (integration project No. 119); Russian Ministry of Industry, Science, and Technologies (grant No. 43.073.1.1.1501, subcontract No. 28/2003); NATO (grant No. LST.CLG.979816).

LOCAL SECONDARY STRUCTURE MAY BE A CRITICAL CHARACTERISTIC INFLUENCING TRANSLATION OF UNICELLULAR ORGANISMS mRNA

Yu.G. MATUSHKIN[1,2]*, V.A. LIKHOSHVAI[1,2,3], A.V. KOCHETOV[1,2]

[1]*Institute of Cytology & Genetics, Siberian Branch of the Russian Academy of Sciences, prosp. Lavrentieva 10, Novosibirsk, 630090 Russia, e-mail: mat@bionet.nsc.ru;* [2]*Novosibirsk State University, ul. Pirogova 2, Novosibirsk, 630090 Russia;* [3]*Ugra Research Institute of Information Technologies, ul. Mira 151, Khanty-Mansyisk, 628011 Russia*

* *Corresponding author*

Abstract Detection and study of the factors changing the efficiency of gene expression is essential for both a better understanding of the key stages of translation and design of artificial systems with a prespecified expression level. It is known that recognition of the translation start site (TSS) depends on the contextual features of neighboring RNA regions. Patterns of codon compositions and local secondary structures of the protein coding regions in genes of 74 unicellular organisms, whose complete genomic sequences had been determined, were studied. We divided all the microorganisms into the five major groups that display markedly different patterns in terms of prevalence of selection for codon-frequency bias or selection against local self-complementarity. We found the specific secondary structure in the TSS region near AUG codon. We found a significant negative correlation between the predicted secondary structure stability and predicted expression level in the TSS region in some organisms. We assume that stability of the secondary structure in the region of translation start site can be an important factor in TSS recognition.

Key words: expression efficiency, translational signals, translation start site (TSS), secondary structure, translation elongation stages, codon usage frequencies

1. INTRODUCTION

Efficacy of gene expression is one of the most general characteristics measuring biological activity of protein-coding genes.

Translation of mRNA in unicellular organisms is one of the most energy-consuming stages of the gene expression process. For example, up to 50% material and energy resources may be spent for translation in an *E. coli* cell. Consequently, increase in efficiency of translation machinery operation might play the role of a long-term factor of evolutionary selection.

The stage of elongation may be a target of such optimization. The elementary act of elongation – attachment of an amino acid residue to a growing polypeptide chain—comprises three successive stages, namely: placement of the charged isoacceptor tRNA in the ribosome A site; transpeptidation, and translocation. In general, each of these three elongation stages may be limiting for the polypeptide growth.

In some organisms, the efficacy of protein synthesis is well correlated to codon usage frequencies within genes encoding these proteins (Sharp and Li, 1986; Shields and Sharp, 1987; Shields *et al.*, 1988; Sharp and Devine, 1989; Lloyd and Sharp, 1993). The rate of codon "readability" correlates with concentration of relevant tRNA fraction (Ikemura, 1985; Yamao *et al.*, 1991). In principle, this approach allows for estimating gene expression efficacy by calculating the frequency of codon occurrence rates in a gene. For some organisms, this approach has been realized (Li and Luo, 1996). On the contrary, in some other organisms, this regularity is not obvious (Andersson and Sharp, 1996; Lafay *et al.*, 1999).

We are relating the efficiency of isoacceptor aminoacyl-tRNA placement to the codon usage frequencies, while the efficiency of translocation with the number of local secondary structures. Since the modern concepts prevent from relating the transpeptidation stage to the mRNA context directly, we omitted it from consideration in this work. We are determining the relative contributions of codons and local secondary structures to the efficiency of translation elongation for 74 organisms whose complete genomes have been sequenced. This allows us to select most informative characteristics of nucleotide compositions of their protein-coding regions to recognize adequately the efficiency of gene expression. Here, recognition of genes encoding ribosomal proteins as highly expressed genes was the criterion of adequacy.

On the other hand, the important feature influencing mRNA translation is the context of translation start codon (Yun *et al.*, 1996; Kozak, 1999). This influence either may be local in the case of prokaryotes due to a direct binding of ribosomal subunits to the start codon (Likhoshvai, 1992; Kolchanov and Lim, 1994) or, according to the scanning model of eukaryotic translation initiation (Kozak, 1999; 2002), postulating that 40S subunit of the eukaryotic ribosome binds to the 5'-untranslated region, may involve an essential part of this region.

It was found that nonoptimal context (pyrimidine at –3 position upstream of AUG) decreased both the translation initiation and mRNA translation efficiencies of eukaryotes. However, analysis of nucleotide sequence databanks has shown that a large part of eukaryotic mRNAs display poor context of the start codon (Suzuki and Ishihara, 2000; Peri and Pandey, 2001). We analyzed structural characteristics of pro- and eukaryotic mRNAs in the region of translation start site.

2. METHODS AND ALGORITHMS

Sequence data. Coding DNA sequences (CDS) for 74 unicellular organisms and CDS with 600-nucleotide-long 5'- and 3'-end extensions for Saccharomyces cerevisiae and S. pombe were extracted from GenBank using the Feature Table information. The sample of yeast full-size mRNAs was extracted from the EMBL databank (http://www.embl-heidelberg.de/). Full-size 5'UTRs were selected from the entries containing description of mapped transcription start sites and complete coding regions. This resulted in a set of mRNA 5'UTRs of 171 nonredundant (<60% identity with CDS) yeast genes. Finally, we used in the analysis 98 of the 171 genes characterized by a single transcription start site.

Earlier (Likhoshvai and Matushkin, 2002a), we developed the elongation efficiency index (EEI), a quantitative measure allowing the elongation rate of mRNA depending on its nucleotide composition to be assessed. This measure takes into account both the codon usage frequencies and local mRNA secondary structures.

The average time spent by a ribosome for one elongation act was calculated using the equation: $\mathrm{EEI} = u_1 T_a + u_2 T_e$.

Here, T_a, accurate to the proportionality coefficient, has a meaning of the average time required for isoacceptor aminoacyl-tRNA to be placed in the ribosome A site and is calculated as $T_a(i) = \sum_{j=1}^{n_i} \beta_{\delta(i,j)} \Big/ n_i$, $\beta_\delta = \dfrac{\sum_{m=1}^{C} \sqrt{\alpha_m}}{\sqrt{\alpha_\delta}}$,

where α_δ has a meaning of the usage frequency of codon δ of the genetic code C within a certain gene subset.

The second component $T_e(i)$, accurate to the proportionality coefficient, has a meaning of the average time spent by a ribosome for translocation and is calculated as $T_e(i) = t_{\min}\,(1 - p(i)) + t_{\max}\, p(i)$, where $t_{\min}$ ($t_{\max}$) is the minimal (maximal) conditional time of translocation, respectively, and $p(i)$, the probability of t_{max} realization, calculated according to the equation

$$p(i) = \int_0^{LCI(i)} \frac{k^{n+1} x^n}{G(n+1)} e^{-kx} dx, \; k = m/\sigma^2, \; n = (m/\sigma)^2.$$

Here, m and σ are the expectation and variance, respectively, of a random positive value with a distribution density of $\frac{k^{n+1} x^n}{G(n+1)} e^{-kx}$, where $G(n+1)$ is a gamma function and LCI(i) is local complementarity index. We used two methods for calculating the local complementarity index.

LCIζ form. This index in its meaning is calculated number of the perfect inverted repeats that meet certain conditions and are capable of forming hairpins. The averaged number of complementary regions disregarding the energy of secondary structure formation is calculated using equation (1) over the region with a length m_i for a fixed translation frame (here, m_i equals the triple number of codons contained in the ith gene plus 55 nucleotides from its 3'-end):

$$LCI\varsigma(i) = \frac{\sum_{m=1}^{m_i - 2s_{max} - l_{max} + 1} \left\{ \sum_{s=s_{min}}^{s_{max}} \left[\sum_{l=l_{min}}^{l_{max}} \zeta\left(\mathrm{con}(m, m+s-1), \overline{\mathrm{con}(m+s+l-1, 2m+2s+l-2)}\right) \right] \right\}}{m_i - 2s_{max} - l_{max} + 1}, \quad (1)$$

where con(i, j) is the gene context between ith and jth nucleotides; $\overline{\mathrm{con}(i,j)}$, the corresponding complementary context between ith and jth nucleotides ($i \leq j$); and ζ(context1, context2) = 1, if the words context1 and context2 are identical, otherwise ζ(context1, context2) = 0. The length of accountable inverted repeat falls between s_{min} and s_{max}; the distance between accountable inverted repeats falls between l_{min} and l_{max} (here, $s_{min} = s_{max} = 3$, $l_{min} = 3$, and $l_{max} = 50$).

LCIψ form. Unlike the previous index, not only the number of potential loops is calculated, but also their energetic stability. In equation (1), ζ is substituted with ψ(context1, context2)—the energy of secondary structure potentially formed by a perfect repeat found (here, $s_{min} = 3$, $s_{max} = 6$, $l_{min} = 3$, and $l_{max} = 50$ are used for LCIψ form). The energies of secondary structures were calculated conventionally (Turner *et al.*, 1988).

The following indices were used for calculations:

a) $u_1 > 0$, $u_2 = 0$, when *only the codon composition* was taken into account;
b) $u_1 = 0$, $u_2 > 0$, when *only the local complementarities* were taken into account *disregarding* (ζ *form*) the secondary structure *energies*;
c) $u_1 = 0$, $u_2 > 0$, when *the local complementarities* were taken into account *regarding* (ψ *form*) the secondary structure *energies*;
d) $u_1 > 0$, $u_2 > 0$, when *codon composition and local complementarities* were taken into account *disregarding* (ζ *form*) the secondary structure *energies*; and
e) $u_1 > 0$, $u_2 > 0$, when *codon composition and local complementarities* were taken into account *regarding* (ψ *form*) the secondary structure *energies.*

Thus, five variants of EEI (elongation efficiency index) were used in the calculations.

To discover the factors critical for elongation in each particular unicellular organism, we ordered its genes according to the decrease in values of each of the five indices described. Then, we analyzed the order of genes encoding ribosomal proteins in the five produced ordered lists of all the genes of each organism, assuming that the ribosomal genes belonged to the group with the highest expression (Kochetov, 1998).

Consequently, the larger is the shift of ribosomal genes from the center of the ordered list of genes, the more reliable is the conclusion on the effect of the characteristic observed (which is taken into account in the corresponding index) on the efficiency of expression. The shifts and their statistical significances were calculated as described in (Likhoshvai and Matushkin, 2002). A separate group of organisms displaying the maximal shift with respect to each particular index variant was formed (Table 1). The significance of the shift for each organism amounted to >0.999.

3. RESULTS AND DISCUSSION

3.1 Analysis of elongation stage

We divided 74 microorganisms into the five major groups that display markedly different patterns in terms of the prevalence of selection for codon-frequency bias or selection against local self-complementarity (Table 1).

The first group contains 32 organisms; the critical step for their elongation efficiency is the time required for isoacceptor aminoacyl-tRNA to be placed in the ribosome A site. Thus, only the codon composition is important, while secondary structures have no effect on the process.

Table 1. Resulting groups of unicellular organisms

Group	Organisms	Number
Group A (*codon compositions of genes are critical for elongation*)	*A. tumefaciens* C58, *B. halodurans* C-125, *B. subtilis*, *B. melitensis*, *C. muridarum*, *C. trachomatis*, *C. pneumoniae*, *C. pneumoniae* AR39, *C. pneumoniae* J138, *E. coli* K12, *E. coli* O157 H7, *E. coli* O157 H7 EDL933, *H. influenzae*, *L. innocua*, *L. monocytogenes* EGD-e, *M. loti*, *M. leprae*, *P. multocida*, *S. typhi*, *S. typhimurium* LT2, *S. meliloti*, *S. aureus* Mu50, *S. aureus* N315, *S. pneumoniae* R6, *S. pneumoniae* TIGR4, *S. pyogenes*, *S. pyogenes* MGAS8232, *Synechocystis* sp. PCC6803, *V. cholerae*, *Y. pestis*, *S. cerevisiae*, and *S. pombe*	32
Group ζ (*the number of local complementarities disregarding the energy is critical for elongation*)	*B. burgdorferi*, *Buchnera* sp. APS, *C. jejuni*, *H. pylori* 26695, *H. pylori* J99, *M. genitalium*, *M. pulmonis*, *U. urealyticum*	8
Group ψ (*the number of local complementarities and the energy of secondary structures are critical for elongation*)	*P. aeruginosa* PA01	1
Group Aζ (*codon composition and the number of local complementarities disregarding the energy are critical for elongation*)	*A. aeolicus*, *C. rescentus*, *C. perfringens*, *C. acetobutylicum*, *D. radiodurans* R1, *F. nucleatum* ATCC 25586, *L. lactis*, *M. tuberculosis* H37Rv, *M. tuberculosis* CDC1551, *M. tuberculosis* H37Rv, *M. pneumoniae*, *Nostoc* sp. PCC 7120, *R. conorii* Malish 7, R. *prowazekii*, *T. maritime*, *T. pallidum*, *A. pernix* K1, A. *fulgidus*, *Halobacterium* sp. NRC-1, *M. thermoautotrophicum*, *M. jannaschii*, *M. kandleri* strain AV19, *M. acetivorans* strain C2A, *P. aerophilum*, *P. abyssi*, *P. horikoshii*, *S. solfataricus*, *S. tokodaii*, *T. acidophilum*, *T. volcanium*	30
Group Aψ (*codon composition, number of local complementarities, and the energy are critical for elongation*)	*N. meningitidis* MC58, *N. meningitidis* Z2491, *X. fastidiosa*	3

The second group comprises eight organisms, for which the translocation step is critical. The translocation rate here decreases with the increase in the degree of local complementarities; however, the secondary structure energies are insignificant. This means that the amount of local complementarities disregarding the energy is critical for elongation.

Only one organism, *P. aeruginosa* PA01, forms the third group. Its elongation efficiency is also determined at the translocation stage; however, the energy of local secondary structures is an important factor influencing the rate of ribosome movement. Therefore, both the amount of local complementarities and the energy of secondary structures are critical for elongation. The stage of tRNA attachment is not critical for the organisms belonging to the second and third groups; in this respect, they are opposite to the organisms of the first group.

The fourth group contains 30 organisms, the fifth, only three. In these organisms, both elongation stages considered influence the elongation efficiency: the codon composition and the amount of local complementarities disregarding the energy are critical for elongation. However, the elongation efficiency of the organisms from the fourth group, similarly to the second, depends only on the presence of complementary regions, whereas the energy of secondary structures remains insignificant.

In the fifth group, as in the third, the elongation efficiency also depends on the energies of secondary structures. In this group, the codon composition, the amount of local complementarities, and their energy are critical for elongation.

Certain common patterns are evident from the results listed in Table 1.

First, for 65 organisms (groups A, Aψ, and Aζ), the codon composition plays an essential role in determining the elongation efficiency *per se*, (group A), taking into account the number of local complementarities disregarding the energy (group Aψ), or regarding both the number of local complementarities and the energy of hairpins. This result complies with the generally accepted view on the role of nucleotide context of the mRNA coding part in determination of translation elongation efficiency.

However, we obtained an unanticipated result: for 42 organisms of the 74 analyzed (groups ζ, ψ, Aζ, and Aψ), an important role of the secondary structure in translation elongation was demonstrated.

Several variants of the effect of secondary structure were discovered. In the case of group ψ, this was the number of local complementarities disregarding the energy; in the case of group ψ, the number of local complementarities and the energy of secondary structures; for group Aζ, the number of local complementarities disregarding the energy but taking into account the effect of codon composition appeared essential; and as for group

Aψ, both structural characteristics—the number of local complementarities regarding the energy and codon composition—were evidently essential.

This result means that not only the nucleotide context of mRNA coding part, but also the secondary structure of the coding region exert their effects on translation machinery of unicellular organisms at the stage of elongation.

In addition, we have demonstrated that all the archaebacteria studied fall into group Aζ. The majority of related eubacterial species also fall into the same group. The eukaryotic organisms (*S. cerevisiae* and *S. pombe*) analyzed belong to one group, A, as well. Thus, evolutionary related species display a trend of clustering together. The results obtained suggest several biological interpretations.

This produces an impression that organisms had used different strategies while optimizing the elongation stage during their evolution. The organisms belonging to the most numerous group A have evolved such efficient mechanisms of mRNA physical movement through the ribosome (or ribosome movement along mRNA) that the translocation stage ceased to be limiting (if it ever was). The selection of these organisms towards increasing the elongation efficiency was achieved through optimizing the codon compositions of their genes.

On the contrary, the data obtained suggest that the translocation stage in the organisms belonging to groups ζ and ψ is sensitive to the local hindrances formed, in particular, due to local complementarities. However, the placement of tRNA in the ribosome A site is either equally efficient for all the codons or proceeds in parallel with the process removing the hindrances ahead of the moving ribosome (preparatory stage of translocation), and this process is slower. Thus, with reference to translation optimization, removal of hindrances becomes the critical process (it is the factor noticed by the natural selection), whereas mutations in the codon composition appear neutral (the natural selection overlooks them).

In the rest groups, both stages influence essentially the elongation efficiency, suggesting consequently that both characteristics—the codon composition and local complementarities—have been optimized.

Groups ζ and Aζ, on the one hand, and groups ψ and Aψ, on the other, suggest that the corresponding organisms might utilize different mechanisms for their ribosomes to overcome the hindrances represented by local secondary structures at the stage of translocation. We may hypothesize that in the case of group ζ and Aζ organisms, encounter of a ribosome with a hindrance triggers a mechanism that spends a predetermined batch of resources (time or energy) to remove all the hindrances within the mRNA region of a certain length. On the contrary, the corresponding mechanism of group ψ and Aψ organisms is somehow capable of estimating the hindrance "capacity" and spends the proportional time (or, possibly, energy) for its removal. Note that the total number of organisms displaying the sensitivity of elongation to the secondary structure energies appeared insignificant, amounting to only four.

3.2 Analysis of 5' gene regions of mRNAs

3.2.1 Bacteria: *Escherichia coli*

To verify functional significance of the LCI distribution, we estimated correlations between LCI in 5' region and the expression level evaluated by EEI (taking into account the codon bias; Likhoshvai and Matushkin, 2002). Overall, 4277 CDS with a length of at least 300 nucleotides were used in calculating LCI profile of *E. coli* K12. To calculate the correlation between LCI and EEI, we divided all the genes into 42 groups, each containing 100 genes (the last group comprised 77 genes), in the ascending order of their EEIs (the result was insensitive to the partition method used). Then, we calculated the mean EEI values and averaged LCI profiles for the remaining genes of all the groups. Thus, we obtained a vector of EEI values of all the sequences (each sequence has its own EEI value) as well as a vector of LCI values of all the positions of the sequence set (one LCI value for each sequence). At the next stage, we calculated the Pearson's correlation coefficients between the vector of EEI values and the vector of LCI values for each nucleotide and the significance of the correlations found using Student's test.

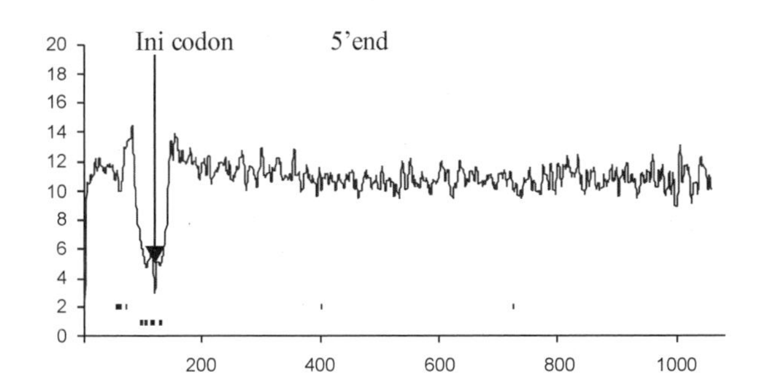

Figure 1. Positions whereat LCI value correlates significantly (at least 99.9% according to Student's test) with EEI value of gene are indicated under the profile. Regions of positive correlations are indicated at level +2 of the ordinate; of negative correlation, at level +1.

A deep "well" in the LCI profile (Figure 1) in the region of initiation codon complies with the steric size of ribosome. A continuous region of negative correlation is observed only at the translation initiation site with the relative coordinates [–12; +40]. Position 1 is the first nucleotide of AUG codon. The significance threshold was exceeded only in this region, Positive correlation is significant in the regions [–7; –2], [+3; +8], [+12; +21], and [+25; +33].

3.2.2 Eukaryotes: *Saccharomyces cerevisiae*

The LCI profiles along 5'UTRs (300 nucleotides upstream and downstream of AUG) are demonstrated in Figure 2. It is evident that LCI declines from about –200 nt to the start AUG codon and increases from +1 to +50 nt. We estimated the correlation between LCI in this region and the expression level evaluated by EEI (taking into account the codon bias; Likhoshvai and Matushkin, 2002) as we did it for *E. coli*. We divided all the genes into 12 groups according to increase in this index. The first group comprised 500 genes with the lowest values of EEI. The 500 genes displaying next levels of the EEI values composed the second group, and so on. The twelfth group contained 304 genes exhibiting the maximal indices (the result was insensitive to the partition method used). Then, we discarded from these groups all the genes containing less than 400 nt in their open reading frames, calculated the mean EEI values, and averaged LCI profiles for the remaining genes of all the groups.

At the next stage (similarly to *E. coli*), we calculated Pearson's correlation coefficients between the vector of EEI values and the vector of LCI values for each nucleotide and the significance of the correlations found using Student's test.

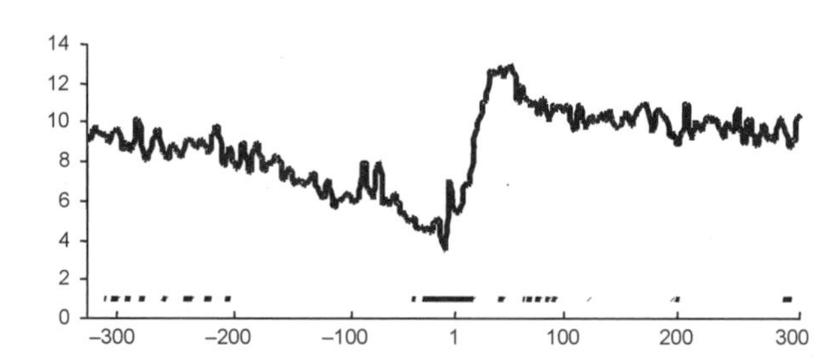

Figure 2. LCI profile of 5'-region of *S. cerevisiae* mRNAs: the abscissa, 5'-region (distances in nt relative to the initiation codon [+1; +3]); the ordinate, values of LCI; bold intermittent line, positions with significant correlations ($P < 0.01$) between LCI and EEI values; calculations were made taking into account the perfect hairpins with stems of 3–6 nt and loops of 3–50 nt.

The longest continuous region displaying a significant negative correlation ($P < 0.01$) insensitive to the parameters used for calculating LCI was the fragment [–19; +13] of 5'UTR (Figure 2). In this case, the significance level decreased to 98%, and this fragment expanded to [–20; +25].

The LCI value in the protein coding regions displays a trend of positive correlation with the expression level. However, it is possible that this trend to a considerable degree is determined by codon selection. Numerical experiments on generation of random sequences with a certain codon composition confirm this hypothesis (data not shown).

3.2.3 Eukaryotes: *Schizosaccharomyces pombe*

We calculated the LCI profile for *Schizosaccharomyces pombe* through averaging over all the coding sequences with a length of at least 300 nt (3988 CDS). Correlation analysis of the vectors of EEI and LCI was performed as described in Section 3.2.1. The results are shown in Figure 3.

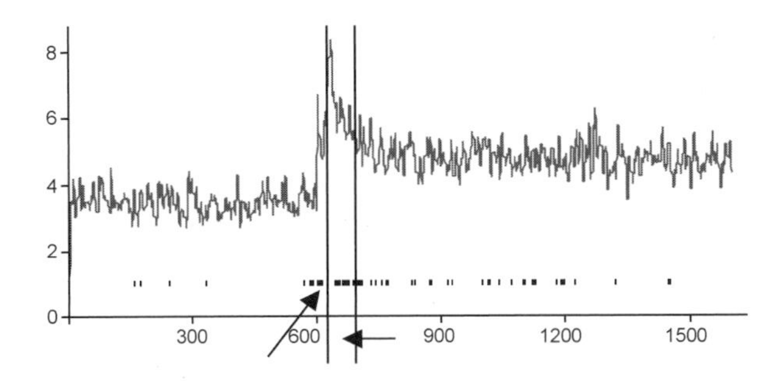

Figure 3. The coordinate +1 corresponds to the first nucleotide of initiation codon (position 601 on the abscissa axis. Left arrow indicate the region with a significant negative correlation, [–1; +9]; position +7 displays the maximal significance (>99.99). Right arrow indicates the regions with a significant positive correlation—[+44; +55], [+61; +80], and [+86; +107]; position +67 displays the maximal significance (>99.9999).

4. CONCLUSION

We have demonstrated that traces of evolutionary optimization of translation elongation are detected in all the organisms analyzed. These traces are found at the levels of gene codon compositions, local secondary mRNA structures, or both at once. All the unicellular microorganisms analyzed fall into the five following groups: (1) the group comprising 32 organisms with the elongation efficiency determined only by the codon composition; (2) the group of 8 organisms with the codon composition inessential for the elongation efficiency; on the contrary, the elongation is limited by the degree of local secondary structures; however, the energies of these local secondary structures play no role; (3) the group consisting of only one organism—*P. aeruginosa* PA01—with no effect of codon composition on elongation efficiency controlled by both the local secondary structures and their energies; (4) the group of 30 organisms where both the codon composition and local secondary structures are essential for determining the elongation efficiency; however, the energies of local structures are insignificant; and (5) the group of 3 organisms where all the three factors—codon composition, local secondary structure, and their energies—are essential.

It was demonstrated experimentally that the secondary structure decreased the translation activity of 5'UTR by slowing the ribosome movement, whereas no strict evidence on the negative influence of hairpins located in CDS or 3'UTR on mRNA expression was found (Kozak, 1999; Niepel *et al.*, 1999). We analyzed the distribution of LCI (local complementarity index), reflecting the potential base pairing of nucleotides along the borders of CDS with 5'UTRs. We assumed that the secondary structure could represent an additional signal marking the transition between coding and mRNA 5'-untranslated parts of yeast and other organisms. We found that there was a marked shift in LCI at the border of 5'UTR and CDS. This difference in potential secondary structure can represent an additional signal increasing the efficiency of AUG recognition. Moreover, there is a significant negative correlation for *S. cerevisiae*, *S. pombe*, *E. coli* K12 between LCI and EEI values, possibly reflecting selection for a more optimal translation start site in highly expressed mRNAs.

The results on LCI allow the region (approximately from –20 to +25) to be outlined where the absence of stable hairpins could be important. It agrees with the scanning model for eukaryotic translation as well as with the ribosome size of prokaryotes and locations of codons exposed in ribosome A site with respect to its external surface.

ACKNOWLEDGMENTS

The work was supported in part by the Russian Foundation for Basic Research (grants Nos. 01-07-90376-в, 02-07-90355, 03-07-96833-p2003, 03-01-00328, 02-04-48802, 02-07-90359, 03-04-48506, 03-04-48829); the Siberian Branch of the Russian Academy of Sciences (integration project No. 119); Russian Ministry of Industry, Science, and Technologies (grant Nos. 43.073.1.1.1501 and 43.106.11.0011); grant PCB RAS (No. 10.4); NATO (grant No. LST.CLG.979816); the U.S. Civilian Research & Development Foundation for the Independent States of the Former Soviet Union (CRDF) No. NO-008-X1.

COMPUTER ANALYSIS OF miRNA–mRNA DUPLEXES AND ITS APPLICATION TO PREDICTING POSSIBLE TARGET mRNAS OF *ARABIDOPSIS*

N.A. OMELYANCHUK[1]*, D.G. VOROBIEV[1,2]
[1]*Institute of Cytology & Genetics, Siberian Branch of the Russian Academy of Sciences, prosp. Lavrentieva 10, Novosibirsk, 630090 Russia, e-mail: nadya@bionet.nsc.ru; denis@bionet.nsc.ru;* [2]*Novosibirsk State University, ul. Pirogova 2, Novosibirsk, 630090 Russia*
* *Corresponding author*

Abstract Prediction of possible targets for microRNAs (miRNAs) in complete genomic sequences is a difficult task due to potential bulges and loops occurring in miRNA–mRNA duplexes. Thermodynamic calculations of the free energy may assist in solving the problem of searching for such potential duplexes. The program GArna-duplex allows the duplexes of various degrees of complementarity to be detected and the free energy of stable duplexes to be assessed quantitatively. Analysis of several cases of formation of miRNA–mRNA duplexes experimentally detected so far demonstrated that (1) GArna-duplex was capable of detecting both the duplexes able to guide mRNA cleavage and those inhibiting translation and (2) the free energy of duplex formation correlated with their function. Search for potential targets for 24 *Arabidopsis* miRNAs demonstrated that the miRNAs differed in both the number of potential duplexes they can form and the least free energy of binding with transcripts. Target mRNAs forming most stable duplexes with the miRNAs in question were predicted.

Key words: microRNA, miRNA, target mRNA, RNA duplex, free energy

1. INTRODUCTION

Activity of the genes is regulated at the levels of transcription, mRNA splicing, translation, and protein degradation. Recently, a new component was added to this regulatory network—the regulation realized by microRNAs (miRNAs). The miRNAs are RNAs usually with a length of 20–22 nucleotides whose complementary binding to mRNA either inhibits its translation (Olsen and Ambros, 1999) or causes its cleavage (Llave *et al.*, 2002; Kasschau *et al.*, 2003; Tang *et al.*, 2003). MicroRNAs are transcribed as a long transcript (pri-mRNA); in the nucleus, the transcript subsequently gives rise to miRNA precursor (pre-miRNA) with a length of 70 nucleotides and a characteristic stem–loop structure (Lagos-Quintana *et al.*, 2001; Lau *et al.*, 2001; Lee *et al.*, 2002). There, pre-miRNAs interact with proteins producing miRNP complex to be transported into the cytoplasm, where mature miRNAs are formed with the help of Dicer, a member of ribonuclease III family, a DEAD/DEAH box helicase (Ketting *et al.*, 2001; Knight and Bass, 2001; Mourelatos *et al.*, 2002).

The miRNA–protein complex is similar or may be even identical to RNA-induced silencing complex (RISC), guiding mRNA cleavage while interference (Hutvagner and Zamore, 2002; Mourelatos *et al.*, 2002). When miRNAs guide mRNA cleavage, they act as the siRNA in the RISC complex; moreover, this cleavage is also inhibited by micrococcal nuclease, which cancels the RISC activity and may be inhibited by RNA-silencing suppressor, P1/HC-Pro (Kasschau *et al.*, 2003; Tang *et al.*, 2003). So far, miRNAs have been discovered in *Caenorhabditis elegans*, *Arabidopsis thaliana*, *Drosophila melanogaster*, human, and mouse (Lagos-Quintana *et al.*, 2001; 2002; Lau *et al.*, 2001; Mourelatos *et al.*, 2002; Reinhart *et al.*, 2002). Prediction of potential target mRNAs for miRNAs is among the topical problems of the modern biology. The majority of miRNAs fail to display a precise complementarity to the target mRNA binding sites; nevertheless, a partial noncomplementarity does not prevent from formation of miRNA–mRNA duplexes (Wightman *et al.*, 1993; Brennecke *et al.*, 2003; Kasschau *et al.*, 2003; Tang *et al.*, 2003). Unfortunately, detection of target mRNAs by the alignment software is incapable of covering the entire range of potential targets, as this software is more adapted to detecting long complementary regions (perfect duplexes).

Thus, it was of interest to apply the program GArna-duplex (available at http://wwwmgs.bionet.nsc.ru/mgs/programs/2dstructrna/), allowing duplexes with various loop sizes and localizations within a duplex to be searched for and free energy of stable duplexes to be quantitatively assessed, to searching for target mRNAs. The program succeeded in detection of all the sites discovered experimentally for both mRNA cleavage and translation inhibition. We assessed quantitatively the free energies of duplexes for both cases and demonstrated that the free energy of stable duplex correlate with the function of miRNA. We predicted the potential targets for 24 *Arabidopsis* miRNAs whereat most stable duplexes are formed. The prediction results obtained were compared with earlier predictions of potential targets for *Arabidopsis* miRNAs based on search for complementary regions (Park *et al.*, 2002; Rhoades *et al.*, 2002; Bartel B. and Bartel D., 2003).

2. METHODS AND ALGORITHMS

We used the program GArna-duplex from the package GArna (Titov *et al.*, 2002) when searching for potential sites for miRNAs. The program GArna-duplex searches for potentially stable RNA duplexes using an approximate yet very fast algorithm for finding duplexes; this algorithm is based on the steepest descent to the free energy space. In the algorithm, the duplex is represented with a list of perfect stems, ListUsed, formed by a pair of sequences considered. The algorithm comprises the following procedures. (1) Calculation of all the possible stems for a pair of sequences S_1 and S_2 (this stage makes the complexity of algorithm linear with reference to the product of the lengths of these sequences, L_1L_2) and adding them to the list of all the stems not involved in the duplex formation, ListFree. Formation of stems is allowed only between the sequences, not within the same sequence. Initially, the list of duplex-forming stems ListUsed is empty. (2) Energy is calculated for each stem from ListFree (using thermodynamic parameters described by Turner *et al.*, 1988). (3) Choosing of the stem h_i from ListFree that decreases maximally the energy of duplex. (4) Transfer of the stem h_i from ListFree to ListUsed. (5) Calculation of free energy of the duplex formed (6) to check whether the free energy of the duplex decreased. If it decreased, the procedure is iterated from step 3; if not, calculation of energy of the duplex is ceased. The output is the ListUsed comprising the components of the duplex searched for. This approach allows the pairs of RNA sequences potentially capable of forming stable duplexes to be chosen for further analysis in more detail, taking into account distinctions in the hydrogen bond energies of the pairs GC, AU, and GU; energy of stacking; and destabilizing energy of loops.

The miRNA sequences were extracted from the database Rfam (http://www.sanger.ac.uk/Software/Rfam/mirna). Sequences of the experimentally confirmed target sites for calculation of the energy of stable miRNA–mRNA duplexes were taken from the corresponding publications. To test the operation of GArna-duplex, the mRNA sequences carrying the targets for miRNA discovered experimentally were extracted from GenBank (http://www.ncbi.nlm.nih.gov/Entrez/index.html). Sequences of all the *Arabidopsis* transcripts from the TAIR database (http://www.arabidopsis.org/) were used to predict potential targets for miRNAs.

3. RESULTS AND DISCUSSION

3.1 Free energy of stable miRNA–mRNA duplexes correlates with the fate of mRNA after duplex formation

Using GArna-duplex, we calculated free energies of the duplexes whose constituent mRNA's fate has been determined experimentally. The results of this analysis are listed in Table 1. Coexpression of *Arabidopsis* miR171 miRNA with two *Arabidopsis* mRNAs *SCL6-III* and *SCL6-IV* in tobacco leaf tissues results in cleavage of both mRNAs in the region of duplex formation. However, no cleavage of mRNA occurred upon coexpression of miR171 with *SCL6-IV* carrying three mutations in the miRNA-binding site (Llave *et al.*, 2002). The free energies of miR171–*SCL6-III* and miR171–*SCL6-IV* duplexes amount to –40.3 and –40.6 kcal/mol, respectively; of the miR171 duplex with *SCL6-IV* carrying mutations at the binding site, to –23.5 kcal/mol. The *Arabidopsis* miRNAs miR165 and miR166 differ in a single C→U transition (Reinhart *et al.*, 2002). In *Drosophila* embryo lysate, an siRNA duplex where one strand carried either miR165 or miR166 sequence caused cleavage of mRNA of *Arabidopsis* gene *PHV*; however, when *PHV* mRNA had a G→A mutation in the binding site, no cleavage was observed (Tang *et al.*, 2003).

Energies of the miR165–*PHV* and miR166–*PHV* duplexes amount to –40.0 and –36.6 kcal/mol, respectively; in the case of mutation in the miRNA-binding sites, to –33.5 and –30.1 kcal/mol. Thus, the increase in free energies by 6.5 kcal/mol caused canceling of the mRNA cleavage.

The cleavage of mRNA resulting from formation of miRNA–mRNA duplexes may be demonstrated by RNA ligase-mediated 5' RACE with untreated RNA extracts followed by sequence analysis of the cloned PCR product. This allowed ten miRNA–mRNA duplexes to be discovered that caused mRNA cleavage in *Arabidopsis* plants (Kasschau *et al.*, 2003). These duplexes displayed the following free energies indicated in parenthesis: miR160–*ARF17* (–45.6 kcal/mol), miR160–*ARF10* (–39.4 kcal/mol), miR164–*CUC1* (–38.1 kcal/mol), miR164–*CUC2* (–37.8 kcal/mol), miR172–AT5G60120 (–35.5 kcal/mol), miR172–*AP2* (–34.5 kcal/mol), miR172–AT2G28550 (–32.9 kcal/mol), miR167–*ARF8* (–32.8 kcal/mol), miR172–AT5G67180 (–32.7 kcal/mol), and miR156–AT5G43270 (–32.7 kcal/mol). The largest difference between free energies of the duplexes formed by the same miRNA (miR160–*ARF17* and miR160–*ARF10*) equals 6.2 kcal/mol. In this case, the growth in free energy by the value shown failed to change the influence of duplex on mRNA cleavage, whereas the increase in free energy by 6.5 kcal/mol in the duplexes miR165–*PHV* and miR166–*PHV* canceled this function.

In *Caenorhabditis elegans,* binding of the miRNAs *lin-4* and *let-7* to mRNA targets do not change the mRNA level but rather inhibit translation of these mRNAs. The sequence of *C. elegans lin-4* miRNA is complementary to seven sites in the 3'UTR region of gene *lin-14;* however, the complementarity is not precise and the corresponding duplexes have loops and bulges. Deletion of all these sites or their part removes or weakens, respectively, the negative regulation of *lin-14* translation (Wightman *et al.*, 1993; Olsen and Ambros, 1999). The free energies of duplexes formed by *lin-4* and the 3'UTR of *lin-14* amount to –15.3, –12.9, –20.4, –15.5, –16.6, –11.4, and –25.1 kcal/mol, respectively. One site in the 3'UTR of *lin-28* gene is complementary to *lin-4* miRNA; the allele with this site deleted displays a gain-of-function phenotype (Moss *et al.*, 1997). Two sites in the 3'UTR of *lin-41* gene are complementary to *let-7* miRNA. Expression of *lin-41::gpf* transgene decreases when expression of *let-7* miRNA is increased (Slack *et al.*, 2000). The free energies of these duplexes also do not exceed –25 kcal/mol. Translation inhibition on the background of an unaltered mRNA level was also demonstrated for the *bantam* miRNA bound to one of the five sites localized to the mRNA 3'UTR of *Drosophila melanogaster hid* gene. Deletion of either one or four sites also reduced negative regulation of translation (Brennecke *et al.*, 2003). Free energies of the duplexes formed by *bantam* in *hid* 3'UTR amount to –21.0, –14.9, –15.9, –23.5, and –15.7 kcal/mol, respectively. Human miR30 miRNA bound to an artificially synthesized target, where the site with imperfect complementarity to the miRNA in question was repeated four times in 3'UTR, blocked translation of the reporter gene, however, had no effect on the level of target mRNA (Zeng *et al.*, 2002). According to our calculations, the free energy of this duplex amounts to –29.2 kcal/mol.

The consolidated data listed in Table 1 suggest that the free energies of the duplexes capable of guiding mRNA cleavage are predominantly below –30 kcal/mol. When the free energies of the duplexes formed are exceeding –29.2 kcal/mol, these duplexes may fail to cause mRNA cleavage; however, in certain cases may inhibit translation. Presumably, after duplex stability, the second in the list of important factors in this process is interaction of the duplex with proteins of RISC-like complex, cleaving mRNAs (Hutvagner and Zamore, 2002; Mourelatos *et al.*, 2002; Kasschau *et al.*, 2003; Tang *et al.*, 2003). Conceivably, miRNAs may display a different affinity for this protein complex. For example, only eight of the ten *Arabidopsis* miRNAs studied are accumulated to increased concentrations in the presence of the protein factor P1/HC-Pro, a viral suppressor of RNA silencing, which is supposed to interfere with assembly or activity of RISC-like complex

(Kasschau *et al.*, 2003). It may be indeed the differences in miRNA affinities for protein complex cleaving mRNAs that could explain the following contradiction: the duplexes miR172–AT2G28550, miR167–*ARF8*, miR172–AT5G67180, and miR156–AT5G43270, displaying free energies of –33 to –32.7 kcal/mol, guide mRNA cleavage, whereas the duplex miR165–*PHV* carrying a mutation in the miRNA binding site, displaying the free energy of –33.5 kcal/mol, fails to cause the cleavage.

Table 1. Free energies (ΔG) of the duplexes exerting experimentally confirmed effect on the target RNAs

ΔG, kcal/mol	Species and duplex
mRNA cleavage	
From –45.6 to –34.5	*A. thaliana,* miR171–*SCL6-III*, miR171–*SCL6-IV* (Llave *et al.*, 2002); miR165–*PHV*, miR166–*PHV* (Tang *et al.*, 2003); miR160–*ARF17*, miR160–*ARF10*, miR164–*CUC1*, miR164–*CUC2*, miR172–AT5G60120, miR172–*AP2* (Kasschau *et al.*, 2003)
From –32.9 to –32.7	*A. thaliana*, miR172–AT2G28550, miR167–*ARF8*, miR172–AT5G67180, miR156–AT5G43270 (Kasschau *et al.*, 2003)
No mRNA cleavage	
From –33.5 to –30.1	*A. thaliana*, miR165–*PHV**, miR166–*PHV** (Tang *et al.*, 2003)
–23.5	*A. thaliana*, miR171–*SCL6-IV** (Llave *et al.*, 2002)
No mRNA cleavage; binding of miRNA to mRNA inhibits translation of target mRNA	
–29.2	*H. sapiens*, miR30–artificial target (Zeng *et al.*, 2002)
From –25.1 to –11.4	*C. elegans*, *lin-4–lin-14-1*, *lin-4–lin-14-2*, *lin-4–lin-14-3*, *lin-4–lin-14-4*, *lin-4–lin-14-5*, *lin-4–lin-14-6*, *lin-4–lin-14-7* (Wightman et al., 1993); *lin-4–lin-28* (Moss *et al.*, 1997); *let-7–lin-41-1*, *let-7–lin-41-2* (Slack *et al.*, 2000)
From –23.5 to –14.9	*D. melanogaster*, *bantam–hid-1*, *bantam–hid-2*, *bantam–hid-3*, *bantam–hid-4*, *bantam–hid-5* (Brennecke *et al.*, 2003)

* Mutation in miRNA-binding site of mRNA.

Nevertheless, the role of affinity for protein complex should not be exaggerated as well, since it is known that protein complexes may have insufficiently high affinity for duplexes; for example, the RISC-like complexes of remote species, such as tobacco and even *Drosophila*, are capable of cleaving *Arabidopsis* miRNA–mRNA duplexes (Llave *et al.*, 2002; Tang *et al.*, 2003). GArna-duplex succeeded in detecting all these sites in the above-listed duplexes of *A. thaliana*, *C. elegans*, and *D. melanogaster* sequences stored in GenBank. This gave us the grounds to apply this software for predicting the targets capable of guiding mRNA cleavage for the known 24 miRNA of *Arabidopsis*.

3.2 Prediction of potential targets for *Arabidopsis* miRNAs

Overall, GArna-duplex predicted 20,864 potential miRNA–mRNA duplexes with a free energy below –25.1 kcal/mol and 1062 potential duplexes with a free energy below –30 kcal/mol for the 24 known miRNAs of *Arabidopsis*. The miRNAs differ in the number of potential targets, the range of free energies of the duplexes with these targets , and the energies of most stable duplexes (Table 2). The number of potential targets at ΔG below –25.1 kcal/mol varies from 18 for miR158 to 7985 for miR179. Presumably, some miRNAs regulate activities of many genes, whereas others are highly specific and regulate activities of a limited number of genes. A certain dependence of the number of duplexes on the maximal miRNA length is evident, as the maximal number of duplexes (3759 and 7985 at ΔG below –25.1; 242 and 512 at ΔG below –30 kcal/mol) was predicted for miR177 and miR179, whose lengths amount to 27 and 32 nt, respectively (Table 2). Nevertheless, the number of potential duplexes depends also on the nucleotide context, since the number of duplexes with a free energy below –30 kcal/mol varies from 1 to 37 in the case of miRNAs with an equal length (21 nt); with a free energy below –25.1 kcal/mol, from 24 to 1086.

As we searched all the *Arabidopsis* transcripts for potential targets, this allows us to infer that each miRNA in the genome has its own limit of duplex stability. All the 24 miRNAs studied fall into three distinct groups with reference to the free energy of the most stable duplex they form. The first group comprises miR160 with its most stable duplex with $\Delta G = -45.6$ kcal/mol and six other miRNAs, whose most stable duplexes display the free energies from –40.6 to –37.9 kcal/mol. The second group, 15 miRNAs, forms most stable duplexes with free energies of –36.1 to –31.8 kcal/mol. The members of the third group, miR158 and miR161, form duplexes exhibiting the least free energies: the former with the mRNAs for reverse transcriptase AT1G10160 (–28.3 kcal/mol) and mitotic checkpoint protein AT1G49910 (–28.1 kcal/mol) and the latter with mRNAs for eight pentatricopeptide repeat-containing proteins (–30.6 to –28.2 kcal/mol).

Thus, the first group comprises seven miRNAs capable of forming the most stable duplexes, namely, miR160, miR164, miR165, miR166, miR171, miR177, and miR179. The free energies of most stable duplexes formed by miRNAs of the second group—miR156, miR157, miR159, miR162, miR163, miR167, miR168, miR169, miR170, miR172, miR173, miR174, miR175, miR176, and miR178—have intermediate values. The third group contains two miRNAs—miR158 and miR161—forming the most stable duplexes with maximal free energies. However, as for the third group miRNAs, it is yet impossible to predict the functions of these duplexes.

Table 2. Number of potential duplexes within different energy ranges and free energies of the most stable duplexes

miRNA	Maximal miRNA length (nt)	Number of duplexes*		Free energy of the most stable duplex	Number of predicted targets	Of them, predicted earlier
		ΔG below –25.1	ΔG below –30			
miR156	21	108	11	–33.8	11	10
miR157	21	58	10	–33.0	10	9
miR158	20	18	0	–28.3		
miR159	21	189	5	–33.0	4	4
miR160	21	851	37	–45.6	3	3
miR161	21	43	1	–30.6		
miR162	21	75	2	–33.9	2	1
miR163	24	178	7	–36.1	7	5
miR164	21	1086	32	–39.4	9	5
miR165	21	565	21	–40.0	5	4
miR166	21	436	22	–38.0	12	1
miR167	21	269	6	–33.7	6	2
miR168	21	1068	24	–33.7	24	1
miR169	21	301	6	–33.0	6	2
miR170	21	38	3	–35.5	3	3
miR171	21	167	6	–40.6	3	3
miR172	21	47	5	–35.5	5	4
miR173	23	77	4	–31.8	4	
miR174	24	1845	53	–35.1	52	1
miR175	21	24	3	–33.3	3	2
miR176	27	1405	42	–34.0	42	
miR177	27	3759	242	–37.9	85	
miR178	20	272	8	–32.3	8	
miR179	32	7985	512	–39.8	63	

* Variants of splicing were considered as a separate target mRNA only when binding sites and free energies of duplexes were different.

To predict the potential targets that form duplexes with the least free energies (most likely, guiding mRNA cleavage) with miRNAs from both the first and second groups, we considered only the duplexes with a free energy below –30 kcal/mol. For the first group, an additional criterion was used. It is based on calculations of experimentally confirmed duplexes with the miRNAs of this group and requiring that the difference between the free energy of the predicted duplex and the least energy predicted for this miRNA does not exceed 6.2 kcal/mol. Thus, we predicted 367 targets for 22 miRNAs except for the third group miRNAs (Table 2); of them, 60 targets were predicted earlier (Rhoades *et al.*, 2002; Park *et al.*, 2002; Bartel B. and Bartel D., 2003).

Capabilities of GArna-duplex in predicting targets for miRNAs are illustrated by the below examples. Our calculations confirmed the prediction of ten and nine mRNAs for SPB and SPB-like proteins as targets for miR156 and miR157, respectively. Of them, nine mRNAs are potential targets for both miRNAs (Rhoades *et al.*, 2002). An additional mRNA—the mRNA for SPB-like protein AT3G15270—is also a potential target for these two miRNAs. The mRNAs for HD-Zip transcription factors AT1G30490 *PHV*, AT2G34710 *PHB*, AT4G32880 Athb-8, AT5G60690 *REV* were also predicted as potential targets for miR165 and AT1G52150, for miR166 (Rhoades *et al.*, 2002). GArna-duplex predicts all these targets for both miRNAs. This prediction found its experimental confirmation for the miR166–*PHV* duplex (Tang *et al.*, 2003). As we predicted, the first four mRNAs for HD-Zip transcription factors listed above form more stable (by 3.4 kcal/mol) duplexes with miR165 compared with miR166, whereas only the mRNA AT1G52150, whose binding site differs from these four in two nucleotides, may form a more stable (by 2 kcal/mol) duplex with miR166. The miR166 has also additional potential targets in mRNAs for glycine-rich protein AT4G36020, homeodomain protein AT4G03250, calcium-dependent protein kinase AT2G46700, succinate-semialdehyde dehydrogenase AT1G79440, CHP-rich zinc finger proteins AT2G13910 and AT1G53340, and an expressed protein AT3G25640. In addition to mRNAs for two auxin response transcription factors, predicted earlier as targets for miR167 (Park *et al.*, 2002; Rhoades *et al.*, 2002), the mRNAs for Tonneau 1b AT3G55005, DNA-binding protein AT3G61310, a hypothetical protein AT1G50800, and an expressed protein AT5G19950 are also potential targets for this miRNA. Moreover, Tonneau 1b is able to form with miR167 the duplexes with a free energy that is by 0.9–1.1 kcal/mol lower compared to that for mRNAs for auxin response transcription factors predicted earlier.

No targets were previously found for miR176, miR178, and miR179, as all the duplexes they formed contained bulges and loops. GArna-duplex predicts a considerable number of potential targets for these miRNAs. Examples of duplexes with new targets found for several miRNAs by GArna-duplex are shown in Figure 1. They include perfect duplexes, the duplexes where G-U pairs are involved in stem formation along with complementary pairs, duplexes with small bulges formed by either mRNA or miRNA, duplexes with a loop, and duplexes carrying several imperfections.

Thus, we demonstrated broad capabilities of the program GArna-duplex in studying of an important aspect of gene activity regulation—miRNA–mRNA interaction—by the example of the available experimental data and prediction of potential *Arabidopsis* targets. Taking into account the free energy of duplexes allowed us to develop certain criteria for the prediction, extended essentially the range of potential targets for certain *Arabidopsis* miRNAs, and permitted targets of the miRNAs lacking yet any forecast targets to be predicted.

Figure 1. Examples of duplexes formed by new targets predicted by GArna-duplex: (*a*) the perfect duplex miR173–AT3G28460 and partially noncomplementary duplex miR156–AT3G15270; (*b*) the duplexes miR172–AT2G39250 and miR163–AT4G24160, where G-U pairs are involved in stem formation along with complementary pairs; (*c*) duplexes with bulges formed by either mRNA (miR164–AT2G47650) or miRNA (miR173–AT4G03010); (*d*) the duplex miR166–AT2G13910 with a loop; (*e*) the duplexes miR167–AT3G55005 and miR164–AT2G01140, combining several imperfections.

ACKNOWLEDGMENTS

The work was supported by the Russian Foundation for Basic Research (grants Nos. 01-07-90376-в, 02-07-90359, 03-07-96833-p2003, 03-07-96833, 03-04-48506-a, 03-04-48829, 03-01-00328, 03-07-06077-мас); grant PCB RAS (No. 10.4); the Siberian Branch of the Russian Academy of Sciences (integration project No. 119); Russian Ministry of Industry, Science, and Technologies (grant No. 43.073.1.1.1501, subcontract No. 28/2003); the U.S. Civilian Research & Development Foundation for the Independent States of the Former Soviet Union (CRDF) No. NO-008-X1; NATO grant No. LST.CLG.979816.

CORRELATIONS BETWEEN SEQUENCE FEATURES OF YEAST GENES FUNCTIONAL REGIONS AND THE LEVEL OF EXPRESSION

A.G. PICHUEVA[1]*, A.V. KOCHETOV[1,4], L. MILANESI[2], Yu.V. KONDRAKHIN[1,3], N.A. KOLCHANOV[1,4]

[1]Institute of Cytology & Genetics, Siberian Branch of the Russian Academy of Sciences, prosp. Lavrentieva 10, Novosibirsk, 630090 Russia, e-mail: anna@bionet.nsc.ru; [2]CNR Institute of Biomedical Technology (CNR-ITB), via Fratelli Cervi, 93, 20090 Segrate (Milan), Italy; [3]Ugra Research Institute of Information Technologies, ul. Mira 151, Khanty-Mansyisk, 628011 Russia; [4]Novosibirsk State University, ul. Pirogova 2, Novosibirsk, 630090 Russia

* *Corresponding author*

Abstract Analysis of contextual features of *S. cerevisiae* noncoding gene regions (5'UTR and core promoter) for the correlation with mRNA ability to support a high translation rate (expressed as a codon adaptation index) suggested characteristics that might modulate the expression efficiency. We found that 5'UTRs of highly expressed mRNAs tend to have a higher content of adenine, lower content of guanine, higher imbalance in the content of complementary nucleotides, and preferences for occurrences of certain dinucleotides. In turn, sequence features of the core promoter tend to have higher content of thymidine, lower content of adenine and guanine, and preferences for certain dinucleotides. Further, we investigated interdependencies between the significant features of noncoding regions and gene expression level by using the method of principal components, regression analysis, and discriminant analysis. It was found that a sample of yeast noncoding regions consisted of three clusters, each described with a specific function allowing for prediction of the level of gene expression. Thus, we demonstrated that (1) noncoding functional regions of yeast genes displayed specific base composition (mono- and dinucleotide contents); (2) core promoter and 5'UTR were characterized by different subsets of features correlating with the level of gene expression; and (3) discriminant analysis of contextual features of noncoding regions showed a principal feasibility to predict gene expression level. It means that the sequence organization of noncoding gene regions is an important element of the structure and regulation of eukaryotic genes.

Key words: yeast mRNAs, translation, context effects, statistical analysis

1. INTRODUCTION

It is known that base composition of eukaryotic gene functional regions (promoter, introns, 5'UTR, 3'UTR, and coding part) differs considerably (Pesole *et al.*, 1997). Some of the region-specific characteristics were found to be of functional importance, e.g. avoidance of AUG triplets and stable hairpins in 5'UTR increased translation initiation efficiency (Kozak, 2002). However, in the vast majority of cases, possible functional role of the intrinsic region features is unclear and may depend on various factors (e.g. different mutation pressure on transcribed or untranscribed regions, genome context, isochores, etc.).

Investigation of such region characteristics may help reveal gene features influencing interaction with cellular expression machinery. In particular, expression rate of eukaryotic genes can be predicted by using this approach (Sharp and Li, 1987; Kochetov *et al.*, 1998; 1999; Pavesi *et al.*, 1999; Davuluri *et al.*, 2000). Comparison of predicted expression levels with microarray, SAGE, or proteomic data may reveal specifically regulated genes (e.g. genes at a certain developmental stage or in a tissue-specific manner).

In this paper, we present the results of analysis of sequence features of *S. cerevisiae* 5'UTRs and core promoters. Yeast genes were chosen because they represent a good model for investigation. It is well known that mRNAs of highly expressed yeast genes are optimized at the level of translation elongation (by preferential usage of synonymous codons corresponding to the major fractions of isoacceptor tRNAs; McCarthy, 1998; Percudani *et al.*, 1999). Previously, we described the correlations between codon adaptation index (CAI), average transcript level, mRNA cytoplasmic stability, weights of translation initiation and termination signals, and some features of gene functional regions (Kochetov *et al.*, 2002). Here, we report evidences on interdependencies between certain contextual features of promoters and 5'UTR and the rate of gene expression.

Since expression involves several consecutive steps (transcription, pre-mRNA processing, mRNA transport into the cytoplasm, translation, and posttranslational modification), the intensity of protein synthesis depends not only on the intensity of transcription, but also on the efficiency of the other processes. According to the limiting step model (McCarthy, 1998; Lesnik *et al.*, 2000), the level of protein synthesis is determined by the least efficient step. This principle underlies most regulatory mechanisms. Hence, a high-level expression of a gene requires optimization of its nucleotide sequence to ensure its efficient interaction with the cell factors. Functional gene regions are responsible for different steps in gene expression, and each region may have a specific context organization.

Sequence features of yeast noncoding regions (mono- and dinucleotide contents) were analyzed by the method of principal components. It was found that the sample analyzed consists of three clusters. Each cluster was characterized with specific eigenvector of features correlation matrix. Further, we used liner regression analysis to obtain the functions describing interdependence between the codon adaptation index (the measure reflecting gene expression level) and corresponding eigenvectors. This demonstrated the possibility to predict codon adaptation index from contextual features of noncoding yeast gene regions (core promoter and 5'UTR). This result supports strongly the basic idea of the rate-limiting stage model (McCarthy, 1998; Lesnik *et al.*, 2000) and provides the background for both modeling functional regions and development of methods for prediction of their functional activities.

2. METHODS AND ALGORITHMS

Sequence data. The mRNA 5'UTR sequences were extracted from the EMBL database. Full-sized 5'UTRs were selected from the entries containing description of mapped transcription start sites and complete coding regions. For the selection of such 5'UTRs from EMBL DNA entries, Feature Table keys «CDS» and one of the following—«mRNA», «precursor_RNA», «prim_transcript», «exon», or «5'UTR»—were used. In this work, we used only the 5'UTRs with experimentally mapped 5'-ends. This resulted in a set of mRNA 5'UTRs of 240 yeast genes. To avoid the bias due to redundant sequence data in the statistical analysis, redundant sequences (CDS homology >70%) were removed. After this procedure, the set comprised 5'UTRs of 171 yeast genes. A sample of promoters (213 sequences spanning 150 nucleotides upstream of the major transcription start site) was also compiled.

Methods. Codon adaptation index (Sharp and Li, 1987) was calculated by the program CodonW 1.3 (written by J. Peden and available at http://www.molbiol.ox.ac.uk/cu). Pearson's linear (LCC) and Kendall's rank tau (TAU) correlation coefficients were used for measurement of correlation between two variables. These coefficients were based on different assumptions (parametric and nonparametric) and thus, the relation between variables was analyzed independently.

Interdependence of sequence features of yeast noncoding regions (mono- and dinucleotide contents) and CAI was analyzed using the method of principal components (Ledermann *et al.*, 1984). To predict gene expression level, the least squares liner regression model (Draper and Smith, 1981) was used. Discriminant features of the clusters were estimated by Fisher discriminant analysis (Ledermann *et al.*, 1984).

3. RESULTS AND DISCUSSION

3.1 Paired correlations between codon adaptation index and promoter and 5'UTR contextual features

It was shown earlier that certain parameters of yeast genes (mRNA level, codon adaptation index, and weights of initiation and termination signals) correlated positively and significantly (Kochetov *et al.*, 2002), agreeing with the rate-limiting stage model (McCarthy, 1998). Correlation coefficients between CAI and base composition of two neighboring regions—promoter and 5'UTR—are shown in Table 1. It is evident that (1) some contextual features of these functional regions correlate significantly with the codon adaptation index and (2) each functional region is characterized by a highly specific subset of such significant features. According to the correlations, an "efficient" 5'UTR (i.e. presumably supporting a high rate of translation initiation) should contain more A and less G and be highly asymmetric in the contents of the complementary nucleotides (G/C and A/U). In addition, some dinucleotides could also be important: it should contain less GpG and UpG and more CpG. Similarly, the core promoter of highly expressed yeast genes tends to have higher content of T, AA, and GG and lower frequency of A, G, AC, and GC. Some of these correlations could be explained by the known experimental data (i.e. lower G content and higher imbalance in the content of complementary nucleotides in 5'UTR could be associated with selection against stable secondary structure and more efficient initiation of translation), whereas functional significance of the other features remains unclear.

Table 1. Correlations between contextual features of promoters and 5'UTRs with codon adaptation index of yeast genes

	Promoter		5'UTR			Promoter		5'UTR	
	LCC	TAU	LCC	TAU		LCC	TAU	LCC	TAU
A	**–0.33**^{3}	**–0.24**^{3}	**0.38**^{3}	**0.26**^{3}	AA	**0.24**^{2}	**0.15**^{2}	–0.04	0.02
G	**–0.32**^{3}	**–0.20**^{3}	**–0.50**^{3}	**–0.34**^{3}	AC	–0.08	**–0.12**^{1}	0.08	0.04
C	0.05	0.06	**0.21**^{1}	0.12	GG	**0.30**^{3}	**0.11**^{1}	**–0.38**^{3}	**–0.28**^{3}
T	**0.46**^{3}	**0.32**^{3}	–0.16	–0.08	GC	**–0.18**^{1}	**–0.10**^{1}	–0.08	–0.11
$I_{(G,C)}$	n.t.	n.t.	**0.38**^{3}	0.11	CG	–0.02	0.03	0.19	**0.15**^{1}
$I_{(A,U)}$	n.t.	n.t.	**0.21**^{1}	**0.14**^{1}	TG	–0.02	–0.06	–0.07	**–0.18**^{2}

Significant correlations are bold-faced; the level of significance is marked by uppercase numbers: $^1P < 0.05$; $^2P < 0.01$; and $^3P < 0.001$; $I_{(G,C)} = |G - C|/(G + C)$, $I_{(A,U)} = |A - U|/(A + U)$; and n.t., not tested.

3.2 Interdependencies between the contextual features of yeast gene noncoding regions and the level of expression

Correlation matrix for all the mono- and dinucleotide frequencies was analyzed by principal component method. The main component revealed was used as an independent variable in linear regression to estimate codon adaptation index. The genes for which CAI was predicted accurately were removed from the sample; they made up the first cluster. New correlation matrix was calculated for the remaining gene set, and the procedure of analysis was consequently iterated. It allowed us to reveal three clusters, each characterized by specific interdependence between significant contextual features (Figures 1, 2).

One can see that 213 core promoters form three clusters (Figure 1; clusters are marked with different type dots). Characteristic of each cluster is a specific linear regression reflecting the interrelation between CAI and the corresponding eigenvector. It was found that the most diverged cluster 3 (marked with crosses) contained core promoters of highly expressed genes. Clusters 1 and 2 are close to each other, and their discrimination is problematic. Similar situation is shown in Figure 2, demonstrating the results of 5'UTR analysis (171 sequences): cluster 3 (marked with crosses) contains 5'UTRs of highly expressed genes, whereas the two other clusters are not so distinct. Comparing with core promoters, 5'UTR-based analysis allowed for a more clear-cut identification of clusters 1 and 2.

Despite the classification of the yeast gene sample into three clusters is stable (Figures 1, 2), it alone cannot be used for an accurate prediction of gene expression rate through analysis of contextual features of noncoding regions. In particular, the same value of a principal component function can correspond to several values of CAI. However, this limitation can be overcome with the aid of additional discriminant analysis.

To illustrate the interrelations between contextual features of noncoding gene regions and the expression level, we developed the composed rule for prediction of highly expressed genes (cluster 3). For this purpose, we used both the characteristics of core promoters and 5'UTRs. The prediction algorithm takes advantage of two functions (D_u and D_p stand for 5'UTR and core promoter, respectively):

$$D_u = -0.000005f_{\mathrm{A}} + 0.004f_{\mathrm{G}} - 0.069f_{\mathrm{C}} - 0.025f_{\mathrm{T}} - 0.025f_{\mathrm{AA}} + 0.125f_{\mathrm{AG}} + 0.164f_{\mathrm{AC}} - 0.213f_{\mathrm{AT}} + 0.342f_{\mathrm{GA}} - 0.064f_{\mathrm{GG}} - 0.281f_{\mathrm{GC}} - 0.549f_{\mathrm{GT}} - 0.128f_{\mathrm{CA}} + 0.164f_{\mathrm{CG}} + 0.231f_{\mathrm{CC}} + 0.026f_{\mathrm{CT}} + 0.364f_{\mathrm{TA}} + 0.391f_{\mathrm{TG}} - 0.129f_{\mathrm{TC}} - 0.074f_{\mathrm{TT}}.$$

$$D_p = -0.000001f_{\mathrm{A}} - 0.004f_{\mathrm{G}} - 0.003f_{\mathrm{C}} + 0.009f_{\mathrm{T}} + 0.009f_{\mathrm{AA}} + 0.023f_{\mathrm{AG}} + 0.061f_{\mathrm{AC}} - 0.628f_{\mathrm{AT}} - 0.353f_{\mathrm{GA}} + 0.062f_{\mathrm{GG}} + 0.119f_{\mathrm{GC}} - 0.290f_{\mathrm{GT}} + 0.005f_{\mathrm{CA}} + 0.062f_{\mathrm{CG}} + 0.151f_{\mathrm{CC}} - 0.193f_{\mathrm{CT}} + 0.039f_{\mathrm{TA}} + 0.221f_{\mathrm{TG}} + 0.405f_{\mathrm{TC}} - 0.309f_{\mathrm{TT}},$$

where f_i are the frequencies of mononucleotides; f_{ij}, preferences in dinucleotide usage. If the value of function D_u will be higher than a threshold $\text{crit}_u = -0.80$, the gene will be classified as a highly expressed and its codon adaptation index will be calculated as $\text{CAI} = -0.0048x + 0.7145$, where x is the value of the principal component function calculated for correlation matrix of 5'UTR contextual features.

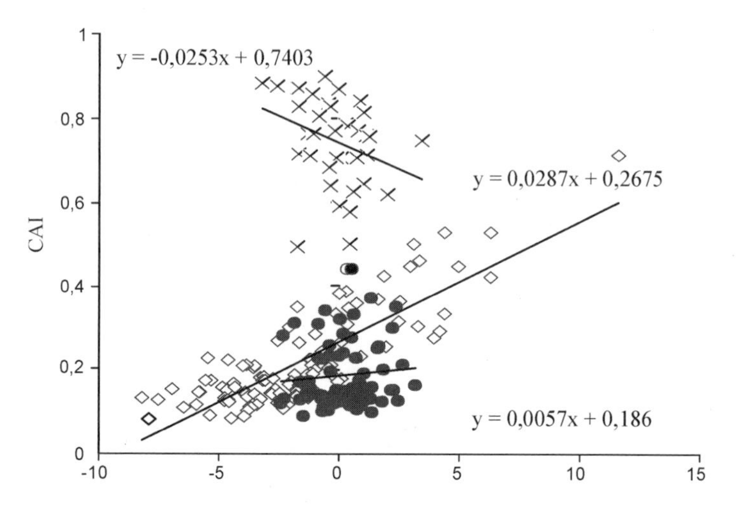

Figure 1. Interdependence (linear regression) between the expression level (CAI) and contextual features of yeast gene core promoters (expressed through the eigenvectors).

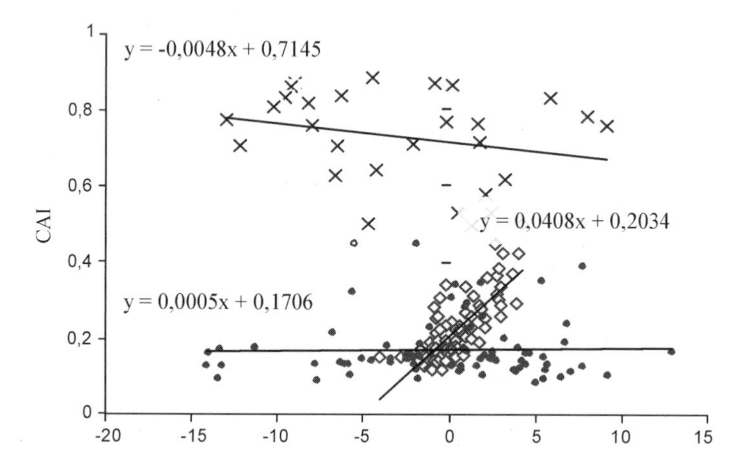

Figure 2. Interdependence (linear regression) between the expression level (CAI) and contextual features of yeast gene 5'UTRs (expressed through the eigenvectors).

In the case of $D_u \leq \text{crit}_u$, function D_p is calculated and compared with a threshold: if $D_p > \text{crit}_p$ (where $\text{crit}_p = -0.26$), the gene is classified as a highly expressed and the codon adaptation index is predicted as $\text{CAI} = -0.0253x + 0.7403$ (where x is the value of the principal component function calculated for correlation matrix of core promoter contextual features).

Finally, if the rule $D_p > \text{crit}_p$ is not true, the gene can not be classified as highly expressed. Note that Fisher's discriminant functions D_u and D_p were found in the process of discrimination of highly expressed genes on the base of contextual features of noncoding regions (core promoters and 5'UTRs). The threshold values crit_u and crit_p were adjusted to minimize a total misclassification error. Function D_u was found to be more informative compared with D_p. However, the role of D_p in the prediction process was also significant: type I (α_1) and type II (α_2) errors of the composed CAI prediction procedure using both D_u and D_p functions amounted to 0.165 and 0.173, respectively. If the procedure of CAI prediction was simplified by omitting function D_p, type II error increased ($\alpha_1 = 0.165$ and $\alpha_2 = 0.230$). An actual possibility to use contextual features of noncoding gene regions to predict their expression level can also be demonstrated by the example of clusters 1 and 2 (Figure 1; circles and boxes; linear regressions $y = 0.0287x + 0.2675$ and $y = 0.0057x + 0.186$). These clusters are very close and their Fisher's discrimination is relatively difficult. Fisher's discriminant function for these clusters was found to be $D = 0.000005f_A + 0.010f_G - 0.001f_C - 0.029f_T - 0.017f_{AA} - 0.464f_{AG} - 0.347f_{AC} - 0.463f_{AT} + 0.094f_{GA} - 0.198f_{GG} - 0.009f_{GC} - 0.251f_{GT} + 0.051f_{CA} - 0.364f_{CG} - 0.220f_{CC} - 0.238f_{CT} - 0.008f_{TA} - 0.121f_{TG} - 0.267f_{TC} - 0.104f_{TT}$. The appropriate threshold was determined to make a misclassification error minimal (if the threshold is equal to –3.75, then only 25% of the genes were misclassified).

4. CONCLUSION

It is known that eukaryotic mRNAs vary in translation initiation efficiency (reviewed in McCarthy, 1998; Kozak, 2002). The translation efficiency of eukaryotic mRNAs associated with their sequence organization is an essential feature of gene expression pattern. It was considered that in yeast, each molecule of well-translated mRNA produced about 4000 protein molecules (Futcher *et al.*, 1999). It is likely that some intrinsic features of mRNA sequence may be essential in supporting high translation intensity.

However, the 5'UTR features affecting translation initiation rate are known rather poorly. Presumably, 5'UTRs of eukaryotic mRNAs possess some contextual features that can influence translatability. In the concept of the rate-limiting stage model (Lesnik *et al.*, 2000), a low efficiency of any consecutive translation stages will be incompatible with a high expression rate. The significant correlations between certain 5'UTR contextual features and gene expression level described in this work may highlight some characteristics essential for high translation initiation efficiency (this assumption is also confirmed by the data reported by Vishnevsky *et al.* and Likhoshvai *et al.* in this issue).

Similarly, preferences in base composition of a core promoter (region upstream of the transcription start site) were found to differ from those of 5'UTR. It was demonstrated that a core promoter played an important role in the control of gene expression (Butler and Kadonaga, 2002). In this research, we analyzed a gene region located upstream of the transcription start site. Thus, it could include TATA box or some other general elements of the core promoter (BRE, Inr, etc.). It may be assumed that contextual features of this region can influence the interaction with basal transcription factors and, thereby, the efficiency of transcription. Significant correlations between the codon adaptation index and contextual features of this region showed that characteristic of highly expressed genes was specific organization of their promoters. This suggests that these deviations in base composition reflect contextual background of the intrinsic promoter DNA conformation and highly expressed genes tend to have more "efficient" promoter region. Anyway, these results may be used to make computational methods allowing gene expression rate to be evaluated.

Interrelations between contextual features of noncoding regions of genes and the level of expression allowed us to develop a method for classification. Such a possibility to predict the level of gene expression taking into account only contextual features of noncoding regions illustrates an adjustment of organization of all the functional regions to support the required expression rate. Further computational and experimental investigations are required to prove the role of contextual features of noncoding regions in the expression process.

ACKNOWLEDGMENTS

This work was supported by the Russian Foundation for Basic Research (grants Nos. 01-07-90376-в, 02-04-48508, 02-07-90355, 03-07-96833-p2003, 03-07-96833, 03-04-48829); grant PCB RAS (No. 10.4); the Siberian Branch of the Russian Academy of Sciences (integration project No. 119); Russian Ministry of Industry, Science, and Technologies (grants Nos. 43.073.1.1.1501, 43.106.11.0011, and Sc.Sh.-2275.2003.4); NATO (grant No. LST.CLG.979816); the U.S. Civilian Research & Development Foundation for the Independent States of the Former Soviet Union (CRDF) No. NO-008-X1.

PART 2. COMPUTATIONAL STRUCTURAL BIOLOGY

LATENT PERIODICITY OF MANY DOMAINS IN PROTEIN SEQUENCES REFLECTS THEIR STRUCTURE, FUNCTION, AND EVOLUTION

A.A. LASKIN[1]*, E.V. KOROTKOV[1], N.A. KUDRYASHOV[2]
[1]Bioengineering Center, Russian Academy of Sciences, prosp., 60-letiya Oktyabrya 7/1, Moscow, 117312 Russia, e-mail: katrin2@biengi.ac.ru; [2]Moscow Engineering & Physics Institute, Kashirskoe shosse 31, Moscow, 115409 Russia
* *Corresponding author*

Abstract We have analyzed many protein sequences for the presence of latent periodicity common for their functionally identical sites. As a result, we found that Rossman-like domains of TPP-binding enzymes, ATP synthases, and pyridoxal phosphate-dependent enzymes as well as many other α/β-proteins, such as dethiobiotin synthases, have latently periodic structure. We also identified the periodicity pattern responsible for formation of parallel β-barrel in a number of enzymes. Possible relationships of regular structures and latent periodicity are also discussed.

Key words: periodicity, repeats, profile analysis, Rossman-like domains, regular structures, molecular evolution, duplication, structure–function relationship

1. INTRODUCTION

It is essential to study repeats and periodicities in genetic sequences, since repetitive elements in sequences often result in regularities in their structures. Manifold duplications of DNA sequence fragments in conjunction with subsequent base substitutions, insertions, and deletions are likely to be the ground for evolution of genes and genomes.

The discovered decomposition of functionally active sites of genetic sequences into repeats is strong evidence that in the past, genes could arise just from repetitions of relatively short DNA fragments. We may also suppose that such structures in protein sequences are involved in selection and stability of the proper conformation of protein globules.

Repeats in genetic sequences are subdivided into tandem and interspersed; in this paper, we focus on approximate tandem repeats, which form a nearly periodic site in a sequence. Various techniques were applied to investigate the relationship between periodicity and structure; as a result, many repeats were assigned their structural meaning (Bornberg-Bauer *et al.*, 1998; Andrade *et al.*, 2000; Katti *et al.*, 2000; Neuwald and Poleksic, 2000). Rackovsky (1998) used amino acid factors (i.e. the averaged values of amino acid indices expressing numerically their physicochemical properties, such as charge or hydrophobicity, or their secondary structure preferences) in his Fourier analysis of amino acid sequences.

That work pioneered in demonstrating that amino acid factors may be used to correlate periodicity and structures, but the author did not succeed in correlating the type of structure to period length and statistical significance of the factors.

The latent periodicity notion and search technique were initially reported by Korotkov and Korotkova (1995) and refined in subsequent works (Korotkov and Korotkova, 1996; Korotkov *et al.*, 1997; Chaley *et al.*, 1999; Korotkov *et al.*, 1999; Korotkova *et al.*, 1999). As a result of the studies performed, we discovered the existence of various types of latent periodicity in numerous amino acid sequences (Korotkov *et al.*, 1999; Korotkova *et al.*, 1999). However, the question of functional significance of the latent periodicities identified and its correlation with protein structures remained open. To a great extent, this resulted from incapability of the information decomposition method (Korotkov and Kudryashov, 2003) of revealing latent periodicity interrupted with insertions and deletions. Therefore, this approach omitted a substantial subset of proteins with certain functional domains, so that no inference about the relationship between latent periodicity and protein functionality could be made.

The papers by Laskin *et al.* (2001; 2002; 2003) presented a pioneer work that shows the existence of latent periodicity type that is common for whole protein superfamilies. We have achieved this result by means of the development of novel mathematical methods and software capable of finding statistically significant latent periodicity of a predefined type in the presence of insertions and deletions. In this paper, we describe the progression of these methods, namely, investigation of periodic patterns with the use of amino acid factor analysis as well as new types of structure- and function-related latent periodicity in amino acid sequences.

2. METHODS AND ALGORITHMS

In this study, we used the latent periodicity database from our previous studies (Korotkov *et al.*, 1999; Korotkova *et al.*, 1999) to search for the presence of periodicity types that were correlated to structural or functional features (we identify periodicity type as "position vs. residue" frequency matrix). The locally optimal method of cyclic alignment (Laskin *et al.*, 2003) was used to identify the protein sequences that possess statistically significant similarity with cyclic profile of a given periodicity type. It is based on WDP, the wraparound dynamic programming (Fischetti *et al.*, 1992). The normalized Z-value was calculated from the local WDP alignment score by simulation of the cyclic alignment with random sequences of the same composition; the formula is:

$$Z = \frac{S_{\text{align}} - E(S)}{\sqrt{D(S)}}, \tag{1}$$

where $E(S)$ and $D(S)$ are the mean and variance of cyclic alignment score. The threshold value of Z was chosen equal to 6.0. Our numerical experiments showed that we were unlikely to observe Z-values greater than 6.0 when analyzing a random test sequence with the same number of symbols as the total number of amino acids in all Swiss-Prot entries.

We also used the techniques of profile optimization by iterative searches of Swiss-Prot data bank (Altschul and Koonin, 1998; Aravind and Koonin, 1999) involving training sequence subsets where appropriate. To determine the structure-related periodicity types, we used amino acid scales and factors. They both numerically define physicochemical, structural, or evolutionary preferences of single amino acids, amino acid factors being combinations of close amino acid scales. The factors were first introduced by Kidera *et al.* (1985); it was shown that 10 properly chosen amino acid factors are enough to reflect 86% of the variability of amino acid properties. For our calculations, there is no difference between scales and factors, so we will further call both of them factors. We took the factor values from the AAindex database (Kawashima *et al.*, 1999), which currently contains 494 amino acid factors; from them, we chose the 30 factors listed in Table 1. As the table indicates, factors 1–4 define α-helix preferences of amino acids; factors 5–7, β-sheet preferences; and factors 8–9, turn preferences. Factors 10–13 are related to amino acid evolution; factors 14–17 characterize polarity; factors 18–21, hydrophobicity; factors 22–26 are related to transmembrane propensity; and factors 27–28, to charge in water surrounding. Notice also that factors 3, 4, 6, 7, 9, 20, and 29 are specific of α/β-protein, while factors 2, 21, and 30, of α-proteins.

Table 1. Selected amino acid factors from the AAindex database

No. of factor	Description in AAindex
1	Normalized frequency of α-helix with weights
2	Alpha-helix indices for α-proteins
3	Normalized frequency of α-helix in alpha/beta class
4	Alpha-helix indices for α/β-proteins
5	Normalized frequency of β-sheet with weights
6	Beta-strand indices for α/β-proteins
7	Normalized frequency of β-sheet in α/β class
8	Normalized frequency for reverse turn with weights
9	Normalized frequency of turn in α/β class
10	Relative mutability factor
11	Relative mutability
12	Relative mutability
13	Side-chain contribution to protein stability
14	Polarity values
15	Polar requirement
16	Mean polarity
17	Polarity
18	Hydrophobicity in folded form
19	Hydrophobicity in unfolded form
20	Normalized hydrophobicity scales for α/β-proteins
21	Normalized hydrophobicity scales for α-proteins
22	Engelman hydrophobicity of nonpolar transbilayer helices
23	Transmembrane regions of non-mt-proteins
24	Transmembrane regions of mt-proteins
25	Turn propensity scale for transmembrane helices
26	Averaged turn propensities in a transmembrane helix
27	Isoelectric point
28	pK of side chain
29	Aperiodic indices for α/β-proteins
30	Aperiodic indices for α-proteins

The factor values were first normalized so that the average value of each factor was equal to zero and variance was equal to unity. This allowed us to define magnitudes of different factors in commensurable quantities:

$$\phi_i' = \frac{\phi_i - E(\phi)}{\sqrt{D(\phi)}}. \qquad (2)$$

Here, φ_i are published and ϕ_i are normalized values of factor φ for amino acid i. Then, we used the periodicity matrix to calculate the factor values for different positions in a period:

$$F_i = \sum_{\text{all i}} p_{i,j}\phi_i^{'}, \tag{3}$$

where $p_{i,j}$ is the frequency of occurrence of amino acid i at position j. We believe that if the observed latent periodicity is caused by periodic distribution of some amino acid properties along the sequence, the factor values at different positions will oscillate. For instance, if the analyzed sequence has cyclic variations of polarity (i.e. consists of alternating polar and nonpolar sites) and these variations caused latent periodicity in this sequence, then the latent period will match the period of these variations and the polarity factor value will be high at some positions in the period and low at other positions. Let us call the magnitude of these oscillations as the magnitude of corresponding factor for the investigated periodicity type, and let us calculate it as the value of the first harmonic in Fourier transformation of the series F_j:

$$A = \sqrt{A_{\sin}^2 + A_{\cos}^2},$$

$$\text{where } A_{\sin} = \sum_{j=1}^{L} F_j \sin\frac{2\pi j}{L}, \quad A_{\cos} = \sum_{j=1}^{L} F_j \cos\frac{2\pi j}{L} \tag{4}$$

(here, L is length of the period, not the sequence).

However, high magnitude of a factor may be caused by a high homology between periods as well. To distinguish these cases, we calculate the magnitude of the "frequency factor" as:

$$\alpha = \sqrt{\sum_{\text{all i}} \left(\alpha_{i,\cos}^2 + \alpha_{i,\sin}^2\right)},$$

$$\text{where } \alpha_{i,\sin} = \sum_{j=1}^{L} p_{i,j} \sin\frac{2\pi j}{L}, \quad \alpha_{i,\cos} = \sum_{j=1}^{L} p_{i,j} \cos\frac{2\pi j}{L}. \tag{5}$$

This value stands for irregularity in distribution of amino acids over different positions in the period. We investigated several previously obtained periodicity matrices with known structural causality (Laskin *et al.*, 2002) and ascertained that 1.5-fold excess of the magnitude of a factor over the magnitude of the "frequency factor" may witness for the corresponding structural causality of this latent periodicity.

We also proposed a measure for the functional causality of latent periodicity. We will call it functionally caused if the results of iterative scanning (Laskin *et al.*, 2003) consist of proteins from a functional group (or a few related groups) by not less than 90%. Functional groups were determined using protein descriptions and Swiss-Prot keywords.

3. RESULTS AND DISCUSSION

Using the techniques described above, we discovered and analyzed more than 20 novel cases of latent periodicity in protein families and gathered the profiles of these periodicity types. Additional information about these cases, including profiles of periodicity and full sets of results of Swiss-Prot scanning (we used Swiss-Prot release 40), can be obtained at http://periodicity.fromru.com/new/.

Analysis of these periodicity patterns performed using magnitudes of amino acid factors allowed us to reveal the structural sense of many novel periodicity types or to put forward a suggestion about types of structures in corresponding protein sites. Let us consider them in more detail.

3.1 Latent periodicity of α/β-domains

Previously, we have investigated the latent periodicity of Rossman-folded NAD-binding sites (Laskin *et al.*, 2003); detailed analysis of the found periodicity types can be found there. Periodicity in these domains turned out closely related to their α/β-structure, one period spanning a helix and a sheet. Amino acid factor analysis showed that factors 1, 3, 4, 5, 6, and 7, defining α- and β-structure preferences and specific of α/β-proteins, were the most expressed; their magnitudes exceeded the frequency factor magnitude by 1.5-fold and more. This expression pattern was proposed to be common for such domains.

Initially (Korotkov *et al.*, 1999; Korotkova *et al.*, 1999), we found latent periodicity with a 25-aa-long period in active sites of two acetolactate synthases, ILVI_BUCAP and ILVI_ECOLI. Iterative search for similar periodicity cases with allowed insertions/deletions showed that this type of periodicity is common for acetolactate synthases as well as for other enzymes that use thiamine pyrophosphate (TPP) as a cofactor. We succeeded in finding latent periodicity in 63 TPP-binding proteins in the absence of false positives (PFam profile PF00205, common for TPP-binding enzymes, identifies 67 proteins in Swiss-Prot). In all cases, we found periodicity at the same protein site, namely, the active TPP-binding site.

It is known that TPP-binding domain has a distinct Rossman-like α/β-structure. As in the case of NAD-binding domain, one period turned out to span a helix and a sheet. The expression patterns for these two periodicity types generally coincide, and α/β-specific factors have the highest magnitude. The analysis of factors in this case clearly affiliates TPP-binding sites with α/β-domains and reveals the structural sense of the latent periodicity found. We obtained similar expression pattern in the case of dethiobiotin synthases, which are also α/β-proteins, but having somewhat different 3D structure that resembles open β-cylinder rather than standard Rossman fold. In this case, the periodicity with a 30-aa-long period was also due to alternation of α- and β-structures. The profile obtained identifies latent periodicity in 15 of the 16 dethiobiotin synthases in Swiss-Prot with no false positives. Since synthesis of dethiobiotin requires ATP, we may suppose that this cyclic structure is essential for the proper binding of this coenzyme.

Latent periodicity with a 32-aa-long period was found in ACC oxidases ACC1_PETHY, ACC4_PETHY, ACC1_LYCES, ACC2_LYCES, alanine aminotransferase ALAM_YEAST, tryptophan synthase TRPB_METTM and aspartate aminotransferases AATC_ORYSA, AAT3_ARATH, and AAT2_ARATH. All these enzymes bind pyridoxal phosphate. The iterative search allowed us to identify 374 protein sequences with the same type of periodicity. The periodic site always lied within the pyridoxal phosphate-binding domain with α/β-structure. Similarly, the period contains one α-helix and one β-sheet. The structure factor analysis confirms that the periodicity is structure-related. Another instance of similar type of periodicity is the periodicity found in active sites of ATP synthases. They are structurally similar to Rossman folds and possess α/β-structure; the period of 28 amino acids spans a helix and a sheet; all 83 proteins in this class contain identified sites of this periodicity.

We also observed similar patterns of magnitudes of factors in a group of methyl transferases (period length, 31) and in many GTP-binding sites (period length, 23). Despite the 3D structures of any proteins in these groups are still not obtained, our analysis allows us to propose that they are α/β proteins or contain α/β domains. This is likely to be true for GTP-binding sites since supersecondary structures of nucleotide-binding sites resemble each other (Baker *et al.*, 1992).

3.2 Latent periodicity of β-helix proteins

Initially, we found a 6-aa-long periodicity in the sequence of ε-subunit of transcription initiation factor E2BE_YEAST and noticed the similarity of periodicity patterns of that protein and acyltransferase LPXA_ECOLI, in which the periodicity is responsible for forming cyclic structural element, the left-handed parallel β-helix. Our iterative searches confirmed the presence of periodicity, similar to acyltransferases, in 10 of the 11 ε-subunit (as well as γ-subunit) EIF-2B sequences. We also found this type of periodicity in glucose-1-phosphate adenylyltransferases, where it was not noticed before (current version of PFam identifies them as containing the so-called context copies of hexapeptide repeat, i.e. not statistically significant themselves, but proposed to be possible, since another co-existing domain is found significant, in this case, the nucleotidyltransferase domain). Latent periodicity is identified in 31 of the 37 proteins of this type; the average length of periodic site is about 78 residues, i.e. nearly 5 turnovers of β-helix (PFam finds only 2 context repeats). For the case of initiation factors, we identify periodic site that spans 104 residues (about 6 turnovers), while PFam identifies only 3 context repeats. The structure factor analysis reveals virtually coinciding magnitude patterns for these three groups (acyltransferases, adenylyltransferases, and initiation factors), suggesting additional evidence that these periodicities form similar structures.

3.3 Function-related latent periodicities

There are also cases where we cannot assign a repeat to a structure element, but the periodic profile clearly identifies a functional group of proteins. An example is periodicity of 22-aa in DNA-binding proteins containing T-box site. The periodic site does not contain cyclic structures, but the structure factor analysis shows some regularity in charge of side chain (factors 27 and 28). Proper distribution of charge is essential for protein to bind to negatively charged DNA. Thus, we may assume that the cyclic distribution of charge, once having appeared as a result of duplications, was then conserved despite of strong conformational rearrangement, since it was critically important for proper function of enzymes. We also observed latent periodicities with high magnitudes of charge-related factors 27 and 28 in other types of DNA-binding proteins.

Additional information about these and other types of latent periodicity found, including color figures of latently periodic regions, can be obtained at http://periodicity.fromru.com/new/.

4. CONCLUSION

Novel techniques for detecting hidden repeats allowed us to see that the latent periodicity is a phenomenon common for many protein families. In a number of cases, it is related to the presence of cyclic structures and has functional meaning; in addition, it may be also be involved in proper folding and stability of globular proteins. The investigation has shown that, as an example, the majority of nucleotide-binding domains have regular structure at the secondary and tertiary levels as well as at the level of primary sequences, where no such regularity was ever demonstrated. The challenge of their α/β structures is that they are less than others conservative from the viewpoint of sequence-structure relationships; i.e. they have the greatest average number of allowable amino acids per position, keeping the structure energetically stable. We believe that latent periodicity is of considerable interest for investigations of evolution and structure. Periodicities in distribution of the properties like charge and hydrophobicity can be of a great value for the proper protein domain folding. We also suppose that duplications are one of the most frequently realized mechanisms in formation of stable protein structure. If that is the case, one could expect latent periodicity to be presented in substantial part of protein domains. The introduced quantities for measuring cyclicity of physicochemical properties along a protein sequence, namely, the magnitudes of structural factors, may help us to understand the structural sense of identified latent periodicities and even predict cyclic-type structures from primary sequences. They can be also applied to classify types of periodicity of different period length from the viewpoint of structures they form up.

LOGICAL ANALYSIS OF DATA AS A PREDICTOR OF PROTEIN SECONDARY STRUCTURES

J. BŁAŻEWICZ[1], P.L. HAMMER[2], P. ŁUKASIAK[3]*
[1]*Institute of Computing Sciences, Poznan Univ. of Technology, Piotrowo 3A, 60-965 Poznan, POLAND;* [2]*Rutgers Center for Operations Research, Rutgers University, 640 Bartholomew Road Piscataway, NJ 08854-8003 New Jersey, USA;* [3]*Institute of Bioorganic Chemistry, Polish Academy of Sciences, Noskowskiego 12/14, 61-704 Poznan, POLAND,*
e-mail: Piotr.Lukasiak@cs.put.poznan.pl
* *Corresponding author*

Abstract Protein secondary structure prediction problem has been known for almost a quarter of century. The idea to create a tool to help molecular biologists was the main reason to choose the new rule-based method—Logical Analysis of Data with its high accuracy in various field of life. The LAD produced overall percentage accuracy of correctly predicted residues Q_3 = 71,6%, and segment overlap measure SOV = 70,9% on the set consist of 126 proteins. The goal of the analysis described in this paper is to create a system that allows for receiving as the output the protein secondary structure, based on its primary structure being an input, and for finding rules responsible for this effect.

Key words: logical analysis of data, protein prediction, protein secondary structure, machine learning

1. INTRODUCTION

Protein secondary structure prediction was one of the first and most important problems faced by computer learning techniques. First generation methods are those making predictions only on the base of information coming from a single residue, either in the form of statistical tendency to appear in an α-helix, β-strand and coil region (Garnier *et al.*, 1978), or in the form of explicit biological expert rules (Lim, 1974). Second generation methods basically apply the connection architecture, taking into account local interactions by means of an input-sliding window with encoding. Values in the output layer discriminate each residue as belonging to one of the three states α-helix (H), β-strand (E) and coil (C). The 3rd generation methods started exploiting the information coming from homologous sequences. This information is processed first doing a multiple alignment between a set of similar sequences and extracting a matrix of profiles (PSSM). The first method incorporating profile-based inputs and going beyond 70% in accuracy was PHD (Rost, 1996). The system is composed of cascading networks. Prediction accuracy can be improved by combining more than one prediction method (Zhang *et al.*, 1992; Cuff and Barton, 1999). Other well-known profile based methods are PSI-PRED (Jones, 1999), which uses 2 neural networks to analyze profiles generated from a PSI-BLAST search, JNet (Cuff and Barton, 2000) and SecPred. An alternative adaptive model is presented in (Baldi *et al.*, 1999). There are other predictors, which are not strictly based on neural network implementations. NNSSP (Salamov and Solovyev, 1995) uses a nearest-neighbor algorithm where the secondary structure is predicted using multiple sequence alignments and a simple jury decision method. The web-server JPred (Cuff *et al.*, 1998) integrates six different structure prediction methods and returns a consensus based on the majority rule. The program DSC (King and Sternberg, 1996) combines several explicit parameters in order to get a meaningful prediction. It runs the GOR3 algorithm (Garnier *et al.*, 1978) on every sequence, to provide mean potentials for the three states. The program PREDATOR (Frishman and Argos, 1997) uses amino acid pair statistics to predict hydrogen bonds between neighboring β-strands and between residues in helices.

2. ALGORITHMS AND METHODS

The goal of the analysis described in this paper is to create a system that allows for receiving as the output the protein secondary structure, based on its primary structure being an input, and to find rules responsible for this effect. Let as set up the subject more formally.

The structure of a protein may be represented hierarchically at 3 structural levels. The primary structure of protein is the sequence of residues in the polypeptide chain and it can be described as a string on the finite alphabet Σaa, with $|\Sigma aa| = 20$.

Let $\Sigma aa = \{A,C,D,E,F,G,H,I,K,L,M,N,P,Q,R,S,T,Y,V,W\}$ be a set of all residues where each letter corresponds to a different residue. Based on the protein chain it is easy to create its relevant sequence of residues replacing a residue (primary structure) in the chain by its code in Latin alphabet. As a result, a word on the residues' alphabet is received. The word *s* is called a protein primary structure on the condition that letters in this word are in the same order as residues in the protein chain are. Let the length of the word s be denoted as $C(s)$ and $A(s, j)$ denote an element of word s, where j is an integer number from the set $[1, C(p)]$. The protein secondary structure refers to the set of local conformation motifs of the protein and schematizes the path followed by the backbone in the space. The secondary structure of protein is built from three main classes of motifs: α-helix (H), β-strand (E) and loop or coil (C). An α-helix is built up from one continuous region in the sequence through the formation of hydrogen bonds between residues in position i and $i + 4$.

A β-strand does not represent an isolated structural element by itself, because it interacts with one or more β-strands (which can be distant in sequence) to form a pleated sheet called β-sheet. Strands in a β-sheet are aligned adjacent to each other such that their residues have the same biochemical direction (parallel β-sheet) or alternating directions (anti parallel β-sheet). α-helices and β-strands are often connected by loop regions, which can significantly vary in length and structure, having no fixed regular shape as the other two elements. Every residue in the sequence belongs to one of the three structural motifs, so the protein secondary structure can be reduced to a string on the alphabet $\Sigma ss = \{H; E; C\}$ being of the same length as the protein primary structure.

A secondary structure is represented here by a word on the relevant alphabet of secondary structures Σss: each kind of a secondary structure has its own unique letter. One can denote this word by d, where the length of word d is equal to the length of word s. Now, one may define the problem as the one consisting in finding a secondary structure of a protein (word d), based on the protein primary structure (i.e. word s). Moreover, for each element $A(s, j)$ one should assign an element $A(d, j)$ in the way that the obtained protein secondary structure r is as close as possible to a real secondary structure of the considered protein. Several standard performance measures were used to assess accuracy of prediction of protein secondary structure. Measure of the three-state overall percentage of correctly predicted residues is usually defined by Q_3 (%) as follows:

$$Q_3 = \frac{\sum_{i \in \{H,E,C\}} \text{number of residues correctly predicted in state } i}{\sum_{i \in \{H,E,C\}} \text{number of residues observed in state } i} \times 100. \quad (1)$$

The segment overlap measure (SOV) (2) (Rost *et al.*, 1994; Zemla *et al.*, 1999) is calculated as shown below:

$$\text{SOV} = \frac{1}{N} \sum_{i \in \{H,E,C\}} \sum_{S(i)} \left(\frac{\text{minov}(s_1, s_2) + \delta}{\text{maxov}(s_1, s_2)} \times \text{len}(s_1) \right) \times 100, \quad (2)$$

where $S(i)$ is the set of all overlapping pairs of segments (s_1, s_2) in conformation state i, len(s_1) is the number of residues in the segment s_1, minov(s_1, s_2) is the length of the actual overlap, and maxov (s_1, s_2) is the total extent of the segment. The Logical Analysis of Data method (Hammer, 1986) has been widely applied to the analysis of a variety of real life data sets classifying objects into two sets. It is not possible to use the original LAD method (Hammer, 1986; Boros *et al.*, 1994; Mayoraz *et al.*, 1995; Boros *et al.*, 1996; Crama *et al.*, 1998) directly for our problem. The first problem lies in the input data representation.

Here one has a sequence of residues but to use the LAD approach one should have a set of observations. Each observation has to consist of a set of attributes and all of them should be in a number format. If all of them are written in binary one can resign from the binarization stage; but this is not the case here, and the binarization procedure must be applied.

The original method is designed only for two classes. In our study, three sets of protein secondary structures have to be considered. Again, a modification of the original method had to be proposed and implemented. Because of a complexity of the algorithm of Logical Analysis of Data (Ekin *et al.*, 1998), it is hard to present all aspects of this method. All important phases one can see in Figure 1.

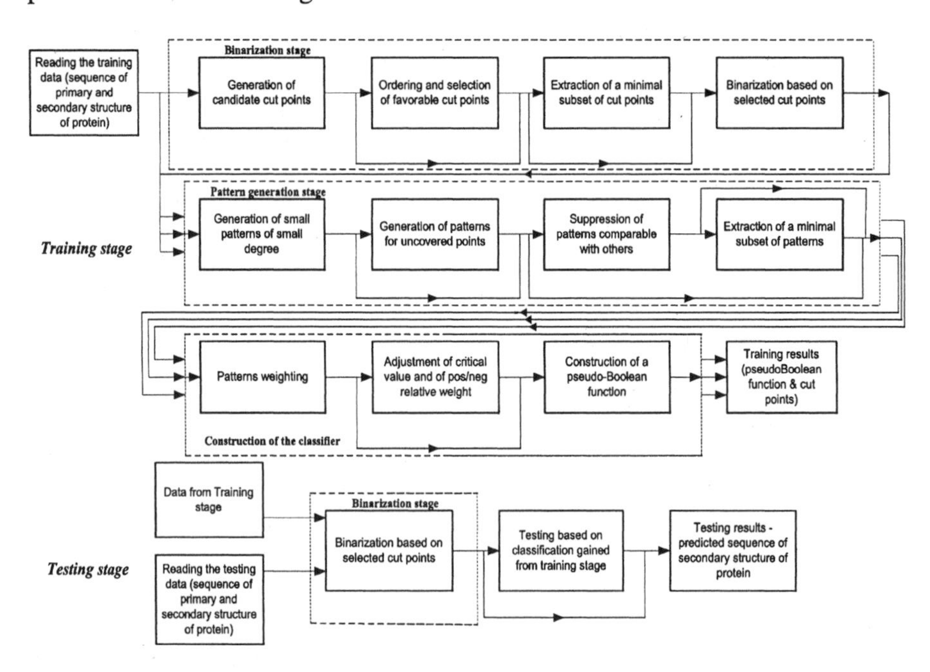

Figure 1. Modified Logical Analysis of Data stages.

2.1 Binarization stage

A data binarization stage is needed only if data are in numerical or nominal formats (e.g. color, shape, etc.). To make such problems useful for LAD one has to transform all data into a binary format. The simplest non-binary attributes are the so-called nominal (or descriptive) ones.

As a result, of this stage, all attributes for each observation are changed into binary attributes. After the binarization phase, all of the observations that belonged to different classes are still different when binary attributes are taken into account.

2.2 Pattern generation stage

Every pattern is defined by a set of conditions, each of which involves only one of the variables. For example if pattern P_1 is defined by (values from hydrophobicity scale (pi-r)): $x_{-3} > -0.705$; $x_{-1} > 0.285$; $x_0 < 0.065$; $x_{+2} < -0.620$, then the meaning is as follows: structure H should appear for a residue situated in position a_0, if simultaneously the value of hydrophobicity scale for residue situated in position a_{-3} is greater than –0.705, for residue situated in position a_{-1} is greater than 0.285, for residue situated in position a0 is smaller than 0.065, and for residue situated in position a_{+2} is smaller than –0.620, respectively. The precise definition of a pattern P_1, involves two requirements. First, there should be no observation belonging to other classes that satisfies the conditions describing P_1, and on the other hand, a huge number of observations belonging to class H should satisfies the conditions describing P_1. Clearly, the satisfaction of condition describing P_1 can be interpreted as a sufficient condition for an observation to belong to class H. The observation is covered by a pattern if it satisfies all conditions describing P_1. For the pattern generation stage, it is important not to miss any of the "best" patterns. Pattern generation procedure is based on the use of combinatorial enumeration techniques that can follow a "top-down" or a "bottom-up" approach. The top-down approach starts by associating to every positive observation its characteristic term. Such a term is obviously a pattern, and it is possible that even after the removal of some literals the resulting term will remain a pattern. The top-down procedure systematically removes literals one by one until arriving to a prime pattern. The bottom-up approach starts with the term that covers some positive observations. If such a term does not cover any negative observation, it is a pattern. Otherwise, literals are added to the term one by one as long as necessary, i.e. until generating a pattern.

Pattern generation used in the experiment described in this paper is achieved by a hybrid bottom-up—top-down approach. This strategy uses the bottom-up approach to generate all the patterns of very small degrees, and then uses a top-down approach to cover those positive observations that remained uncovered after the bottom-up step. During the experiments, it was not allowed to cover an observation belonging to an opposite class. Patterns have been generated until the whole set of all observations has been covered by at least one pattern. All patterns have been generated using breadth first search strategy (for the patterns of up to degree 8) and depth first search strategy (for other patterns).

2.3 Classifier construction stage

For any particular class, there are numerous patterns that cover only observations belonging to that class. The list of these patterns is too long to be used in practice. Therefore, we restricted our attention to a subset of these patterns, called [class_indicator] model (e.g. H model). Similarly, if one studied those observations which not belong to the particular class, one can consider the not H model. A H model is simply a list of patterns associated to the observations which belong only to class H, having the following two properties if an observation is covered by: a) at least one of the pattern from H model, in position a_0, for that observation appears class H; b) none of the pattern from H model, in position a_0 for that observation does not appear class H. Before this stage is performed, every positive (negative) observation point is covered by at least one positive (negative) pattern, and it is not covered by any negative (positive) patterns that have been generated. Based on that it can be expected that an adequately chosen collection of patterns can be used for a construction of a general classification rule. This rule is an extension of a partially defined Boolean function, and will be called below a theory.

A good classification rule should capture all the significant aspects of the phenomenon. The simplest method of building a theory consists in defining a weighted sum (3) of positive and negative patterns and classifying new observations according to the value of the following weighted sum:

$$\Delta = \sum_{k=1}^{r} \omega_k^+ P_k + \sum_{l=1}^{s} \omega_l^- N_l, \qquad (3)$$

where ω_k^+ is non-negative weight for positive pattern P_k (for $1 \le k \le r$), r is the number of positive patterns; ω_l^- is non-positive weight for negative pattern N_l (for $1 \le l \le s$), and s is the number of negative patterns.

An interested reader is referred to (Hammer, 1986; Boros *et al.*, 1997) for a more detailed description of the Logical Analysis of Data method.

As in previous experiments (Błażewicz *et al.*, 2001a, b) at the beginning one constructed 3 binary classifiers one-versus-rest, where one means positive class (e.g. H) and rest means negative class (in that case E, C) denoted: H/~ H, E/~ E, C/~ C. The one-versus-rest classifiers often needs to deal with two data sets with different sizes, i.e. unbalanced training data (Hsu and Lin, 2002), so during experiments one decided to add three additional classifiers denoted: H/E, H/C, E/C. The set of all six classifiers allows for distinguishing the observation between each of two states. However, a potential problem of the one-versus-one classifier is that the voting scheme might suffer from incompetent classifiers. One can reduce that problem by using decision graph (Hua and Sun, 2001) with some modifications as shown in Figure 2.

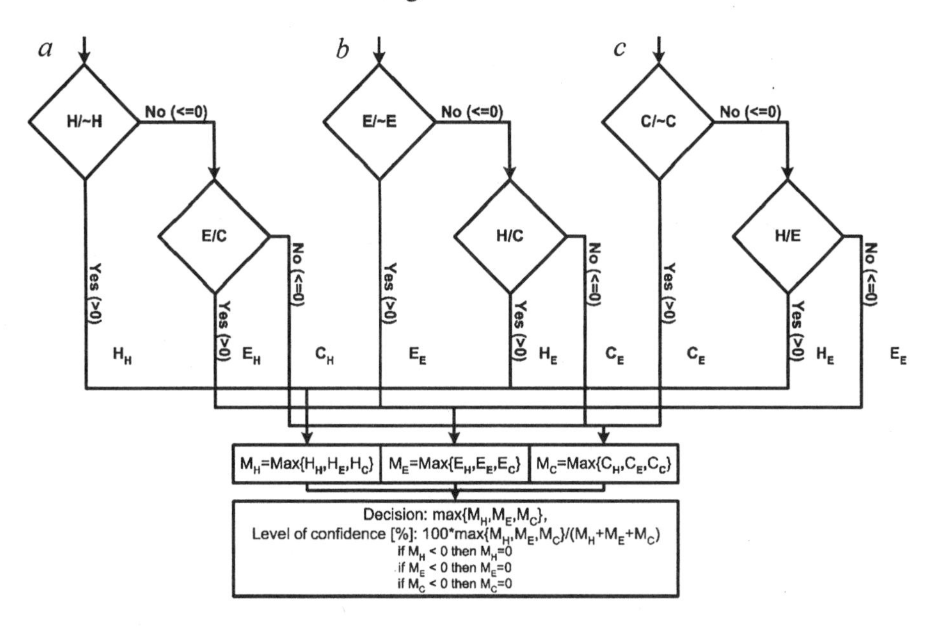

Figure 2. The decision graphs for classifiers; each of them is made up of two binary classifiers: (*a*) classifiers H/~H, E/C; (*b*) classifiers E/~E, H/C; (*c*) classifiers C/~C, H/E.

The protein secondary structure is assigned from the experimentally determined tertiary structure by DSSP (Kabsch and Sander, 1983), STRIDE (Frishman and Argos, 1995) or DEFINE (Richards and Kundrot, 1988). To implement the methods and extract the basics properties of proteins, examples were obtained from the Dictionary of Secondary Structures of Proteins. There are many ways to divide protein secondary structures into classes. Here we used the most popular based on information obtained from DSSP. Data gained from the DSSP set consist of eight types of protein secondary structures: α-helix (structure denoted by H in DSSP), 3_{10}-helix (G), π-helix (I), β-strand (E), isolated β-bridge (B), turn (T), bend (S) and rest (–). The following sets of secondary structures have been created: a) helix (H) consisting of: α-helix (structure denoted by H in DSSP), 3_{10}-helix (G) and π-helix (I); b) β-strand (E) consisting of E structure in DSSP; c) the rest (C) consisting of structures belonging neither to set H nor to set E.

Making a transformation from a protein sequence to the set of observations one has to assume that the main influence on the secondary structure have residues situated in the neighborhood of the observed residue. We took also into account that some n-mers are known always to occur in the same structure in many proteins, while others don't. Certain 4-mers and 5-mers are known to have different secondary structures in different proteins. To fulfill this assumption and avoid naive mistakes, a concept of windows (King and Sternberg, 1990) was used to create a set of observations. It should be done carefully because when the size of window is too short, it may lose some important classification information and prediction accuracy, while a too long size of window may suffer from inclusion of unnecessary noise. One checked the size of window from 3 to 7 as a function of accuracy of prediction (Table 1) for the subset consists of 80 random selected proteins from considered set of 126 proteins, and decided to use for experiments the window of the size 7. Below an example is presented, that illustrates the way a protein chain is changed into a set of observations.

Table 1. Accuracy of each type of classifier for different sizes of window

Type of classifier	Window length = 5	Window length = 7
H/~H	73.15	75.32
E/~E	70.28	73.21
C/~C	64.33	66.90
H/E	75.01	77.80
E/C	70.12	70.91
H/C	69.23	72.12

Let us consider a protein chain called 4gr1 (in PDB). The first and the last fifteen residues in the sequence are shown below:

MKRIGVLTSGGDSPG ... TIDQRMYALSKELSI

For every residue, the corresponding protein secondary structure in DSSP is given as follows:

_EEEEEEESS_TT ... ___HHHHHHHHHH_

One may change this structure into protein secondary structures involving three main secondary structures only in the way depicted below:

CCEEEEEEECCCCCC ... CCCHHHHHHHHHHCC

A window of length 7 generates an observation with 7 attributes (a_{-3}, a_{-2}, a_{-1}, a_0, a_{+1}, a_{+2}, a_{+3}) representing a protein secondary structure corresponding to the residue located in place a_0. Of course, at this moment all values of attributes are symbols of residues. Secondary structures of proteins on the boundaries (first three, and the last three residues) have been omitted and treated as unknown observations (e.g. the first observation can be constructed by residues MKRIGVL and that observation describes the class for residue situated in the middle (residue I)—class E, the next observation is created by window shifted one position to the right, etc.). At the end, a chain consisting of n residues is transformed into set consisting of n – 6 observations. During experiments to develop and test the algorithms, the non-homologous data set proposed by (Rost and Sander, 1993) was applied. They used percentage identity to measure the homology and defined non-homologous to mean that no two proteins in the set share more than 25% sequence identity over a length of more than 80 residues. This set consists of 126 non-homologous proteins that can be obtained from ftp.cmbi.kun.nl/pub/molbio/data/dssp. As an attributes one used subset of the best amino acids' residues physicochemical properties (obtained from previous experiments (Błażewicz *et al.*, 2001a, b)) combined with 3D position of residues in the learning stage.

3. RESULTS AND DISCUSSION

The result obtained during experiments looks interesting (Table 2). Q_3 score calculated based on experiments shows that LAD is 2.0% better than result obtained by PHD, but worse than the second result gained by PHD and NNSP. LAD is also better than methods like PREDATOR and DSC. Only CONSENSUS is much better than LAD. Results for PHD, DSC, PREDATOR, NNSSP and CONSENSUS were obtained from Cuff and Barton (1999), PHD* obtained from Rost and Sander (1993) and Rost *et al.* (1994). LAD* is new method proposed by authors.

Table 2. Comparison between different methods

	Q_3	SOV 94
PHD*	70.8	73.5
PHD	73.5	73.5
DSC	71.1	71.6
PREDATOR	70.3	69.9
NNSSP	72.7`	70.6
CONSENSUS	74.8	74.5
LAD*	71.6	70.9

The SOV score also placed LAD in the fourth position from tested algorithms (after PHD, NNSP and CONSENSUS). As one mentioned before, the window of length 7 was chosen for experiments. One considered 126 proteins, and the sevenfold cross validation test was applied. Average accuracy for each of binary classifiers was over 70%. Only for class C, the accuracy was lower than the rest, but it can be caused by the random conformations of protein secondary structures included in that class. Example of rules is presented in Table 3.

Table 3. Example of rules (horizontal line in a cell means that the value of attribute is not important for that pattern to take decision)

N	A_{-3}	A_{-2}	A_{-1}	a_0	a_{+1}	a_{+2}	a_{+3}	Property
1	>–0.705	—	>0.285	<0.065	—	<–0.620	—	Hydrophobicity
2	<–0.620	<–0.130	—	>1.795	—	>–0.020	—	scale (pi-r)
3	—	>1.745	<0.195	>1.225	>1.795	—	>0.195	class H
1	—	—	<11.705	>14.195	<11.365	>14.765	<11.295	Av. surround.
2	>15.285	—	>12.700	>12.295	—	<11.705	—	hydrophobicity
3	>15.690	<11.395	—	<13.195	>15.285	—	—	class E
1	>10.45	—	—	>9.10	<5.60	>10.45	>9.10	Polarity
2	>11.95	>10.90	—	<10.45	>12.65	—	—	
3	—	>9.80	>8.80	>11.95	—	<6.60	—	class C

4. CONCLUSION

LAD gave results similar to the best standing alone methods used for protein prediction problem in the context of machine learning algorithms. In the molecular biology, context LAD reached the boundary of 70% of accuracy prediction. As one can see, a context is an important factor in determining secondary structure. Also, the protein backbone often folds back on itself in forming a structure, so flexibility is another important attribute. This additional information has not been taken into account during experiments. The sequence information being used is identical to that used by per-residue secondary structure prediction algorithms such as PHD. The PHD algorithm makes use of a global amino acid composition, but the relative importance of the amino acids has not been determined. LAD has determined the most important amino acids' properties and explored rules. To get better than 70% accuracy, LAD should either explore a possibility of parallelization and high performance computing or should be used as a first stage of analysis, together with another method that is able to take into account more detailed understanding of the physical chemistry of proteins and amino acids.

ACKNOWLEDGMENTS

The work is supported by the State Committee for Scientific Research (grant No. 7T11F02621).

STRUCTURE-SPECIFICITY RELATIONSHIP IN PROTEIN–DNA RECOGNITION

H. KONO[1], A. SARAI[2]*

[1]Neutron Research Center and Center for Promotion of Computational Science and Engineering, Japan Atomic Energy Research Institute, 8-1, Umemidai, Kizu-cho, Souraku-gun, Kyoto 619-0215 Japan, e-mail: kono@apr.jaeri.go.jp; [2]Dept. Biochemical Engineering and Science, Kyushu Institute of Technology, 680-4 Kawazu, Iizuka, 820-8520 Japan, e-mail: sarai@bse.kyutech.ac.jp

* *Corresponding author*

Abstract Regulatory proteins bind to specific DNA sequences and play a critical role in controlling gene expression. However, the mechanism of specific binding is not well understood yet. Structural data on protein–DNA complex provides clues for understanding the mechanism of target DNA sequence recognition by regulatory proteins. Such structures indicate that there is no simple one-to-one correspondence between base and amino acid in protein–DNA complexes. We have, therefore, developed statistical potentials based on the complex structures to quantify the specificity of interactions. Previously, we demonstrated that the potentials successfully discriminated target sequences among random sequences. It is interesting to see if they can detect difference in specificity in symmetric/asymmetric and cognate/non-cognate binding in which subtle structural changes are often observed. We demonstrate how such examples are analyzed by the potentials and discuss how they are different in protein–DNA interaction.

Key words: protein–DNA interaction, cooperativity, asymmetric recognition, cognate/non-cognate interaction, DNA deformation

1. INTRODUCTION

Regulation of gene expression in higher organisms is achieved by a complex network of regulatory proteins and their target genes. However, the recognition mechanism is not fully understood. To understand the gene regulation, the mechanism of target recognition by the proteins must be elucidated. Increasing amount of structural data on protein–DNA complex has provided us detailed information about specific interactions between DNA bases and amino acids, and showed that there is no simple code-like rule in protein–DNA recognition (Matthews, 1988). It is thus useful to develop a method to quantitatively evaluate the specificity of interactions.

In our previous work (Kono and Sarai, 1999), we have developed base-amino acid pair statistical potentials based on the amino acid distribution around base using protein–DNA complex structures. The derived statistical potentials were then used to examine the compatibility between DNA sequences and the protein–DNA complex structure in a combinatorial "threading" procedure, showing high discrimination ability between target sequences and random sequences. Transcription factors usually bind to DNA as homodimer, heterodimer, or more complex form, and the structure of DNA and/or proteins is often deformed compared with their free forms.

Such cooperative changes in structure appear to play an active role in the specific recognition. Even homodimeric transcription factors bound to their target DNA show structural asymmetries (Marmorstein *et al.*, 1992; Marmorstein and Harrison, 1994; King *et al.*, 1999), which are usually subtle, making it difficult to determine which regions of proteins are responsible for the difference in specificity. Also, the structures of transcription factors bound to their cognate sequence and non-cognate sequence show subtle structural differences. Here, we describe how our method using the statistical potentials can reveal which interaction produces the differences in specificity in cognate/non-cognate, symmetric/asymmetric and cooperative bindings.

2. METHODS

We briefly explain how the statistical potentials are derived from the protein–DNA complex structures (Kono and Sarai, 1999). There are preferential interactions between amino acid-base pairs though they are not one-to-one interactions (Matthews, 1988). We need to quantify such preferences. Protein–DNA complex structures provide such information. First, the spatial distribution of each amino acid against each of bases is derived from the complex structures. Figure 1 shows the Asn distribution around adenine. A cluster was found at the tip of N7 and N6 atoms. The distribution can be converted into a so-called statistical potential given by the following equation (Sippl, 1990):

$$\Delta E(s) = -k \ln(f_{\text{ab}}(s) / f(s)),$$

where ΔE is potential energy at position s (here $C\alpha$ position is used), $f_{ab}(s)$ is frequency of amino acid a around base b, $f(s)$ is frequency of any amino acid around base b and k is constant. Against adenine, the potential map of Asn shows a distinct difference from that of Asp (Figure 1), which has already suggested the specificity difference between Asn and Asp. These potentials derived for all the base-amino acid combinations were applied for examining the compatibility between DNA sequences and the protein–DNA complex structure by using a combinatorial threading procedure similar to protein sequence–3D structure compatibility.

The sum of the potential energy for a sequence with a given length is regarded as the energy of the protein–DNA complex including the sequence. The compatibility of the sequence in the crystal structure was characterized by Z-score, defined by a difference between interaction energy for the target sequence and an average for random sequences, divided by standard deviation. This normalized quantity serves as a measure of specificity for the protein–DNA complex. Against 52 protein–DNA complexes, we obtained an average Z-score of −2.8 against random sequences (Kono and Sarai, 1999). Here, the Z-score is defined as $(X - m)/\sigma$, where X is the energy of a particular sequence, m is the mean energy of 50 000 random sequences, and σ is the standard deviation.

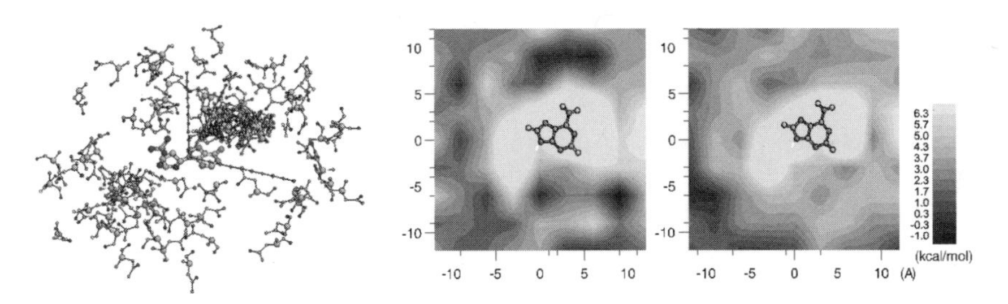

Figure 1. Asn distribution around adenine (left) derived 52 protein–DNA complexes and derived potential map (middle). For comparison, potential map for Asp is also shown (right). In the major group, many Asn are found and form double hydrogen bonds with adenine. Such Asn residues ($C\alpha$ positions) are located at a dark region ((0, 8.5) to (5, 8.5)) in the potential map.

3. RESULTS AND DISCUSSION

We applied the statistical potentials for heterodimer, homodimer, cognate/non-cognate binding forms. Here, we show how sensitively observed structural differences in complex forms are described by the potential energies.

3.1 Cooperativity in heterodimer of MATa1 and MATα2

Transcription factors usually bind to their target sites in cooperation with other factors, thereby enhancing the specificity to DNA sequences. We show such an example in MATa1 and MATα2 heterodimer, which binds to DNA and represses transcription of *HO* gene in yeast. There are two crystal structures available, one is MATa1–DNA complex (PDB code: 1APL) (Wolberger *et al.*, 1991) and the other is MATa1, MATα2–DNA complex (PDB code: 1YRN) (Li *et al.*, 1995). It is interesting to see how their specificity is different. In Figure 2, it is clearly described that the MATα2 (left) from a MATa1/MATα2–DNA complex has more amino acid residues which have favorable interactions with DNA than the MATα2 (right) from a MATα2–DNA complex. The potentials successfully capture the structural changes upon binding of the two proteins to DNA. The Z-scores for these interactions indeed show that the specificity of MATα2 from a MATa1/MATα2–DNA complex is significantly higher than that of MATα2 from a MATα2–DNA complex (Kono and Sarai, 1999).

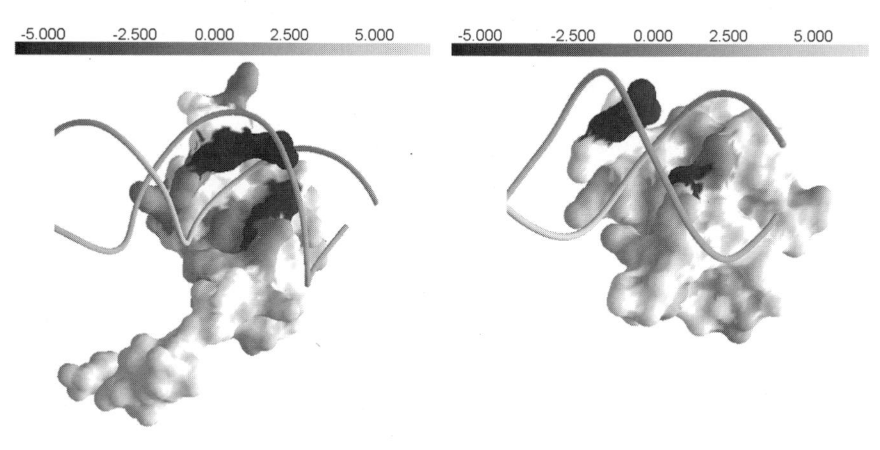

Figure 2. Distribution of statistical potential energy mapped on the molecular surface of MATα2–DNA complexes. MATα2 (left) from MATα2/MATa1–DNA complex (PDB code:1YRN) shows cooperative binding to DNA with more dark region than MATα2 (right) from MATα2–DNA complex (1APL). The Figures 2 to 5 are prepared using GRASP (Nicholls *et al.*, 1991).

3.2 Asymmetry in binding

Transcription factors, especially in prokaryotes, often work in the form of homodimer. Homodimers such as Rel/NF-κB, bZIP homeodomains and helix-loop-helix family members often bind to target DNA sequences in asymmetric manner, leading to quasi-symmetric structures, where identical subunits in sequence adopt similar but different conformations. These examples are of particular interest because comparison of the subtle structural differences among similar contacts within the two halves of protein dimmer provides valuable information about how specificity is determined.

We calculated Z-scores for 23 homodimer protein–DNA complexes belonging to several structural families (Selvaraj *et al.*, 2002). Among the 23 complexes, 20 structures show the Z-score differences ($|\Delta Z| > 0$) in monomers. Especially, 12 complexes have $|\Delta Z|$ of more than 0.5. Such amount of difference is statistically significant by a jackknife-type test. These proteins are good targets to examine the specificity difference in their structures. One group known to bind to asymmetrically to DNA is transcription factors containing a Zn_2Cys_6 binuclear cluster domain such as HAP1, GAL4, and PPR1 (Marmorstein *et al.*, 1992; Marmorstein and Harrison, 1994; King *et al.*, 1999). They bind as a homodimer to a DNA sequence containing two CGC motifs in a tandem orientation. The chain B of PPR1 in the complex shows higher specificity (Z-score = −3.1) than the chain A (Z-score = −1.1). The potential energy map in Figure 3 reveals that the main difference in specificity is due to Lys41, which favorably interacts with DNA in chain B, but not in chain A.

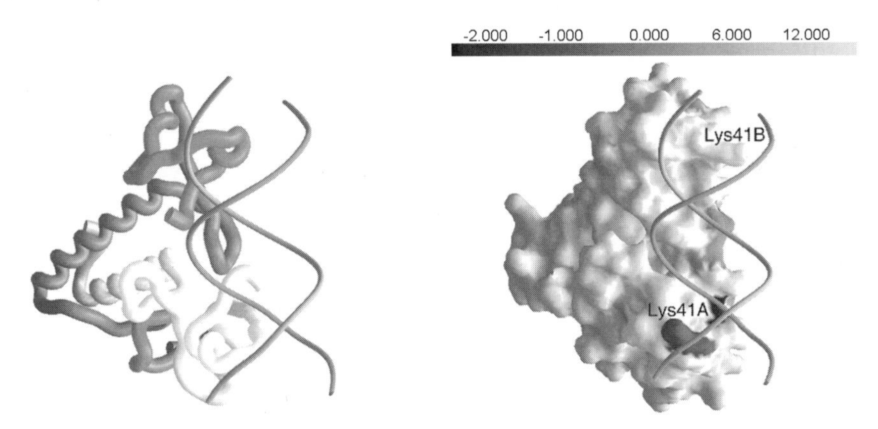

Figure 3. Distribution of statistical potential energy mapped on solvent-accessible surface of pyrimidine pathway regulator structure bound to twofold symmetric DNA sequence (PDB code: 1PY1). In left panel, a darker tube denotes chain A and white chain B. In right panel, more favorable protein–DNA interactions are observed in chain B (down part) than chain A (upper part), especially at Lys41 sticking toward DNA.

3.3 Cognate/non-cognate interaction

The structural database also contains some structures of the same protein bound to different DNA sequences. A transcription factor, NF-κB is such an example and a good test case to examine our potentials because the affinity to different sequences was also experimentally known on this protein.

NF-κB p50 homodimer binds to a duplex oligonucleotide with 11-bp consensus recognition site located in the major histocompatibility complex class I enhancer (Muller *et al.*, 1995). (PDB code: 1SVC). Ghosh *et al.* (1995) used a 10-bp idealized motif related, but not identical, to the natural sites (PDB code: 1NFK). These structures are very similar, but it has been known that NF-κB binds 30 times more strongly to the former than to the latter (Chytil and Verdine, 1996). Consistent with the experimental result, the Z-score for 1SVC was higher than that for 1NFK. The differences in favorable protein–DNA interactions between two structures are shown in Figure 4. The favorable interaction sites shown by darker colors are wider in 11-bp consensus DNA binding (left) than the 10-bp idealized DNA binding (right). Note that the specificity does not always have the relationship with the affinity (thermodynamic preference). However, the result suggests that the derived statistical potentials can detect subtle differences in the specificity caused by subtle structural differences.

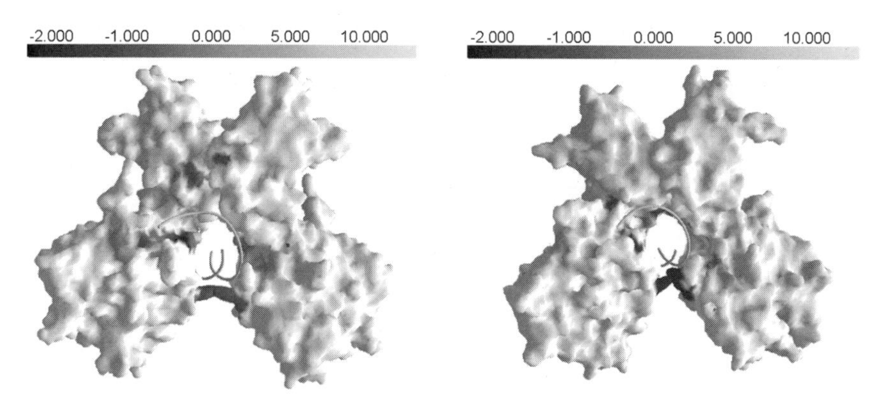

Figure 4. Distribution of statistical potential energy mapped on the structures of NF-κB bound to an 11-bp consensus DNA (left, PDB code: 1SVC) and a 10-bp idealized DNA (right, 1NFK). The left panel has wider dark region than the right one.

Another example of cognate/non-cognate interaction is found in BamHI endonucleases (Newman *et al.*, 1995; Viadiu and Aggarwal, 2000). It is noteworthy that restriction endonucleases have stringent and exquisite sequence specificity while transcription factors in general are tolerate to small changes in target sequences. Thus, one base change in target sequence drastically reduces the cleavage activity. BamHI protein in the cognate binding shown in Figure 5 left, wraps DNA more tightly than one in the non-cognate binding. BamHI in the non-cognate binding has a few favorable interactions and appears to facilitate diffusion along DNA sequence. These differences in protein conformation are reflected by Z-scores obtained for the respective complexes (Z-score = −1.7 for cognate binding, 2.2 for non-cognate binding).

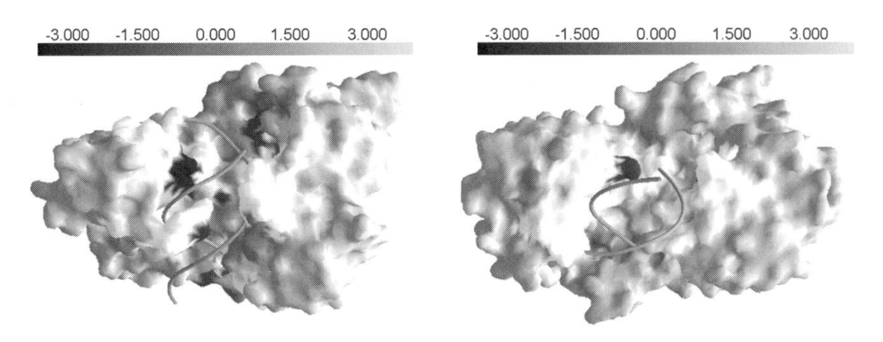

Figure 5. Distribution of statistical potential energy mapped on the structures of BamHI endonuclease bound to cognate DNA (left, PDB code: 1BHM) and non-cognate DNA (right, 1ESG).

4. CONCLUSION

We have demonstrated the ability of the statistical potentials derived from protein–DNA complex structures to detect subtle specificity difference caused by structural changes in cooperative binding, cognate/non-cognate complexes and symmetric/asymmetric complexes. The structural changes are frequently observed in most protein–DNA interactions and are believed to play a critical role in specific recognition of DNA sequences by proteins. Thus the statistical potentials serve as a powerful tool to establish a relationship between structure and specificity to provide insight into the molecular mechanism of protein–DNA recognition. Thus, the statistical potentials can be applied to the prediction of target sequences of transcription factors if genome sequences are used for the threading. The method is also useful to determine which pair of amino acid-base is critical in protein–DNA, enabling one to design amino acid sequences suited for specific target DNA sequences.

ACKNOWLEDGMENTS

This work was supported in part by Grant-in-Aid for Scientific Research 15014233 (H. K.) and 13208037 (A. S.) from MEXT.

MACROMOLECULAR MODELING AS A TOOL FOR EXPANDING BIOINFORMATICS DATABASES

Yu.N. VOROBJEV

Novosibirsk Institute of Bioorganic Chemistry, Siberian Branch of the Russian Academy of Sciences, prosp. Lavrentieva 8, Novosibirsk, 630090 Russia, e-mail: ynvorob@niboch.nsc.ru

Abstract The development and use of efficient computational tools can add to the completeness of bioinformatics databases and our understanding of molecular structure and relationships between structure and function. Reliable macromolecular modeling can provide tools for expanding (experimental) bioinformatics databases beyond the capacity of experimental methods and provide new knowledge. An approach that combines the explicit solvent (ES) and implicit solvent (IS) models has been developed for calculating free energies of protein and nucleic acid conformations. A protocol for calculating context-dependent DNA conformational parameters has been developed.

Key words: protein stability, free energy calculation, molecular dynamics, DNA conformational dynamics, context-dependent DNA conformational parameters

1. INTRODUCTION

Recent progress in macromolecular modeling methods provides increasing evidence that simulated bioinformatics data can be obtained by advanced modeling to fill gaps in databases. When properly designed and used, these methods can aid more accurate determination of three-dimensional (3D) molecular structures, discovery and refinement of structure details, and investigation of internal conformational dynamics under different thermodynamic conditions well beyond the capacity of experimental methods. Macromolecular modeling methods can be efficiently applied in several fields:

1. 3D structure reconstruction of missing structural information on some elements of a macromolecule, i.e. atomic coordinates not defined in the PDB database; e.g. flexible loops on a protein surface;
2. 3D structure determination using conformational restraints based on experimental data;
3. Internal conformational dynamics related to thermal fluctuations;
4. 3D structure prediction—"protein folding" and "nucleic acid folding" problems and stability of various conformations.

Macromolecular modeling methods allow missing bioinformatics knowledge to be "interpolated" and "extrapolated" from the available information. Protein and nucleic acid conformations as well as protein–ligand, protein–protein, and protein–nucleic acid binding are guided by the total free energy of the system and the corresponding forces in solvent. Therefore, the importance of accurate determination of free energies is obvious from the applications. However, the development of an accurate and fast method to calculate the free energy of a macromolecular complex in solution on the basis of strictly statistical mechanical principles is a task of great complexity (Roux and Simonson, 1999). Here, we concentrate on an advanced method for calculating the free energy of a macromolecule in an aqueous solution and consider preliminary results of simulation of context-dependent conformational parameters of a DNA duplex.

2. METHODS AND ALGORITHMS

2.1 Macromolecular free energy in an aqueous solvent

We have developed an approach that combines the explicit solvent (ES) and implicit solvent (IS) models (Vorobjev *et al.*, 1998). The key to success is to obtain accurate representative conformations by simulation using the explicit solvent model, and then to estimate the free energy of the protein–solvent interactions by the implicit solvent model. The ES/IS method suggests a consistent approach, which involves simulation of a set of microscopic structures compatible with the molecular structure of the aqueous solution and a reasonably reliable physical implicit model of the solvent for calculation of the solvation free energy (Vorobjev and Hermans, 1999). The implicit solvation model includes empirical parameters, born radii, to define a molecular surface and dielectric surface interface. Consistent calibration of the born radii parameters in the implicit model of the ES/IS method with free energy calculations by the free energy perturbation method for reference database molecules has been performed to develop the force-field-consistent ES/IS method. Evaluation of the force-field-consistent born radii set was performed for the GROMOS force field. The free energy F_A of a solute molecule in macroscopic conformation A is

$$F_A = < U_m(\boldsymbol{x}) >_A - Ts_{conf,A} + < W(\boldsymbol{x}) >_A, \tag{1}$$

where <>A is an average over the microconfigurations of conformation A; U_m, intraprotein conformational energy; and S_A, the entropy of conformation A. In the ES/IS method, a representative set of microscopic configurations $\boldsymbol{x}_{A,i}$ of a solute in a solvent is generated by MD simulation with the explicit solvent along a relatively short trajectory (50–100 ps). The solvation free energy $W(\boldsymbol{x})$ can be written as a sum of terms for cavity formation, solute–water van der Waals interactions, and electrostatic polarization of the solvent by polar components of the solute. As a result, eq. (1) takes the form

$$F_A = <U_{m,sh}>_A + <U_{m,coul}>_A - TS_{conf,A} + <G_{cav}>_A + <G_{s,vdw}>_A + <G_{pol}>_A, \tag{2}$$

where the intramolecular potential energy U_m is represented as a sum of the short-range (i.e. angle deformation and van der Waals) terms, $U_{m,sh}$, and electrostatic Coulombic interactions, $U_{m,coul}$. Three of the six terms in eq. (2), namely, $<U_{m,sh}>$, $<U_{m,coul}>$, and $<G_{s,vdw}>$, are accumulated as averages during the molecular dynamics simulation: the free energy of the van der Waals interactions between solute and solvent, $G_{s,vdw}$, can be accurately approximated by the potential energy of these interactions:

$$G_{s,vdw} = U_{s,vdw}. \tag{3}$$

The reorganization energy of an aqueous solvent and, consequently, the related entropy are low because of small perturbation of the liquid structure by the short-range and weak van der Waals attraction, as follows from the perturbation theory of liquids (Hansen and McDonald, 1986).

The entropic term, $TS_{conf,A}$, is estimated in a harmonic approximation from the covariance matrix of positional fluctuations along the dynamic trajectory. This term pertains only to conformational fluctuations of the solute molecule. The two remaining terms, the free energy of cavity formation G_{cav} and the free energy of solvent polarization G_{pol}, which are difficult to simulate microscopically, are found by using models in which the solvent is treated implicitly with appropriate physical models, such as a continuum or dielectric continuum. The free energy of formation of a solute-sized cavity in the solvent, G_{cav}, is expressed as a product of the surface area and surface free energy. This functional form follows from the experimental data on hydrocarbon solvation energy and the scaling theory of liquids (Hermann, 1972; Tomasi and Persico, 1994):

$$G_{\text{cav}} \approx \gamma_{\text{mic}} S, \tag{4}$$

where γ_{mic} is the microscopic surface free energy and S, the molecular surface that confines the solvent excluded volume. Here, we estimate the value of γ_{mic} to reproduce experimental hydrocarbon solvation energies with the SPC water model and CEDAR (GROMOS) atom–atom force field. For hydrocarbons, which are hydrophobic, electrostatic interaction with water solvent is expected to be very small; therefore, the free energy of transfer $G_{\text{n,gw}}$ is the sum of the free energy of cavity formation G_{cav} and the van der Waals free energy $G_{\text{s,vdw}}$:

$$G_{\text{n,gw}} \approx G_{\text{cav}} + G_{\text{s,vdw}}. \tag{5}$$

The free energy $G_{\text{s,vdw}}$ of interactions of a hydrocarbon solute with the solvent is calculated by explicit simulation of hydrocarbons in SPC water along the 200–400-ps trajectories. The calculation results for three different sets of radii for C and H varying within 0.15 Å are shown in Figure 1. We conclude that the approximation of the free energy G_{cav} by eq. (4) is excellent and little sensitive to small parameter variations.

Taking into account that the SIMS and FAMBE methods are more effective, if CH_n groups are treated as united atoms, we found the optimal radius of CH_n ($n = 1, 2, 3, 4$) groups equals 2.21 Å and $\gamma_{\text{mic}} = 76$ cal/Å^2. This value is in good agreement with the estimation $\gamma_{\text{mic}} \approx \gamma_{\text{macro}} \langle S_{\text{macro}}/S_{\text{micro}} \rangle = 67$ cal/Å^2, where the ratio estimated for the molecular surface and flat macroscopic surfaces $\langle S_{\text{macro}}/S_{\text{micro}} \rangle = 0.69$ and the macroscopic surface tension of the liquid–gas interface $\gamma_{\text{macro}} = 102$ cal/Å^2 (Lide, 1997).

The free energy of solvent polarization, G_{pol}, is found by approximating the solvent and macromolecular volume as a continuum dielectric and solving the corresponding electrostatic Poisson equation to calculate the solvent polarization charge density on dielectric interface by the FAMBE method (Vorobjev and Scheraga, 1997).

The protein–solvent dielectric interface surface is approximated by a smooth molecular surface and is calculated by the SIMS method (Vorobjev and Hermans, 1997). Validity of the continuum dielectric approximation is assayed via simulation of the polarization free energy of charging by a slow-charging method (Vorobjev and Hermans, 1999).

An optimal set of atomic radii, which defines the solute–solvent dielectric surface, is derived for the ES/IS model by fitting the FAMBE free energies of solvent polarization to the values calculated by slow-charging simulations. The slow-charging simulation was performed for a set of 40 charged groups in dipeptides in the SPC water solvent. The FAMBE solvent polarization free energy calculated with the optimal atom radii set shows a good correlation with the results of the slow-charging method.

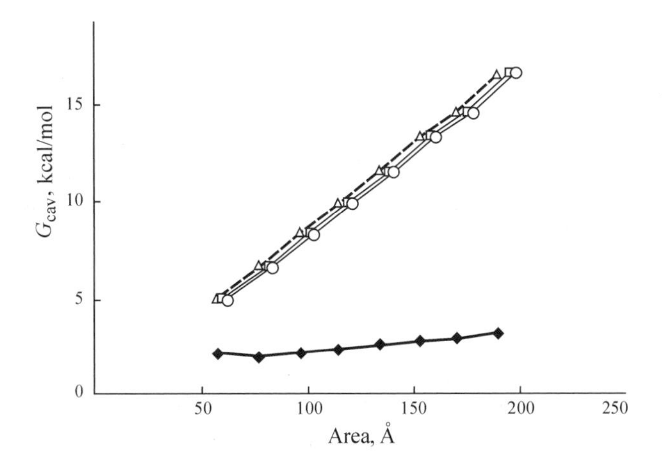

Figure 1. Free energy of cavity formation G_{cav} in hydrocarbons: ○, □, and Δ indicate the calculated G_{cav} for different values (2.1–2.3 Å) of born radii of united CH_n atoms and ♦ shows gas-to-water transfer energies $G_{h,gw}$ of hydrocarbons (methane, ethane, *n*-butane, propane, *n*-pentane, *n*-hexane, and *n*-octane), obtained experimentally (Ben-Naim and Marcus, 1984).

3. RESULTS

3.1 Stability of protein decoys

The accuracy of the ES/IS method was demonstrated by comparing hundreds of non-native conformations ("decoys") of several proteins from the Park–Levitt decoy set and models presented at CASP3 with their native structures. It was found that the latter had lower free energies for all the proteins (Vorobjev and Hermans, 2001; Figure 2). The required molecular dynamics simulations of proteins were performed with the CEDAR (GROMOS) force field.

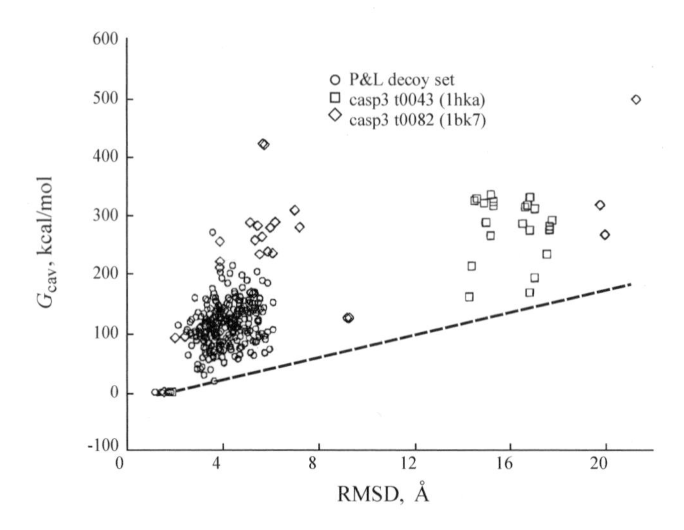

Figure 2. Total excess free energy of decoys versus RMSD from the corresponding native structure: dash line is the minimum discrimination.

3.2 Context-dependent DNA properties

This advanced modeling method can be used as a computational tool to produce new bioinformatics knowledge unavailable from experimental data. An example is the context-dependent internal conformational dynamics of DNA fragments. Protein-binding sites of genomic DNA should have appropriate conformations and conformational mobilities for the proper interaction with cell proteins. Therefore, the context-dependent conformational properties of a DNA sequence can serve as a natural set of descriptors to characterize the probability of interaction between a site and proteins (Ponomarenko *et al.*, 1999).

The available context-dependent static conformational parameters of DNA derived from experimental 3D structures of a DNA fragment in the crystal state are neither complete nor accurate because of the distortion effect of crystal packing and detectable conformational differences in the crystal and in the solution for some sequences (Dornberger *et al.*, 1999). The context-dependent conformational properties of DNA sequences can be obtained in a solvent environment via molecular dynamics simulations. The covariance fluctuation matrix C (Vorobjev *et al.*, 1998) is given by the equation

$$C_{ij} = \langle(x_i - \langle x_i\rangle)\,(x_j - \langle x_j\rangle)\rangle_A, \quad (6)$$

where $x(t)$ are atomic coordinates along the MD trajectory. If a conformation is stable, then the molecular motion along the MD trajectory can be approximated as quasiharmonic to give a statistical description of actual anharmonic intramolecular motions over a large number of microstates of the molecule. The eigenvectors define the normal modes, while the eigenvalues λ_i of the (mass-weighted) covariance matrix C define frequencies and positional fluctuations of the normal coordinates. Considering a set of short 10–12-bp DNA fragments with all the possible double and triple sequences at the centers of duplexes and random sequences at the ends, it is possible to investigate context-dependent conformational parameters of pairs, triplets, etc. The helical parameters Ω, τ, and ρ (twist, tilt, and roll; Figure 3) (Saenger, 1984; Lavery and Sklenar, 1988) for each base-pair step of a DNA duplex can be calculated from atomic coordinates and the average values and amplitudes of thermal fluctuation. To obtain statistically reliable values of conformational parameters, the MD simulations should be made until convergence of the fluctuation matrix C. The convergence required can be obtained for a simulation of ~ 1 ns in length. The reliability of MD simulations depends on the computational method and force field. The required molecular dynamics simulations of DNA duplexes are performed with the GROMOS96 (URL, 2000), and AMBER6.0 (URL, 2001) MD simulation packages. For charged macromolecules, a proper account of long-range electrostatic interactions, such as the Ewald sum, is very important for the stability and accuracy of MD simulations (McConnel and Beveridge, 2000). We investigated the GROMOS96 and AMBER6.0 MD simulation packages for stability of long simulations. The test MD simulations for a DNA duplex showed that AMBER6.0 with the PME option for the long-range electrostatics package provided a stable MD nanosecond trajectory, while the GROMOS96 package with the generalized reaction field method for taking into account long-range electrostatics gave less satisfactory results.

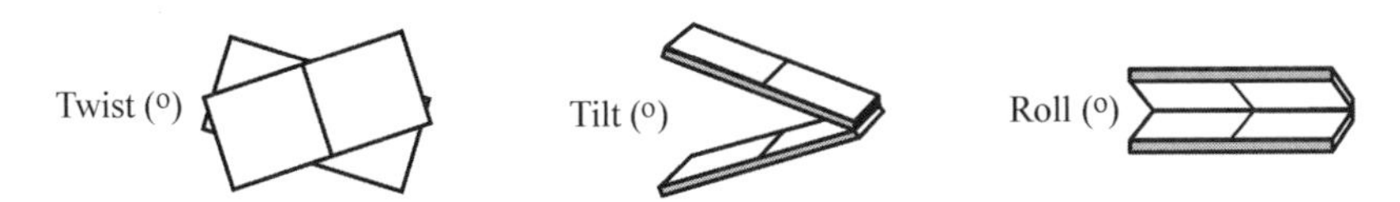

Figure 3. Definitions of DNA helix parameters for a base-pair step.

Preliminary results of simulation of several helix DNA parameters of the DNA duplex d(GCGTTAACGC)$_2$ are shown in Table 1. Four series of 1-ns simulations were performed and analyzed for dinucleotide steps repeated twice. It is evident from Table 1 that a certain dinucleotide step displays little variation in helix parameters when positioned in different parts of the duplex. The helix parameters of different dinucleotide steps have different values, i.e. the simulation is context-dependent. The values of thermal fluctuations of helix parameters (which define conformational flexibility) depend on the dinucleotide step. Complete modeling of the context dependence of helix parameters is in progress.

Table 1. Simulated values of helix parameters of the DNA duplex d(GCGTTAACGC)$_2$

Dinucleotide step (helix parameters*)	Average 1	Average 2	Average 12	Diff 1–2	Diff 1–2/av,%	RMSD
GC (h, Å)	3.352	3.359	3.355	0.007	0.2	
GC (twist, degr)	29.983	29.940	29.961	0.042	0.1	
GC (roll, degr)	1.844	1.559	1.701	0.286	16.8	
GC (tilt, degr)	15.364	16.513	15.939	1.150	7.2	
CG (h)	3.076	3.108	3.092	0.032	1.0	0.25
CG (twist)	36.810	36.875	36.842	0.066	0.2	3.07
CG (roll)	12.913	11.296	12.105	1.617	13.4	3.14
CG (tilt)	5.807	5.972	5.889	0.166	2.8	4.09
GT (h)	3.252	3.248	3.250	0.004	0.1	0.19
GT (twist)	34.453	31.894	33.173	2.559	7.7	2.41
GT (roll)	15.422	11.336	13.379	4.086	0.305	2.71
GT (tilt)	7.930	6.624	7.277	1.306	0.179	8.31
TT (h)	2.871	2.842	2.856	0.029	0.10	0.21
TT (twist)	39.941	40.123	40.032	0.181	0.5	2.25
TT (roll)	12.500	13.221	12.860	0.721	5.6	1.54
TT (tilt)	10.348	8.913	9.631	1.435	14.9	3.41

* Helix parameters are defined as described in (Saenger, 1984; Figure 3).

4. CONCLUSION

We have shown that molecular modeling methods are able to recognize native folds among a large set of non-native compact folds and discover context-dependent conformational data from molecular dynamics computer experiments. This implies that modern modeling methods can be efficiently used to restore experimentally unavailable bioinformatics data and to obtain new unique data on (i) 3D structure reconstruction of the missing structural information, i.e. coordinates of missing atoms that are undefined in a protein data base, e.g. flexible loops on a protein surface (the reconstruction is based on the advances in free energy calculation and its optimization for a protein in an aqueous solvent); (ii) internal conformational dynamics of DNA duplexes resulting from thermal fluctuations, which can be investigated by the molecular dynamics method at a required level of performance, and (iii) unique data on context-conformational dependence unavailable from experiment.

ACKNOWLEDGMENTS

This work was supported by the Russian Ministry of Industry, Science, and Technologies (grant No. 43.073.1.1.1501) and Russian Foundation for Basic Research (grant No. 02-04-48915).

GRAPHIC REPRESENTATION OF EQUILIBRIUM AND KINETICS IN OLIGOPEPTIDES: TIME-DEPENDENT FREE ENERGY DISCONNECTIVITY GRAPHS

A.V. GAVRILOV[1], S.F. CHEKMAREV[2]*
[1]*Novosibirsk State University, ul. Pirogova 2, Novosibirsk, 630090 Russia;* [2]*Kutateladze Institute of Thermophysics, Siberian Branch of the Russian Academy of Sciences, prosp. Lavrentieva 1, Novosibirsk, 630090 Russia, e-mail: chekmare@itp.nsc.ru*
* *Corresponding author*

Abstract This paper introduces time-dependent disconnectivity graphs in terms of free energy that not only exhibits the equilibrium residence probabilities of conformers, but also allows for an insight into kinetics of the system. To characterize a transition state, a "kinetic" free energy (KFE) is introduced, which takes into account the number of transitions through the saddle for a given time τ, thus compensating for the deficient degree of freedom in the transition complex. The time τ enters into KFE as a parameter, and with its increase, KFE barriers between minima become lower. Correspondingly, the free energy surface deforms with τ, and a series of the surfaces for increasing τ shows how equilibrium in the system is attained with time—first, in certain groups of the structures; then, between these groups; and finally, totally in the system. The approach is illustrated for a solvated alanine tetrapeptide.

Key words: peptide, all-atom model, confinement molecular dynamics, free energy surface, equilibrium, kinetics

1. INTRODUCTION

The potential energy surface (PES) of a many-body system completely defines the behavior of the system under specific conditions. Therefore, the knowledge of the PES landscape is an essential element in studying a many-body system. Recently, considerable attention was given to a graphic representation of PESs. Berry and Kuntz (1995) suggested using for this purpose a "minimum–saddle–minimum" sequences, and Becker and Karplus (1997) introduced the disconnectivity graphs. In both cases, the critical points of PES—local minima and saddles—were used to represent PES, but the way of representation was different: the minimum-saddle-minimum sequences presented various pathways connecting the low-lying minimum (or minima) with higher minima, and the disconnectivity graphs indicated the minima that were connected by pathways lying below a certain energy threshold.

Critical points are essential elements of PES, but they do not provide all the information necessary for understanding equilibrium properties and kinetics of a system. For this purpose, the geometry of PES, at least in the vicinity of these points, should be taken into consideration. For example, a high-lying basin may be wider than a low-lying one, so that its statistical weight and thus, the probability of finding the system in it, will be higher. A characteristic example is given by solvated alanine tetrapeptide: while the α-helix conformer presents a ground-state atomic configuration, the system predominantly resides in the basin for the extended β-strand conformer, whose conformation entropy is higher (Krivov *et al.*, 2002).

The entropy is responsible for the geometrical factor. Hence, the free energy surface (FES) of the system should be considered instead of PES. For a basin on PES associated with the ith minimum, the free energy is

$$F_i = -k_{\mathrm{B}} T \ln Z_i, \tag{1}$$

where $Z_i = \frac{1}{h^{3n}} \frac{1}{h_i^{\mathrm{PG}}} \int_i \exp(-H_i / k_{\mathrm{B}} T) d\boldsymbol{r}^{(3n)} d\boldsymbol{p}^{(3n)}$

is the partition function. Here, k_{B} is the Boltzmann constant; T, temperature; n, number of atoms in the system; h, Planck constant, h_i^{PG}, the order of the point group; and $d\boldsymbol{r}^{(3n)}$ and $d\boldsymbol{p}^{(3n)}$, vectors of the atomic coordinates and momenta, respectively. $H_i = \sum_1^n \boldsymbol{p}_j^2 / 2m + U_i$ is Hamiltonian of the system for the ith basin, where m is the atomic mass and U_i, potential energy. Equilibrium residence probability for the ith basin is then written as

$$p_i^{\mathrm{eq}} = \frac{Z_i}{\sum_i Z_i} = \frac{\exp(-F_i / k_{\mathrm{B}} T)}{\sum_i \exp(-F_i / k_{\mathrm{B}} T)}. \tag{2}$$

As for saddles, definition of their free energy is more complicated. In the simplest case of a saddle of the first order, the corresponding atomic structure presents a transition state (TS), which has one degree of freedom less than the system itself. The partition function of TS is density of the states along the reaction coordinate, and not the number of states as Z_i for the minimum. Therefore, these partition functions for TSs and minima are related to the regions of phase space of different dimensionalities, and thus, they are inconsistent. Correspondingly, their free energies are inconsistent.

One way to solve this problem is to use a definition employed in the Gibbs energy diagrams. According to it, the free energy of the TS connecting minima i and j can be written as

$$F_{ji} = F_i - k_B T \ln(hk_{ji} / k_B T), \quad (3)$$

where F_i is free energy of the ith minimum and k_{ji}, the rate constant for transitions from minimum i to minimum j. Middleton *et al.* (2001) used this kinetic definition to group the minima on PES, while Krivov and Karplus (2002) and Evans and Wales (2003) employed it to build FESs.

In this paper, we use a similar definition of TS free energy, except that for compensation of the deficient degree of freedom we take into account the number of transitions through the saddle for a given time τ. The time τ enters into KFE as a parameter, and with its increase, KFE barriers between the minima become lower. Correspondingly, the free energy surface deforms with τ, and a series of the surfaces for increasing τ shows how equilibrium in the system is attained with time—first, in certain groups of conformers; then, between these groups; and finally, totally in the system.

2. "KINETIC" FREE ENERGY OF A TRANSITION COMPLEX

To introduce kinetic free energy (KFE), let us turn to the transition states theory (TST). According to the Rice–Ramsperger–Kassel–Marcus theory (Forst, 1973), the rate constant for transitions from the basin i to basin j is given by the expression

$$k_{ji} = (k_B T / h) Z^*_{ji} / Z_i, \quad (4)$$

where Z_i and Z_{ji} are the partition functions around the minimum and transition complex, respectively.

Following the early TST (Eyring *et al.*, 1980), define the partition function of the system in the vicinity of TS as

$$Z_{ji} = \left(\sqrt{2\pi m^* k_B T} \delta / h\right) Z^*_{ji}, \quad (5)$$

where the factor $\sqrt{2\pi m^* k_B T}\delta/h$ is the partition function for a one-dimensional box of the length δ, which is oriented along the reaction coordinate (m^* is an effective mass related to system's motion along the reaction coordinate). Assume further that the box lies to the reactant basin side with the end attached to the transition state, and its length increases with the time τ as $\delta = v\tau$, where $v = \sqrt{k_B T / 2\pi m^*}$ is the average velocity of the ballistic motion of the representative point of the system in the box. Then, substituting δ into (5) yields

$$Z_{ji} = (k_B T\tau/h) Z^*_{ji}.$$

Comparing this equation with equation (4), we finally obtain for Z_{ji}

$$Z_{ji} = k_{ji}\tau Z_i.$$

In turn, according to (1), the free energy of TS is written as

$$F_{ji} = F_i - k_B T \ln(k_{ji}\tau). \tag{6}$$

This equation shows that the definition of TS free energy (3) used by Krivov and Karplus (2002) and Evans and Wales (2003) corresponds to

$$\tau^* = h / k_B T, \tag{7}$$

which at $T = 300$ K is equal to 160 fs.

From equation (6), it follows that at $\tau \le \tau_{ji}$, where

$$\tau_{ji} = 1/k_{ji}, \tag{8}$$

$F_{ji} \ge F_i$, so that F_{ji} can be considered as a barrier that should be overcome when passing from minimum i to j. Moreover, the height of this barrier decreases with τ, making transitions easier and easier, which is consistent with this interpretation of F_{ji}. At $\tau > \tau_{ji}$. F_{ji} becomes lower than F_i, and F_{ji} losses the physical meaning of a barrier. However, this circumstance is not crucial in the context of building of FESs, rather it requires a proper interpretation. The time τ_{ji} (8) is an average time required to pass from minimum i to minimum j.

3. SOLVATED ALANINE TETRAPEPTIDE: RESULTS AND DISCUSSION

For a model protein, we consider solvated alanine tetrapeptide at a room temperature ($T = 300$ K). Simulations were performed with the CHARMM29 program (Brooks *et al.*, 1983), using the polar hydrogen parameter set for peptides and proteins (param19; Neria *et al.*, 1996) and ACS implicit solvation model (Schaefer *et al.*, 1998). To simulate conditions at a constant temperature, the Langevin dynamics, friction coefficient of $\alpha = 64$ ps^{-1}, and time step of $\Delta\tau = 1$ fs were used.

To calculate the rate constants, we employed the confinement technique (Chekmarev and Krivov, 1998; Chekmarev, 2001), and more specifically, the method of successive confinement (Krivov *et al.*, 2002). Quenching the system at regular intervals of $\tau_{\text{quench}} = 10^3$ time steps, the molecular dynamics trajectory of the system was successive confined to various basins on PES. This allowed us to calculate both the energies of the minima $U_{\min,i}$ and the transition probability matrix $Q = \{Q_{ji}\}$, the Q_{ji} component of which presents the probability that the system, being in basin i, is found in basin j at the subsequent quenching after a time interval τ_{quench}. Solving the equation $\mathbf{P}^{\text{eq}} = Q \cdot \text{P}^{\text{eq}}$, the equilibrium residence probabilities $\mathbf{P}^{\text{eq}} = \{p_i^{\text{eq}}\}$ were calculated, and then, using equation (2), the relative values of F_i were found. The rate constants were estimated as $k_{ji} = Q_{ji}/\tau_{\text{quench}}$. To found the transition states between the minima, the TRAVEL algorithm (Fischer, Karplus, 1992) was employed. Following this way, a fragment of PES was explored that contained 179 conformers connected by 374 TSs; all the conformers are related to an all-*trans* form of the molecule. The total length of the run was 1.5×10^8 time steps. During this time, the system made approximately 150 transitions from each basin to the others.

Figure 1 shows a disconnectivity graph for the system in terms of potential energy (PES graph). This graph shows which minima are connected, directly or indirectly, by a pathway lying below the current energy threshold (U_{thr}). Vertical lines correspond to conformers, with the unattached ends of the lines indicating the potential energy of the minima, $U_{\min,i}$, and the horizontal lines linking the isomers connected by the pathways lying below the corresponding value of U_{thr}. Both $U_{\min,i}$ and U_{thr} are plotted along the ordinate. The statement "directly or indirectly" implies that if $U_{ij} < U_{\text{thr}}$ and $U_{jk} < U_{\text{thr}}$, then, irrespective of whether $U_{ik} < U_{\text{thr}}$ or not, the minima i and k are considered to be connected (U_{ml} is the potential energy of the TS connecting the minima m and l). If several pathways between two minima exist, the lowest one is chosen. It is seen that PES consists of several clusters of minima, with the lowest one containing an α-helix and extended β-strand conformers.

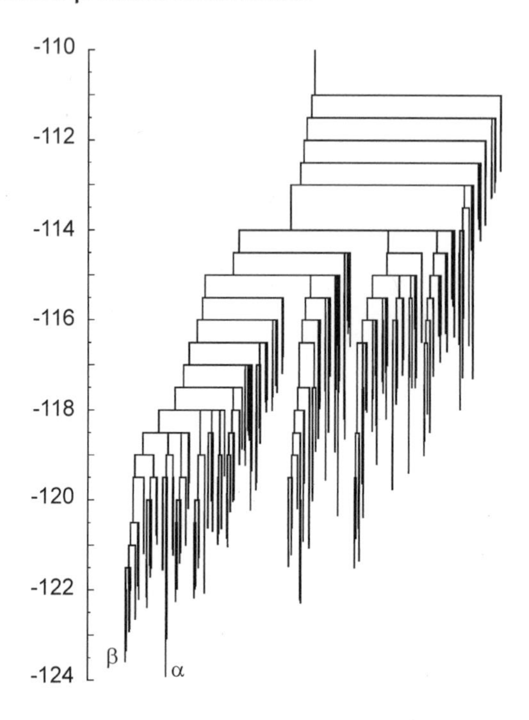

Figure 1. Potential energy disconnectivity graph for the solvated alanine tetrapeptide at 300 K (the energy scale in kcal/mol).

Now, turn to the disconnectivity graphs in free energy terms (FES graphs). A series of the graphs for several values of the elapsed time τ is shown in Figure 2, starting from $\tau = \tau^*$ (7). Each of the graphs is build similar to PES graph, except that the values of F_i and F_{ji} are used instead of $U_{\min,i}$ and U_{ji}, respectively. Energies are counted from the lowest value of F_i. In addition, for large values of τ, some modifications are made (see below). First, as is well seen from panels *a*, *b*, and *c* of Figure 2, FES graph is characteristically different from PES graph (Figure 1): the minima lowest in potential energy are not necessarily lowest in free energy—this depends on the contribution of configuration entropy, that is, of geometry of the basin. In particular, the extended β-strand conformer, which is higher in potential energy than the α-helix, is lower in free energy. As τ increases, the KFE barriers lower, and the horizontal lines linking the minima shift downward. The space between the lines is kept constant, since τ does not affect the relative values of r_{ji}. Correspondingly, the linking lines approach the minima, and, at a certain value of τ (approximately at $\tau = 10^4$ fs; panel *c* of Figure 2), some minima cross the lines.

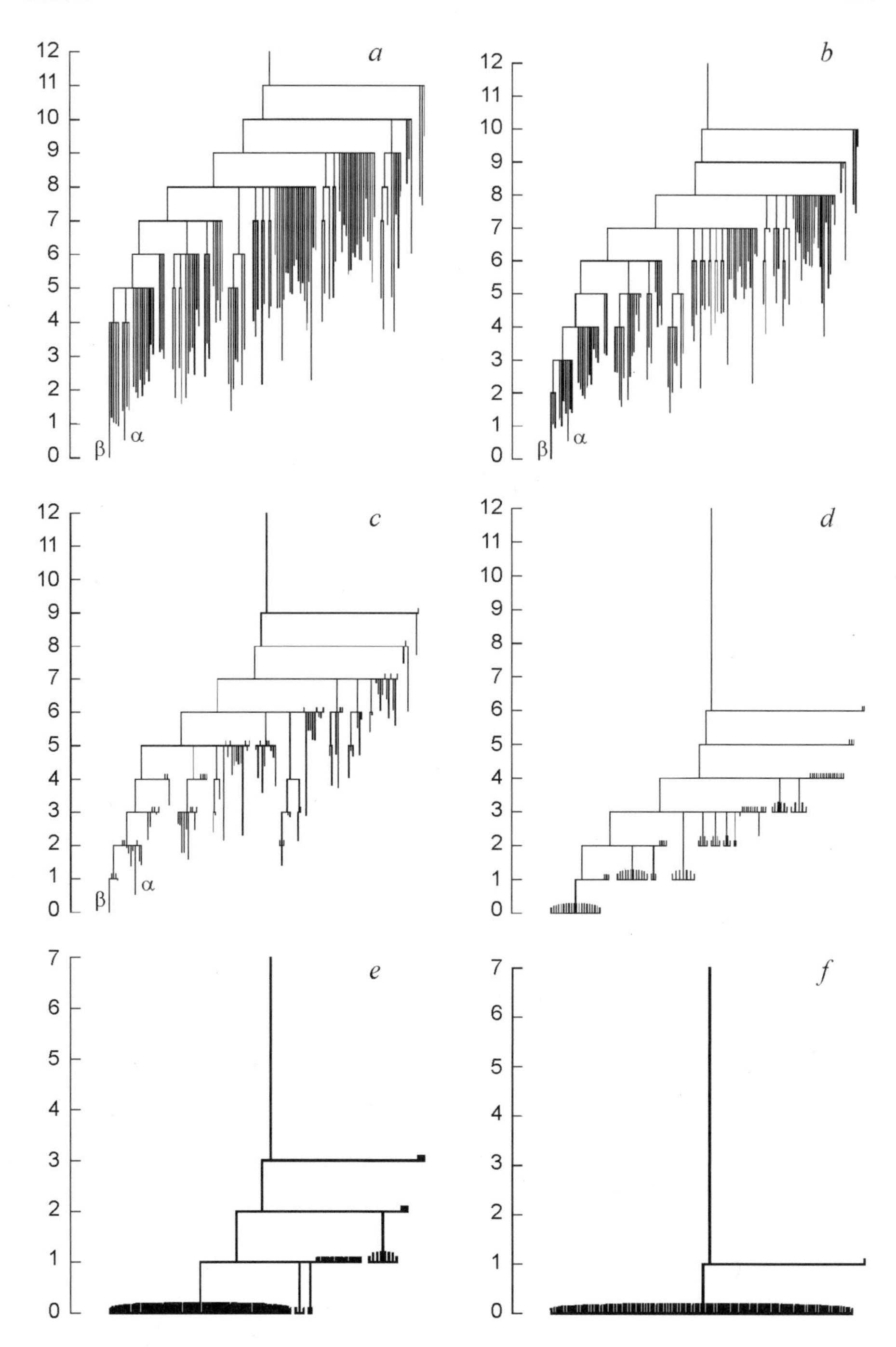

Figure 2. Free energy disconnectivity graphs for solvated alanine tetrapeptide at 300 K: (*a*) r = $1.6 \times\cdot 10^2$ fs; (*b*) $\tau = 1\cdot\times 10^3$ fs; (*c*) $\tau = 1\cdot\times 10^4$ fs; (*d*) $\tau = 1\cdot\times 10^6$ fs; (*e*) $\tau = 1\cdot\times 10^8$ fs; and (*f*) $\tau = 1\cdot\times 10^{10}$ fs (the energy scale in kcal/mol).

For a minimum i, this moment formally corresponds to the condition that τ coincides with the minimum value of τ_{ji} (8) for all the direct or indirect transitions connecting this minimum with the others in a given cluster. From physical point of view, this means that the time τ is sufficient, in average, to provide a satisfactory exchange between a given minimum and the others. Further, when all the minima in a cluster meet this condition (i.e. they all cross the linking line), local equilibrium in the cluster is anticipated. To simplify the graphs, the minima that have crossed the linking line are indicated by bars directed upward; the unattached ends of them do not longer indicate the values of F_i.

At further increase of τ, the KFE barriers continue to decrease, and clusters of minima coalesce into larger ones (panels *d* and *e*). Eventually, virtually all the minima group into a single cluster (panels *f*), which indicates that the equilibrium has been reached. We thus can infer that at a room temperature, the equilibrium in a given domain of PES, which represents all-*trans* conformations of the molecule of alanine tetrapeptide, is attained at approximately 10 μs.

4. CONCLUSION

We have introduced time-dependent disconnectivity graphs in terms of free energy. These graphs exploit a kinetic definition of the transition state free energy (KFE), which takes into account the number of transitions through the transition state for a given time τ, thus compensating for the deficient degree of freedom in the transition complex. The time τ enters into KFE as a parameter, and with its increase, KFE barriers between the minima become lower. Correspondingly, the free energy surface deforms with τ, and a series of surfaces for increasing τ shows how equilibrium in the system is attained with time. Due to this, the time-dependent disconnectivity graphs do not only exhibit the equilibrium residence probabilities for the conformers, but also give an insight into kinetics of the system.

ACKNOWLEDGEMENTS

This work was supported by the Russian Foundation for Basic Research (grant No. 02-03-32048); INTAS (grant No. 2001-2126); and Siberian Branch of the Russian Academy of Sciences (grant No. 119).

SOLVENT ELECTROSTATIC SCREENING IN PROTEIN SIMULATIONS

S.M. SCHWARZL, S. FISCHER*, J.C. SMITH
Interdisciplinary Center for Scientific Computing, Computational Molecular Biophysics, Im Neuenheimer Feld 368, 69120 Heidelberg, Germany, e-mail: sonja.schwarzl@iwr.uni-heidelberg.de; stefan.fischer@iwr.uni-heidelberg.de
* *Corresponding author*

Abstract The treatment of electrostatic interactions is one of the key issues in protein simulations. In particular, solvent effects need to be considered. Several methods have been developed that address this problem, involving a trade-off between accuracy and speed. The current work discusses a method that aims at approximating Poisson–Boltzmann interaction energies with a conventional Coulomb potential by introducing scaling factors that are used to reparametrize the partial atomic charges. Thus, solvent effects can be incorporated into simulations while retaining maximal speed and ease of implementation.

Key words: solvent screening, continuum electrostatics, Poisson–Boltzmann equation

1. INTRODUCTION

Protein simulations performed at an atomic-detail level are based on a molecular mechanics force field. Such force fields include energy terms typically representing stretching of covalent bonds, bending of valence angles, twisting of dihedral angles and nonbonded interactions such as van der Waals terms and electrostatics. The electrostatic interactions are calculated as the Coulomb interaction energy between partial atomic charges that are determined for small building blocks in vacuum and transferred to the macromolecule (Brooks *et al.*, 1983; Cornell *et al.*, 1995).

Since proteins are surrounded by water in their native environment, it is important to not only compute the interactions within the molecule but also account for solute: solvent interactions. The influence of the solvent is twofold: (1) Electrostatic effects (Bashford and Case, 2000). The electrostatic field due to partial atomic charges of the macromolecule polarizes the bulk solvent, orienting on average the water molecules. The resulting electrostatic field generated by the sum of all the individual water dipole fields is the so-called reaction field. Due to this reaction field, the effective electrostatic interaction between any two charges in the protein is diminished in solution relative to vacuum, an effect that is known as solvent screening. Moreover, a charge may also interaction with its own reaction field, an interaction that does not exist in vacuum. The corresponding energy is known as the self-energy. (2) Hydrophobic effect (Sharp *et al.*, 1991). This effect results from the necessity to form a cavity in the bulk solvent, in which the protein can be placed. Since water molecules must be displaced from this cavity, and moreover, the hydrogen-bonding pattern among the water molecules is disturbed, this cavity formation is energetically unfavourable. However, this is partially compensated by favourable van der Waals interactions between the solute and the solvent. The present paper is concerned solely with the solvent screening effect.

A straightforward way of including solvent effects into protein simulations is to explicitly add water molecules. When using molecular dynamics simulations, these can reorient and automatically generate the proper reaction field. This technique is frequently applied, even to simulations involving systems as large as the ATP synthesising machine F_1ATPase (Böckmann and Grubmüller, 2002). However, when using minimization-based algorithms such as in ligand binding studies (Caflisch *et al.*, 1997) or during the calculation of minimum energy reaction paths (Fischer *et al.*, 1993) explicit bulk water molecules would "freeze" and thus interfere with the process under investigation.

To overcome the above difficulty, implicit solvent representations have been developed, in which the solvent is modelled as a highly polarizable continuum whose electrostatic properties are described by a dielectric permittivity, ε. The electrostatic potential can then be described by the Poisson–Boltzmann equation that relates the electrostatic potential, $\Phi(\vec{r})$ to the charge density, $\rho(\vec{r})$ (Honig and Nicholls, 1995):

$$\nabla\varepsilon(\vec{r})\nabla\Phi(\vec{r}) - \kappa'\Phi(\vec{r}) = -4\pi\rho(\vec{r}), \qquad (1)$$

where κ' is the Debye–Hückel screening constant that is related to the ionic strength of the solution.

This second-order differential equation (1) must be solved numerically. This is usually done on a grid using a finite-difference scheme. However, this is also computationally demanding. Therefore, simpler models have been developed that aim at yielding results with an accuracy comparable to

Poisson–Boltzmann calculations but using an analytical formula to allow rapid evaluation. Examples of these are the Generalised Born model (Still *et al.*, 1990; Qiu *et al.*, 1997) and the Analytic Continuum Electrostatic method (Schaefer and Karplus, 1996). Although it has been shown that such methods accurately describe the behaviour of peptides (Schaefer *et al.*, 1996) their applicability to large macromolecules is under debate (Calimet *et al.*, 2001).

An even simpler and quite old implicit solvation method is the use of a distance-dependent dielectric permittivity. However, this has no rigorous physical base and, indeed, changes the Coulomb interaction energy from a $1/r$ dependence to a $1/r^2$ dependence.

In summary, there is a clear gap between methods that are accurate but computationally demanding and methods that are computationally fast but inaccurate. To fill this gap we have developed a method that approximates the electrostatic interaction energies in solution as calculated with the Poisson–Boltzmann approach in such a way that a simple Coulomb potential can still be used.

2. METHODS

The screening from the surrounding solvent results in apparent weaker electrostatic interaction E_{ij}^{solv} between charges q_i and q_j:

$$E_{ij}^{\text{solv}} = \frac{q_i}{\lambda_I} \frac{q_j}{\lambda_J} \frac{1}{r_{ij}}, \tag{2}$$

where r_{ij} is the distance and λ_I is a scaling factor for all atoms belonging to a group *I*. The protein is split into side-chain groups and backbone groups such that the local multipoles on chemical groups are preserved. The scaling factors are calculated from group-specific screening constants ε_I:

$$\lambda_I = \sqrt{\varepsilon_I}, \tag{3}$$

which in turn are determined as the ratio

$$\varepsilon_I = \frac{\sum\limits_{J,\varepsilon_{IJ}>0} |E_{IJ}^{\text{vac}}|}{\sum\limits_{J,\varepsilon_{IJ}>0} |E_{IJ}^{\text{solv}}|}. \tag{4}$$

E_{IJ}^{vac} is the electrostatic interaction energy between groups I and J in vacuum as determined with the Coulomb potential, and E_{IJ}^{solv} is the corresponding energy in solution as determined from a Poisson–Boltzmann calculation. The sum extends over all groups J within a specified cutoff distance around I and for which E_{IJ}^{vac} and E_{IJ}^{solv} have the same sign. This cutoff is the same as used in the subsequent simulations. It must be considered in the sums of Eq. (4) because it implicitly constitutes a form of shielding.

In the example given here, the electrostatic interactions were smoothly brought to zero by multiplying by a cubic switching function between 6 Å and 12 Å (Brooks *et al.*, 1983). Poisson–Boltzmann interaction energies were obtained using the finite difference scheme as implemented in CHARMM version 28 (Im *et al.*, 1998). A focusing approach with a final grid spacing of 1.0 Å and an ionic strength of 145 mM was used. The dielectric boundary was smoothed from –0.75 Å to +2.25 Å across the van der Waals surface.

To check the applicability of the resulting scaling factors, interaction energies obtained with the Coulomb law using reparametrized charges E_I^{shield} are plotted against the reference interaction energies obtained from Poisson–Boltzmann calculations E_I^{solv}. Here, the interaction between a group I and the rest of the protein is considered, in contrast to the way the scaling factors were derived, where group: group interaction energies were used.

Calculations were performed for the molecular motor protein myosin in its OPEN conformation (PDB code 1mmd, Fisher *et al.*, 1995). Surface water molecules were removed whereas internal water molecules were treated explicitly.

3. RESULTS AND DISCUSSION

The scaling factors obtained range from 1.07 to 7.62 with an average of 1.93. Due to the procedure according to which the scaling factors have been determined a systematic deviation between E_I^{shield} and E_I^{solv} may occur. To compensate this, an empirical correction factor γ was introduced that is determined as the square root of the slope of a linear least squares fit of E_I^{shield} versus E_I^{solv}. The correction factors were determined independently for side-chain and backbone groups, resulting in values γ_{SIDE} = 0.87 and γ_{BACK} = 0.71. Both are smaller than one indicating that the reparametrization procedure without this correction slightly overestimates screening.

After the correction, the charge scaling factors λ_I range from 0.81 to 6.63 with an average of 1.53. The electrostatic interaction energies between each group and the rest of the protein are shown in Figure 1. Energies were calculated using a Coulomb potential with (i) unmodified charges E^{vac}, (ii) a distance-dependent dielectric permittivity E^{rdiel}, and (iii) with reparametrized charges E^{shield}. These results are compared to Poisson–Boltzmann interaction energies E^{solv}.

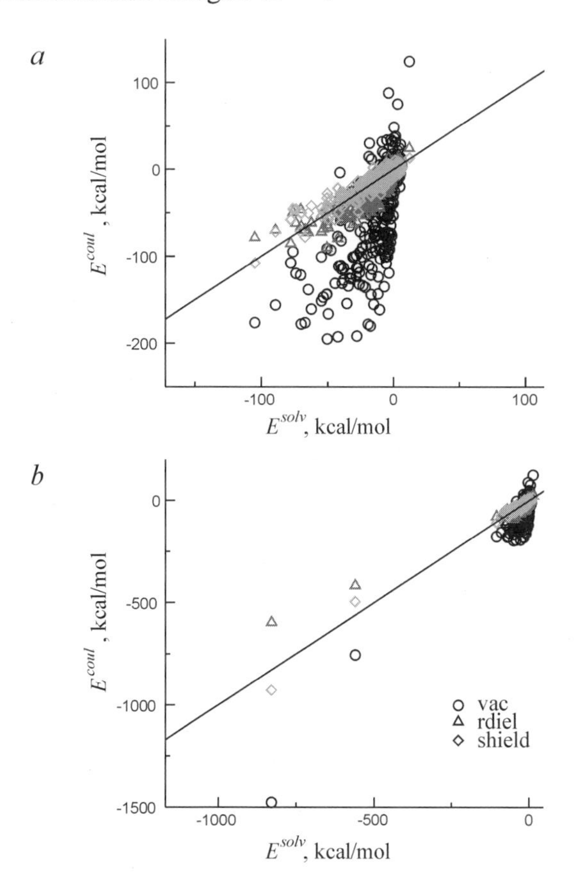

Figure 1. Scatter plot of electrostatic interaction energies between each group of myosin and the rest of the protein calculated with a Coulomb potential, (i) unmodified charges (vac), (ii) distance-dependent dielectric permittivity (rdiel), and (iii) reparametized charges (shield) versus interaction energies computed from Poisson–Boltzmann calculations. The two plots differ only in the scale shown. The outliers in panel B are ATP^{4-} and Mg^{2+}.

In case (i) only weak correlation can be seen. This is reflected by a root mean square error (RMSD) of 34.1 kcal/mol. Case (ii) shows a clearer correlation with a RMSD of 9.9 kcal/mol. However, a deviation from the diagonal is observed. The usage of reparametrized partial atomic charges (case iii) leads to a clear correlation throughout the range of interaction energies observed with an RMSD of 5.5 kcal/mol. In particular, the interactions of Mg^{2+} and ATP^{4-} are well-behaved in case (iii) in contrast to their strong deviation from the Poisson–Boltzmann energies in cases (i) and

(ii) (see Figure 1*b*). Since highly charged moieties contribute most to the overall electrostatic energy it is important to use a method that yields adequate interaction energies not only for weak interactions but especially for strong interactions.

Charge reparametrization allows computation of screened electrostatic interaction energies with a Coulomb potential, therefore ensuring high speed of the calculations. Since it is not necessary to introduce explicit bulk water molecules, the new method may be used in conjunction with minimisation-based calculations. Moreover, because the only modification is the scaling of the partial atomic charges, no interface programming between different modules of simulation packages is necessary. Charge reparametrization models the electrostatic screening effect of the bulk solvent; any structural or functional water molecules may be included explicitly.

Limitations may occur when the process under investigation involves major conformational rearrangements. In this case, the conformation-dependent scaling factors may have to be evaluated for more than one conformation.

ACKNOWLEDGMENTS

This work was funded in part by the Boehringer Ingelheim Fonds.

PDBSITESCAN: A PROGRAM SEARCHING FOR FUNCTIONAL SITES IN PROTEIN 3D STRUCTURES

V.A. IVANISENKO[1,2,3]*, S.S. PINTUS[1,2], D.A. GRIGOROVICH[1], L.N. IVANISENKO[3], V.A. DEBELOV[4], A.M. MATSOKIN[4]
[1]*Institute of Cytology & Genetics, Siberian Branch of the Russian Academy of Sciences, prosp. Lavrentieva 10, Novosibirsk, 630090 Russia, e-mail: salix@bionet.nsc.ru;* [2]*Novosibirsk State University, ul. Pirogova 2, Novosibirsk, 630090 Russia;* [3]*State Research Scientific Center of Virology and Biotechnology Vector, Koltsovo, Novosibirsk oblast, 630559 Russia;* [4]*Institute of Computational Mathematics & Mathematical Geophysics, Siberian Branch of the Russian Academy of Sciences, prosp. Lavrentieva 6, Novosibirsk, 630090 Russia*
* *Corresponding author*

Abstract An Internet-oriented program PDBSiteScan making structural alignment of the sites contained in the database PDBSite with 3D structure of a user-specified protein was developed. The program automatically selects all the variants where the maximal distances with respect to N, $C\alpha$, and C atoms of the fragment's residues in 3D structures of the protein studied and the site do not exceed a user-specified value. An indispensable condition in this realization of the method was identity of the types of amino acids for superposed residue pairs. PDBSiteScan may be used for recognition of functional sites in protein 3D structures and is available at http://wwwmgs.bionet.nsc.ru/mgs/systems/fastprot/pdbsitescan.html.

Key words: PDBSite, functional site recognition, 3D structure, structural alignment, best superposition, hydrolase family, human protein p53, mutants

1. INTRODUCTION

Various methods for recognition of functional sites in primary structure of proteins are now well developed, for example, the database Prosite contains patterns of numerous sites (Falquet *et al.*, 2002). However, many functional sites, such as metal-binding sites and catalytic centers of enzymes, cannot be distinguished well only on the basis of their primary structure (Gregory *et al.*, 1993; Wallace *et al.*, 1996; Yang *et al.*, 2002). These sites in proteins are formed by the amino acid residues that are brought together in the 3D structure but located at certain distances in the primary structure. Recognition of discontinuous sites requires the data on their 3D structure. Earlier, we have developed a database of 3D structures of functional sites, PDBSite (Ivanisenko *et al.*, 2002).

In this paper, the program PDBSiteScan intended for search for potential functional sites in the proteins compiled with PDB (Bernstein *et al.*, 1977) and displaying both the similarity in their 3D structure and full identity of their amino acid residues with the real sites contained in the database PDBSite is described. PDBSiteScan allows both the sites continuous in polypeptide chain and discontinuous sites to be searched for.

2. METHODS AND ALGORITHMS

The method for recognizing functional sites was based on a search for the fragments in 3D structure of proteins displaying structural similarity and identity of the amino acids with sites from the database PDBSite. The known methods for protein structural alignment (Gibrat *et al.*, 1996; Eidhammer *et al.*, 2000) are inadequate for solving this problem, as these methods involve aligning of the 3D structures represented by continuous polypeptide chains.

The realized algorithm of structural alignment was based on consecutive expansion of the set of compared residues of the site and protein studied using exhaustive search with restrictions for their formation. Similarly to Pennec and Ayache (1998), only N, $C\alpha$, and C atoms of the residues, whose positions determine directions of side chain radicals of amino acids in space, were considered. The names of amino acid residues, their order within the site, and x, y, and z coordinates of the atoms in question were taken from PDBSite.

The algorithm operation comprises the following steps. Every possible triplet of amino acid residues in various regions of protein 3D structure was considered as a potential nucleus for the site required. Each triplet was compared in space with the first three residues of the site from PDBSite. In the case of satisfactory structural similarity, the appropriate sets of the residues compared were extended by a consecutive addition of one residue until their number appeared equal to the number of residues in the site. The decision about addition of a residue to the set was made basing on estimation of the structural similarity. In addition to estimating the structural similarity basing on 3D comparison of amino acid residues, identity of their types was also checked. The result of algorithm operation was all the found sets of amino acid residues of the protein with a given order of comparison to the residues of the site that satisfied similarity criteria for the pair site–protein. The structural similarity was estimated upon the best superposition of 3D structures by calculation of the maximum distance mismatch (MDM) of the atoms in space:

$$\text{MDM} = \max\left\{\sqrt{\left(x_i^s - x_i^p\right)^2 + \left(y_i^s - y_i^p\right)^2 + \left(z_i^s - z_i^p\right)^2}\right\}_{i=1}^{n}, \tag{1}$$

where x_i^s, y_i^s, z_i^s are the coordinates of the *i*th atom of the site; x_i^p, y_i^p, z_i^p are the coordinates of the *i*th atom of protein fragment; and *n*, the number of atoms for the corresponding list of residues. The order of atoms for each amino acid residue of both the site and protein was specified as *N*, *Cα*, and *C*. Thus, the residues were matched by superposing the atoms of only the same type.

The best superposition of 3D structures was performed as described by Kabsch (1976). A brief description of the method is given in Appendix.

2.1 Program PDBSiteScan

PDBSiteScan is an Internet-oriented program integrated with the database of 3D structures of protein functional sites—PDBSite. PDBSiteScan is available at http://wwwmgs.bionet.nsc.ru/mgs/systems/fastprot/pdbsitescan.html.

The input data for PDBSiteScan is the file with 3D structure of the protein analyzed in PDB format and the value of maximum distance mismatch. The output data of the program is the table containing the information about each pair site–protein fragment whose maximum distance mismatch does not exceed a preset value. Another option is site–protein structural alignment as a PDB file that allows for visualizing it with standard tools.

3. RESULTS AND DISCUSSION

3.1 Recognition of the catalytic centers of enzymes

Let us consider the application of the program by the example of recognition of the catalytic sites in hydrolase family. Recognition of such sites only from the primary structure is impossible since they are short (consist of three residues) and discontinuous (the residues are remote from one another on the primary sequence). Prosite lacks any patterns for their recognition (Falquet *et al.*, 2002). Two samples of nonhomologous proteins (homology is less or equal to 35%) were generated, each containing 23 proteins. The positive sample included hydrolases with known catalytic sites (Table 1), whereas the negative sample contained the proteins lacking any catalytic centers of hydrolases. The samples of nonhomologous proteins were formed using the program available at http://www.fccc.edu/research/labs/dunbrack/culledpdb.html). Only the proteins with known 3D structures were considered.

Table 1. Catalytic sites of hydrolase family proteins used for testing PDBSiteScan

PDB Id	Catalytic site	PDB Id	Catalytic site
1A2Z	80E, 143C, 167H	1CV8	24C, 120H, 141N
1AUO	114S, 168D, 199H	1CVL	87S, 263D, 285H
1B2L	138S, 151Y, 155K	1E6U	107S, 136Y, 140K
1B6G	124D, 260D, 289H	1E6W	155S, 168Y, 172K
1BDB	142S, 155Y, 159K	1EA5	200S, 327E, 440H
1BIF	256H, 325E, 390H	1ELV	460H, 514D, 617S
1BIO	57H, 102D, 195S	1H2W	554S, 641D, 680H
1BN7	117D, 141E, 283H	1H4W	57H, 102D, 195S
1BRT	98S, 228D, 257H	1HAZ	57H, 102D, 195S
1BS9	90S, 175D, 187H	1QJ1	57H, 102D, 195S
1CUJ	120C, 175D, 188H	1QJ4	80S, 207D, 235H
1SVP	141H, 163D, 215A		

Type I and II errors of recognition of catalytic sites in these proteins at different MDM threshold values were estimated using these two samples. The recognition was performed by automatic search for the 3D fragments similar to catalytic sites compiled in PDBSite in the proteins analyzed. Overall, PDBSite contains 297 catalytic sites of the hydrolase family. A site was considered recognized if the program found a structural homology, provided that the amino acids were coinciding, between the 3D fragment of the protein analyzed corresponding to a real site and at least one of the hydrolase sites from PDBSite at certain search parameters specified. Before the recognition testing, the catalytic site of the protein analyzed was deleted from PDBSite. The dependences of type I and II errors in recognition of catalytic sites on MDM threshold values are shown in Figure 1. The type I error was calculated as a fraction of unrecognized catalytic sites at a specified MDM threshold:

$$E_1(\mathrm{MDM}) = \frac{n(\mathrm{MDM})}{N_p}, \tag{2}$$

where n(MDM) is the number of unrecognized catalytic sites in the positive sample $N_p = 23$. The type II error was calculated as

$$E_2(\mathrm{MDM}) = \frac{k(\mathrm{MDM})}{N_n}, \tag{3}$$

where k(MDM) is the number of proteins with at least one falsely recognized catalytic site in the negative sample $N_n = 23$.

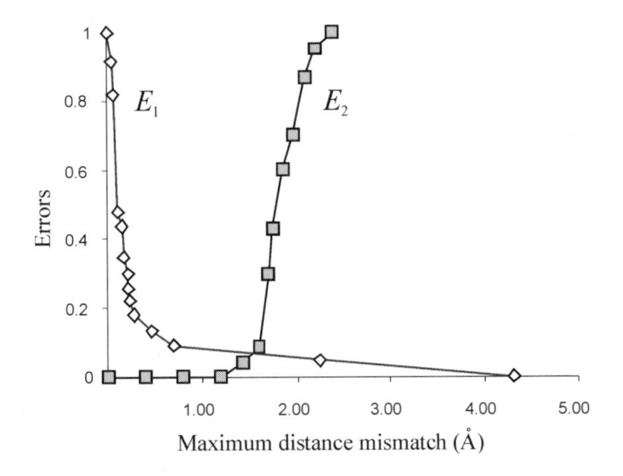

Figure 1. Dependence of type I and II errors of recognition of the catalytic sites in the sample of hydrolases on MDM threshold values: light diamonds mark the curve for type I error (E_1); black squares, for type II error (E_2).

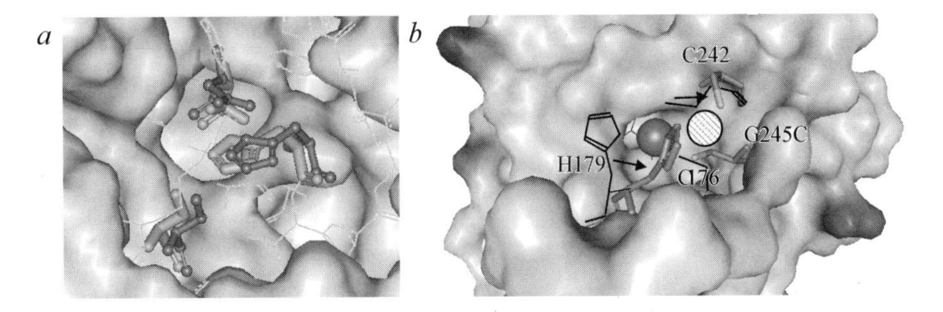

Figure 2. Site recognition by PDBSiteScan: (*a*) image of spatially aligned residues of the catalytic center of protein 1ELV and a catalytic site from PDBSite (Id 1BQYB) exemplifying result of PDBSiteScan opertaion. The residues of 1ELV catalytic center of the protein are shown as stick model; the residues of 1BQYB site, as stick and ball model; (*b*) image of zinc-binding site of the human p53 DNA-binding domain (PDB Id 1gzh): fine lines show the residues of normal p53 zinc-binding site; stick model, residues of the additional zinc-binding site generated due to the mutation G245aC. The structure of the latter site is obtained by structural alignment of this fragment of p53 protein with the zinc-binding site of cytidine deaminase (PDB Id 1AF2). Arrows indicate the displacement directions of residues to the position whereat the new site binds zinc ion; dark ball, zinc ion bound by the normal site; and hatched circle, zinc ion bound by the potential site. The images are obtained using ViewerLite.

It is evident from Figure 1 that the recognition displays a good accuracy. For example, the number of positively recognized sites exceeds 80% at the MDM threshold value slightly lower than 1 Å, while the number of falsely recognized sites equals zero. Structural alignment of the catalytic site in a protein from the positive sample (1ELV) with a catalytic site from PDBSite, obtained as a result of recognition, is shown in Figure 2*a*.

3.2 Analysis of mutant forms of human p53 protein

The role of p53 protein in regulation of the cell cycle is widely known. This protein is capable of halting cell division or causing apoptosis (Lee *et al.*, 1995). Mutations in human p53 gene are thought to be implicated in cancer diseases. The 3D structure of the DNA-binding domain of human p53 protein was taken from PDB (Id 1gzh); data on the mutations of human p53 protein involved in cancer development, from Swiss-Prot (Id P53_HUMAN; Boeckmann *et al.*, 2003). Overall, 181 amino acid substitutions in p53 associated with various human tumors were analyzed. The 3D structures for each mutant p53 form were obtained as follows. As the positions of N, $C\alpha$, and C atoms accounted by PDBSiteScan are invariant to the types of substitutions, the PDB file for a mutant was generated simply by replacing the name of the substituted amino acid with that of the amino acid substituting it in the initial PDB file for p53.

The analysis of p53 mutant forms allowed us to find out a potential zinc ion binding site that emerged due to a point mutation (Figure 2*b*). The new site is localized to the pocket of the already existing normal zinc-binding site. According to PDB (Id 1gzh), the normal zinc-binding site includes residues C176, H179, C238, and C242. The mutation G245→C generates the new site formed of H179, C242, and C245, overlapping with the normal site and displaying a structural similarity to the binding site of Zn^{2+} cytidine deaminase (PDB Id 1AF2). Consequently, this suggests a competition of the normal site with the newly appeared site for binding zinc. As is known, binding of Zn^{2+} is necessary for normal interaction of p53 with DNA (Cho *et al.*, 1994). The results of analysis suggest that it is impaired zinc binding by the normal site due to the competition with the site generated by the mutation G245→C that may underlie the development of tumors associated with this mutation.

4. CONCLUSION

The proteins having fragments of 3D structure similar to functional sites of other proteins frequently possess similar biological properties (Branden and Tooze, 1991). Over six thousand various protein sites have been accumulated in the database PDBSite, and this information is being constantly replenished with new data (Ivanisenko *et al.*, 2002). The database compiles the data on the ligand-binding sites, catalytic active centers of various enzymes, biochemically modified sites, etc. Identification of the sites displaying high structural similarity to the sites from PDBSite in protein 3D structures allows new properties of these proteins to be discovered. Further, we are planning to perform a number of modifications of the algorithm to accelerate the comparison of 3D structures. Preprocessing of the database on coordinates of atoms will be accomplished and geometrical hashing will be used. We also plan to increase the site recognition accuracy by taking into account their 3D environment. Furthermore, a database of 3D patterns of biologically important sites considering the variation of amino acids at different positions within the site will be developed on the basis of PDBSite (Bork, Koonin, 1996).

ACKNOWLEDGMENTS

The work was partially supported by the Russian Foundation for Basic Research (grants Nos. 01-07-90376-в, 03-07-96833-p2003, 03-07-96833, 03-07-06079-мас); Russian Ministry of Industry, Science, and Technologies (grant No. 43.073.1.1.1501, subcontract No. 28/2003); grant PCB RAS (No. 10.4); the Siberian Branch of the Russian Academy of Sciences (integration projects Nos. 119 and 65); the U.S. Civilian Research & Development Foundation for the Independent States of the Former Soviet Union (CRDF) No. NO-008-X1.

APPENDIX

The problem of best superposition of two ordered sets $P = \{P_1, P_2, \ldots, P_N\}$ and $Q = \{Q_1, Q_2, \ldots, Q_N\}$ of points from R^3 consists in definition of the rigid-body transformation (x_0, T): $R^3 \rightarrow R^3$ providing minimum of the functional

$$F = \frac{1}{2}\sum_{n} w_n \left(T(P_n - x_0) - Q_n\right)^2 \qquad (1)$$

under conditions that

$$\sum_{k} t_{ki}t_{kj} - \delta_{ij} = 0\,, \qquad (2)$$

where x^0 is the burgers vector; $T = (t_{ij})$, rotation matrix; and δ_{ij}, elements of identity matrix. The method described by Kabsch (1976), based on a direct solution of the problem of minimization of a given functional, was used to determine the rotation matrix. To solve this problem, a symmetric matrix of Lagrange multipliers $L = (l_{ij})$ is introduced to define the auxiliary function

$$M = \frac{1}{2}\sum_{i,j} l_{ij}\left(\sum_{k} t_{ki}t_{kj} - \delta_{ij}\right) , \tag{3}$$

which is added to F to produce the Lagrangian function

$$\mathrm{G} = \mathrm{F} + \mathrm{M}. \tag{4}$$

To find the minimum of function F with restrictions (2), we determine the minimum of functional (4). For this purpose, the matrix $R = (r_{ij})$ is constructed, where

$$r_{ij} = \sum_{n} w_n p_{ni} q_{nj} . \tag{5}$$

Preliminary, we get rid of the problem of shift by transferring the coordinate system of each set of points to their center of mass. For the matrix being the product $R^T R$, where R^T is the transposed matrix R, the eigenvalues μ_k and corresponding eigenvectors, forming the matrix $\boldsymbol{a_k}$, are found. Further, values of the matrix $\boldsymbol{b_k}$ are calculated using the following equation:

$$\boldsymbol{b_k} = \frac{1}{\sqrt{\mu_k}} \boldsymbol{R a_k}. \tag{6}$$

Values of the required matrix T are calculated as products of the elements of matrices $\boldsymbol{b_k}$ and $\boldsymbol{a_k}^T$:

$$t_{ij} = \sum_{k} b_{ki} a_{kj} . \tag{7}$$

If one of the eigenvalues of the matrix $R^T R$ is zero, for example, $\mu_3 = 0$, than the corresponding columns of matrices $\boldsymbol{a_k}$ and $\boldsymbol{b_k}$ are constructed using the other two columns:

$$\boldsymbol{a_{3+}} = \boldsymbol{a_1} \times \boldsymbol{a_2} \boldsymbol{b_3} = \boldsymbol{b_1} \times \boldsymbol{b_2}. \tag{8}$$

A GENETIC ALGORITHM FOR THE INVERSE FOLDING OF RNA

I.I. TITOV[1,2]*, A.Yu. PALYANOV[2]
[1]*Institute of Cytology & Genetics, Siberian Branch of the Russian Academy of Sciences, prosp. Lavrentieva 10, Novosibirsk, 630090 Russia, e-mail: titov@bionet.nsc.ru;*
[2]*Novosibirsk State University, ul. Pirogova 2, Novosibirsk, 630090 Russia, e-mail: palyanov@bionet.nsc.ru*
* *Corresponding author*

Abstract Solution of the inverse folding problem of biopolymers is important both for understanding the structure codes and for the design of nanoscale devices (nanotechnology). We developed a genetic algorithm for RNA inverse folding. Our algorithm can find sequences folding into a given secondary structure (as earlier developed Vienna RNA Secondary Structure Package, Hofacker *et al.*, 1994) and as well, it can optimize RNA structure properties (what is not implemented in Vienna Package). We applied the algorithm for finding the sequences that can form hypothetical RNA structures (a cube and a square lattice). We estimated the algorithm efficiency on test cases of finding thermodynamically stable and unstable RNA secondary structures. Our algorithm calculates one out of few best sequences among 10^{10}–10^{11}. Then we searched for the sequences with mutationally robust secondary structures. We found that nucleotide content of this sequences deviates from uniform to the same direction as nucleotide content of natural RNAs.

Key words: RNA, inverse folding, structure design, pseudoknot, genetic algorithm, nanotechnology

1. INTRODUCTION

Because of a specificity of interactions and development of the manipulation techniques, biopolymers are amongst the most promising materials for design of nanoscale devices with predictable properties (templates for molecular electronics, sensors, springs, etc.). The natural evolution could hardly exhaust all the possible sequences and, following the common principles of molecular biology, one can create the novel structures of interest (Seeman, 1985).

Most RNAs and proteins fold spontaneously into their unique shapes. Although the question of how this ability is coded in a symbolic sequence has been extensively studied for several decades, it still remains vague. The reconstruction of a sequence from a spatial structure (i.e. the inverse problem of folding) can help to understand the rules of folding. Combinatorial complexity of the problem and degeneracy of folding code (typically, there are many sequences folding into the same shape) justify the heuristic approaches. A practical way to solve the inverse problem is to refer to the Nature and conduct an evolution in a tube. Recently the similar computational procedure for the inverse folding of RNA has been developed (Hofacker *et al.*, 1994). It uses an adaptive walk towards target structure in sequence space. Our paper describes a more general approach, based on genetic algorithm (Goldberg, 1989), for the inverse folding problem of RNA. Besides finding RNA sequences matching a given structure, our algorithm allows structural characteristics of RNA to be optimize. We apply the algorithm to design the RNA sequences that fold into a given structure or into a structure with specified structural properties (stability, instability, mutational robustness).

2. METHODS AND ALGORITHMS

2.1 Solution representation

The solution of the RNA inverse folding problem is RNA sequence, optimal for structural requirement included into algorithm. This requirement can be a folding of RNA into given structure, and also structural properties of the molecule (thermodynamical or mutational stability etc.). In the GA, each solution (individual) is characterized by the set of its elements (genes). In our algorithm, nucleotide positions of the RNA sequence were treated as genes. Each gene can be in one of four states (A, U, G, and C).

2.2 Energy model

In this work, we used the nearest neighbor energy model. For calculation of RNA energy and structure we use the subprogram—algorithm GArna (Titov *et al.*, 2002). GArna finds one of the most stable secondary structures of RNA. This algorithm has the same typical time complexity as dynamical Zuker's algorithm, so it can be used as a subprogram even for time-expensive calculations. The energy parameters were taken from Turner and co-authors (1988). To estimate the nonrandomness of the sequence obtained we have calculated deviation of its secondary structure's energy relative to secondary structure energies of random RNAs with the same nucleotide content—Z-score (Titov *et al.*, 2002).

2.3 Genetic algorithm

The genetic algorithm for solving the inverse folding problem included the following steps:

1. Generation of the initial population from sequences of random composition.
2. Stochastic selection of individuals from the population according to their fitness values (individuals with better fitness survive with higher probability).
3. Recombination of randomly chosen pairs of parents. Offsprings fill the vacancies in population (formed after step 2) until restoration of the initial number of individuals.
4. Point mutations, deletions and insertions at random positions, implemented as conventional algorithms of a stochastic search (Metropolis *et al.*, 1953).
5. Repeat from Step 2 until algorithm convergence.

Fitness value depended exponentially on a score optimized, as in the work of Titov *et al.* (2002). The choice of the algorithm parameters (frequency of mutations and recombinations, temperature of selection, population size, etc.) also followed Titov *et al.* (2002) in general. For different tasks, we optimized the following structure characteristics (Figure 1):

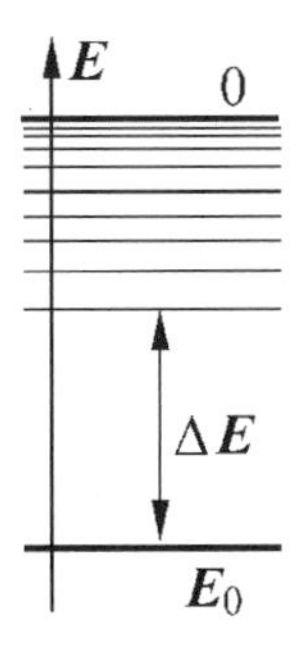

Figure 1. Qualitative picture of RNA secondary structure energy spectrum.

a) the thermodynamical gap ΔE between target structure and suboptimal foldings (Sec. 3.1);
b) the energy of RNA secondary structure E_0 (Sec. 3.2.1);

c) the negenergy of RNA secondary structure $-E_0$ (Sec. 3.2.2);
d) mutational stability of the structure (Sec. 3.2.3).

3. RESULTS AND DISCUSSION

3.1 Artificial structures

The following examples of application of the algorithm belong to the field of biotechnology. Accurate selection of a nucleotide sequence is important for the product yield. Due to a small size of the nucleotide alphabet, a straightforward stabilization of the target structure by increasing the GC content might cause formation of alternative structures with the similar energies. Thus, the most appropriate way is to maximize the thermodynamic probability of the target structure, i.e. to widen its thermodynamic gap ΔE. The stability of RNA secondary structure relates to its fast folding because they both depend on secondary structure parameters (helices and loops energies) in similar way (Mironov *et al.*, 1985). Therefore, for the majority of RNAs with wide thermodynamical gap one should expect fast folding into the ground state.

3.1.1 Calculation of the RNA sequences forming a square lattice

Square DNA lattices were obtained earlier at Seeman's laboratory (Seeman, 1985). Let us briefly describe their experimental procedure. At first, short DNA chains were synthesized, four of these chains forming a 4-arm branched junction—an elementary lattice cell. The nucleotide chain composition was selected to avoid alternative pairing. The arms had "sticky ends", which bound the junctions into a lattice by complementary interactions. The subsequent linking by ligases created covalent links between the elementary cells.

For the calculations, we have taken 4 oligonucleotide sequences and fixed reciprocal complementarity in the way that they formed a junction. Then we calculated the energy of the junction and the energy of the most stable structure formed by these oligonucleotides. Genetic algorithm maximized the gap between these two energies. For methodological purposes, we varied the lengths of the oligonucleotides beginning from sequences forming a junction with half-turn helices. Finally, even in this case the gap was significant ($\Delta E = 16.3$ kcal/mole per junction, what corresponds to Boltzmann factor of 10^{12}). The energy of the junction itself was equal to –45.0 kcal/mol. Oligonucleotide sequences forming a junction with half-turned double helices are shown in Figure 2. It is likely impossible to obtain a lattice with a shorter period.

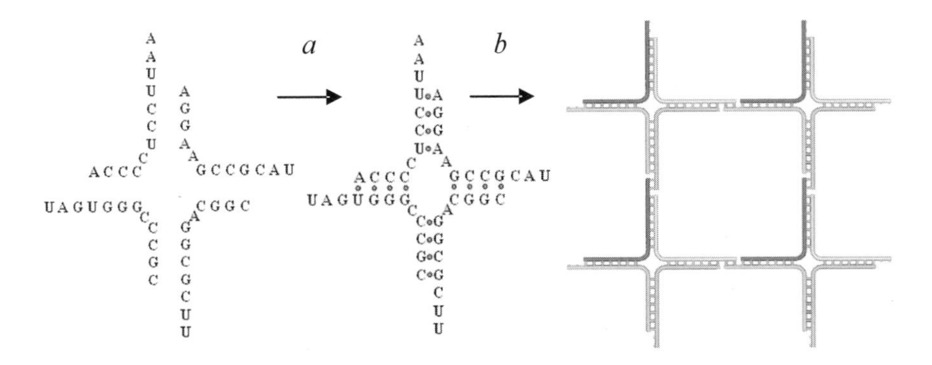

Figure 2. Four RNA oligonucleotides (*a*) folding into a junction with sticky ends; (*b*) the junctions form a square lattice.

3.1.2 RNA cube design

Previously, a DNA cube was made from several chains (Chen and Seeman, 1991). It would be interesting to design a cube formed by spontaneous folding of a single RNA strand. The secondary-structure trefoil could transform into the tertiary cube-like structure through pseudoknot interactions of the loops (Figure 3). Previously single or double pseudoknots were observed in natural RNAs.

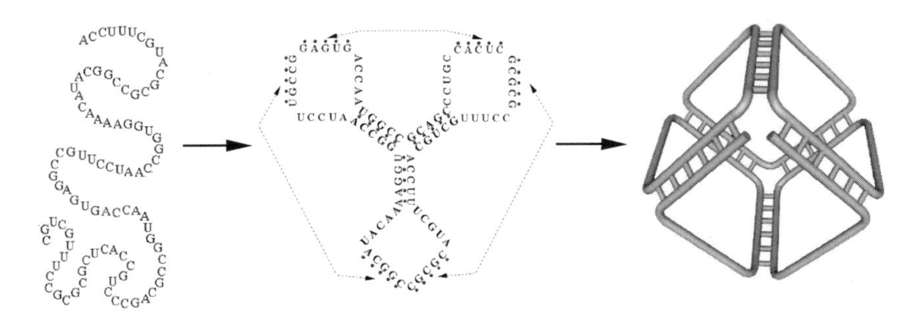

Figure 3. RNA oligonucleotide folding into the trefoil with potential pseudoknot contacts. The formation of pseudoknots gives a cube-like structure of RNA (artist view).

A search for an RNA cube sequence was similar to Section 3.1.1, when we maximized the width of the energy gap ΔE. The search was performed only among sequences with length of 90 nt and with fixed complementarity corresponding to target structure. Calculated sequence is shown in Figure 3. Since the thermodynamic table is not available for pseudoknots, we used the secondary structure parameters. In this approximation, the energy of the target structure was equal to –60.7 kcal/mole. Basing on thermodynamical data (Puglisi *et al.*, 1988), we suggest that our procedure overestimated pseudoknot stability by few kcal/mole. At the same time the thermodynamic gap obtained is too wide (15.5 kcal/mol per pseudoknot or 46.4 kcal/mol total) to be crossed by the energy destabilization neglected. Nevertheless, we plan to take into account sterical restrictions and test our results with molecular force calculations.

It is interesting to speculate on possible applications of such a cube. A reversible folding used to control cell functions of the most biomolecules. The RNA-cube binding the other molecules can assemble protein complexes (placed on the cube sides) or hold a ligand inside. RNA splicing controlled by conformational switches could be another application of a cube. It can be realized by placing catalytic groups for cutting RNA into two loops of the trefoil and into the third loop—a catalytic group for joining RNA chains cleft.

3.2 Optimization of RNA secondary structure characteristics

Our algorithm is able to optimize RNA structural characteristics. To illustrate this property the algorithm was tested on the following simple examples. RNA sequences that we got in these tests can be generated with a few rules. The reason is that they encode "simple messages" about the secondary structure: "the most stable structure" (Sec. 3.2.1) and "the most unstable structure" (Sec. 3.2.2).

3.2.1 Optimization of RNA secondary structure stability

The sequence with the most stable secondary structure is manually predictable: it forms a perfect hairpin composed of two homopolymer (G_n and C_n) fragments with optimal size loop. To estimate the optimization quality of the algorithm we have tested how the solutions found can be close to optimal at different lengths of sequences. Our algorithm gave close to optimal solutions, especially for short oligonucleotides. The results of the calculations for length of 50 are shown in Table 1. This test allowed estimating optimization quality of our algorithm. Let us define optimization quality as a ratio of the number of all possible sequences to the number of sequences with more stable secondary structure than we got. This ratio was found equal to 10^{10}–10^{11} for 30 nt sequences. The sequences that we got had more stable secondary structures and were significantly more nonrandom than natural (Table 1).

Table 1. The results of optimization of RNA structural characteristics by our algorithm

Characteristic	RNA sequence	Energy	Z-score
Instability of secondary structure with 50% G and 50% C (length = 54)	CGGGGGGGCCCGCGCCCGCCGCCGCCGCC GGCGCGCCCGCCGCGGGGGGCCGC	–17.1	4.1
	GCGGGGCCGCCGGGCCCGCGCGGGGG GCCGCCGCCGCCGCCGCCGCCGCCGG	–18.1	3.9
	CCGCCGCCGGCCGGCCGCCGCCCGGGGG GGGGGGCCGCGCCGCGCGCCGCGCCG	–19.6	3.5
	GGCGGCGGGCCGCGGGCGGCCCCCCCCC CGGCCCCGGGCGGCGCGGCGCGGGC	–18.4	3.8
	CGGCGGGCCGCGCGCGCCCGGCCCCCCC CCCGGCGGCGGCGGGCGGCGGCGGG	–17.0	4.2
Stability of secondary structure (length = 50)	GGGCCACGCCCCCCGGGGGCCCCUAGUG GGGCCCCCGGGGGGCGUGGCCC	–59.3	–11.0
	GCGGGGACGCGGGCGCGCCGGGCGGAUG CCCGGCGCGCCCGCGUCCCCGC	–58.4	–9.8
	GGGCCGGGCGGCCCCCCGCCCUGGCAGC CAGGGCGGGGGGCCGCCCGGCC	–56.3	–8.6
	GCGUCGGCGCCCCCCCCCAGGCCGUGAG GCCUGGGGGGGGGCGCCGACGC	–57.8	–11.1
	GGGGGCCUGGGGGGCGCCCCGGAAACCG GGGCGCCCCCCAGGCCCCCUCA	–53.8	–10.5
Mutational robustness (length = 100)	UAUCGGGGUGACCCUGGGGCGCGCAAGC GCCCGAAUGACCCGCGGCCCCUGGACCG AAGCCGGUUCGGUCCAGGGGCGCGGGUC GGGCAGGGUCACCCUG	–78.7	–11.4
	GCCUCGCACGGGGCCUUGCACUCCGUAG GCUGGUCCGGGGUAAGAUACCCCGGACC GGCCUGUGCAAGGCCCCGUGCGAGGCGG GGGCUGGGAACAGCCC	–86.6	–14.4
	UUGUGAGCCGUGGUGGUCAUACUGACCA CCUGGGCUCACUGGUCCCGCGACCGACC CUCUGGCCAUAUCAGAUGGCCAUCGAGG GUCGGUCGCGGGACCA	–77.7	–15.3
	UGGCGGGGUAUAGUCAAGAGUCCCGGCG CUAGUGCGCCGGGACUCUAAUGGCUAUG CCCCGCCGAGCCCCCAGCGGCGUGGCCGC CACGCCGCUGGGGGA	–85.9	–14.7
	GUGGCUGUGCCCGGGCGACGUGCUGGGU UCUGUACCGAGAACCGGUACAGAACCCA GCAUGUCGCCCGGGCACAGCCACUGCCU GGCCGCACGGCCGGGC	–93.4	–17.73

For each characteristic 5 sequences are shown, with energy and Z-score for each.

3.2.2 Selection towards an unstable secondary structure of RNA

The simple solution here is that the best sequences lack complementary partners (A_n, C_n, G_n or U_n). The calculations were performed for different lengths from 30 to 100 nt. Here, we have revealed that there are many sequences of mixed nucleotide content that cannot form a stable structure, i.e. the structure with negative energy. Our algorithm spent fewer cycles to find such solutions than in case of 3.2.1 (Figure 4). This difference of optimization speed agrees with known result that sequences forming stable secondary structures are deficient amongst all the possible sequences (Titov *et al.*, 2002). To make the task more complex, and the solution less obvious, we restricted our search by the sequences of G and C nucleotides with equal contents. This restriction leaves a minor part of the sequence space available for searching, and the sequences evolved slowly and exclusively via nucleotide rearrangement. The final sequence (with length of 54 nt, Z-score = 4.6 and $E_0 = -15.1$ kcal/mol) consisted of triplets $(GCC)_n$ with G_n insertion compensating C (Table 1).

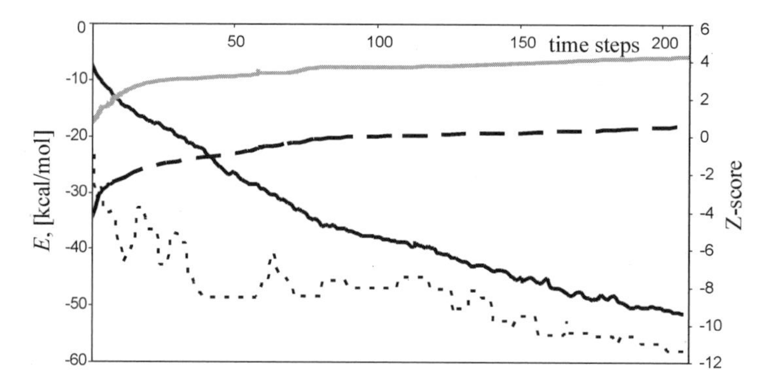

Figure 4. Evolution of RNA secondary structure energy (solid line) and Z-score (short dash line) during selection towards stable secondary structure; evolution of RNA secondary structure energy (long dash line) and Z-score (gray line) during selection towards unstable secondary structure.

3.2.3 Calculation of mutationally robust RNAs

RNAs with mutationally stable secondary structures can be useful for biotechnological experiments that include intensive mutagenesis. Unlike previous examples (Sections 3.2.1, 3.2.2), in this one the resulting RNA sequences were less predictable. Let us describe the procedure in more details. Calculation of the RNA mutational stability included the following steps:

1. Calculation of RNA structure by algorithm Garna.
2. Filling the complementary matrix M_{ij} for calculated RNA structure. M_{ij} has the size [L×L], where L is the number of nucleotides in the RNA. If

nucleotides i and j form the complementary pair in native structure then $M_{ij} = 1$ and otherwise 0.

3. Production of 20 clones of the original RNA and exposing them to random point mutations at randomly selected positions. Number of mutated positions was 7% out of RNA length. This gives us 20 different random mutants of the original RNA.
4. Calculation for each clone its secondary structure and the complementary matrix.
5. Compare original RNA's and each clone's complementary matrices and calculate mutational stability as $\sum_k \frac{2 \cdot \left(\sum_{i,j} \text{abs}\left(M_{ij}^0 - M_{ij}^k \right) \right)}{\sum_{i,j} M_{ij}^0 + \sum_{i,j} M_{ij}^k}$, where $\sum_{i,j} M_{ij}^0$ is the number of complementary pairs in the original RNA's complementary matrix M_{ij}^0, $\sum_{i,j} M_{ij}^k$ is the number of complementary pairs in the kth mutant RNA's complementary matrix M_{ij}^k, and $\sum_{i,j} \text{abs}\left(M_{ij}^0 - M_{ij}^k \right)$ is the number of complementary pairs, not common for original and mutant RNA's matrices.

Mutational stability takes values between 0 and 1, where 1 corresponds to absolutely equal structures and more closer to 0 for more differing structures. It characterizes how strongly the RNA's native structure changes under exposure of mutations. We have tried another scores reflected the structure mutational robustness with similar results. For the comparison, we have preliminarily calculated the nucleotide content for two groups of natural RNAs—16sRNAs and tRNAs (Table 2). These groups turned out to be GC-rich and close to each other by nucleotide content. This fact indicates that structural RNAs have similar evolutionary restrictions. The sequences that we have calculated contained all 4 types of nucleotides, predominantly G and C (Table 2). Although the trends of nucleotide content to the GC-rich area for the artificial and natural RNAs were similar, the artificial sequences were biased significantly stronger. In our computational experiments, we started from uniform content and already at 50–200 cycles of optimization, long before algorithm convergence, the predominance of GC reached the level of natural RNAs. And finally, we have got typical values of Z-score = $-14 \div -17$ for $75 \div 100$ nt artificial sequences (Table 1) in comparison with Z-score = –2 for tRNAs (Rivas and Eddy, 2000). We suggest that, as for structure stability, mutational robustness appears not to be the main factor for RNA sequences evolution.

Table 2. Nucleotide content (normalized to 1) of two classes of natural RNAs and of the calculated sequences

RNA	Number of sequences	A	U	G	C
16s RNA	1102	0.25 ± 0.03	0.21 ± 0.03	0.31 ± 0.03	0.23 ± 0.03
tRNA	535	0.23 ± 0.06	0.19 ± 0.06	0.29 ± 0.06	0.28 ± 0.05
Artificial	10	0.16 ± 0.03	0.16 ± 0.03	0.36 ± 0.03	0.33 ± 0.02

4. CONCLUSION

We have developed genetic algorithm for solving RNA inverse folding problem. Earlier developed Vienna Package (Hofacker *et al.*, 1994) reconstructs RNA sequences folding into the target secondary structure. It stops when the first sequence matching a given structure is found. Solving the same problem, our algorithm maximizes thermodynamical probability of the target structure. Wide thermodynamical gap in energy spectrum can be an important factor of the product yield in biotechnological experiments. In our work, we illustrate this ability by calculation of two artificial structures, interesting for nanotechnology (RNA-junction and a cube from a single RNA strand). We suppose that these structures cannot be miniaturized further; at the same time, wide energy gap should provide high probability of their formation.

The important property of our algorithm is its ability to optimize structural characteristics of RNA. Simple examples are shown in this work: calculation of the sequences forming (a) stable, (b) unstable, and (c) mutationally robust secondary structures.

ACKNOWLEDGEMENTS

The discussions with N.A. Kolchanov greatly improved the manuscript. The research described in this publication was made possible in part by Award No. REC-008 of the U.S. Civilian Research & Development Foundation for the Independent States of the Former Soviet Union (CRDF) No. NO-008-X1; Russian Foundation for Basic Research (grant No. 01-07-90376-в); the Siberian Branch of the Russian Academy of Sciences (integration projects Nos. 119 and 10.4); INTAS grant 2001-2126.

SIGNAL TRANSDUCTION PATHWAYS INITIATED VIA CELL SURFACE RECEPTOR CD150: *IN SILICO* AND *IN VITRO* ANALYSIS

S.P. SIDORENKO*, S.V. MIKHALAP, L.M. SHLAPATSKA,
M.Y. YURCHENKO, Y.M. AKIMOV, V.F. CHEKHUN
Kavetsky Institute of Experimental Pathology, Oncology and Radiobiology, National Academy of Sciences of Ukraine, ul. Vasylkivska 45, Kyiv, 03022 Ukraine,
e-mail: svetasid@onconet.kiev.ua
** Corresponding author*

Abstract Cell surface receptor CD150 represents a novel group of receptors that could transmit both activating and inhibiting signals regulating the cell fate. The CD150 subfamily members are involved in the development and pathogenesis of severe immunodeficiencies, and CD150 serves as a cell receptor for measles virus. CD150 receptor functions are linked with the presence of a paired immunoreceptor tyrosine-based switch motif (ITSM) TxYxxV/I in its cytoplasmic tail. This motif can bind different SH2 domain–containing molecules, including tyrosine and inositol phosphatases, Src-family kinases, and adaptor molecules. Using different experimental and computational biology approaches, we identified the molecules that are interacting with the cytoplasmic tail of CD150 (CD150ct), their binding sites, and possible combinations of binding. An experimental system involving chicken B-cell line DT40 transfected with CD150 and adaptor protein SH2D1A allowed us to clarify the mechanisms of regulation of CD150-initiated signal transduction pathways. We found that ligation of CD150 activates the ERK pathway, which is SH2D1A-independent; however, CD150-mediated Akt phosphorylation depends on the presence of SH2D1A. A 3D-model of CD150ct was constructed. This model was used for docking with the SH2-containing proteins SH2D1A and SHP-2. We found preferential bindings of SH2D1A to pY281 and SHP-2 to pY327 in CD150ct. We also showed that binding of SH2D1A to Y281-containing motif in CD150ct may change the conformation of CD150ct. Docking of the SH2-containing molecules to CD150ct demonstrated possible combinations of binding with the molecules that linked this receptor to signal transduction pathways.

Key words: signal transduction pathways, CD150, SH2-containing proteins, docking of proteins, computer modeling

1. INTRODUCTION

CD150 (IPO-3, SLAM) is a cell surface receptor expressed on activated T and B lymphocytes, dendritic cells, and monocytes. CD150 is a member of the CD2 subfamily of the Ig superfamily that shares homology with CD58 (LFA-3), CD2 (LFA-2, T11), BLAME (B lymphocyte activator macrophage expressed, BCM-like membrane protein), SF2001 (CD2F-10), NTB-A (SF2000, Ly108), CD84 (Ly9B), CD48 (BCM1, Blast-1, OX-45), CS1 (19A24, CRACC), CD229 (Ly9), and CD244 (2B4, NAIL) (Cocks *et al.*, 1995; Fraser *et al.*, 2002; Sidorenko and Clark, 2003). CD150 serves as a receptor for measles virus and other morbilliviruses (Tatsuo *et al.*, 2000) and is involved in pathogenesis of X-linked lymphoproliferative disorder, B-cell non-Hodgkin's lymphoma, and familial hemophagocytic lymphohistiocytosis (Morra *et al.*, 2001). Divergent functions of CD150 are linked with a unique structure of CD150 cytoplasmic tail, especially with the presence of paired immunoreceptor tyrosine-based switch motif (ITSM) TxYxxV/I (Shlapatska *et al.*, 2001). This motif could bind different subsets of signal transduction molecules. However, it is yet unknown how the binding of these molecules is regulated and what signal transduction pathways are initiated via CD150. Using experimental approaches, we identified SH2-containing molecules that bind CD150 receptor in B lymphocytes and found that both Y281 and Y327 in CD150 cytoplasmic tail are essential for binding of both SHP-2 and SHIP. We also showed that CD150 ligation activates the extracellular signal-regulated kinase (ERK) pathway independently of SH2D1A. Another CD150-initiated signaling pathway leading to Akt phosphorylation requires SH2D1A expression. Computer simulation showed that CD150 cytoplasmic tail (CD150ct) is conformationally flexible, and macromolecular docking studies demonstrated preferential binding of SH2D1A and SHP-2 to different ITSM motifs in CD150ct. Computational analysis of CD150ct docking with SH2D1A showed that SH2D1A might function as a molecular switch by changing the conformation of CD150ct. Therefore, ITSM motifs may play key roles in the regulation of CD150-mediated signal transduction pathways in normal lymphocytes as well as in X-linked lymphoproliferative disorder and certain lymphomas.

2. METHODS AND ALGORITHMS

2.1 Experimental systems

For identification of molecules interacting with CD150 and signal transduction pathways initiated via this cell surface receptor, we used the following three different experimental systems. (1) Immunoprecipitation of

CD150 with the monoclonal antibody IPO-3 followed by biochemical identification of proteins that coprecipitate with CD150 (Mikhalap *et al.*, 1999). (2) GST fusion proteins of CD150ct (GST–CD150ct) in non-phosphorylated and tyrosine-phosphorylated forms. The GST fusion protein construct of the cytoplasmic tail of CD150 (GST–CD150ct) was prepared as described in (Mikhalap *et al.*, 1999). Using PCR-based site directed mutagenesis, we made constructs of GST–CD150ct fusion proteins with phenylalanine (F) replacements at tyrosines Y269, Y281, Y307, Y327, and Y281 + Y327. These mutant fusion proteins were used in pull-down experiments for typing of binding sites in CD150ct (Shlapatska *et al.*, 2001). (3) To clarify the signal transduction pathways initiated via CD150, we developed an experimental system involving DT40 chicken B lymphoma cell line transfected with CD150 alone or together with SH2D1A. CD150 cDNA (provided by Dr. G. Aversa, DNAX, USA) was cloned into the pApuro expression vector and SH2D1A cDNA (provided by Dr. Kim Nichols, USA) was cloned into the pcDNA3 vector (Invitrogen, USA).

2.2 Modeling and docking of CD150 cytoplasmic tail

The model of CD150ct was built using a threading approach on FUGUE (http://www.cryst.bioc.cam.ac.uk/~fugue/prfsearch.html) and SAUSAGE (http://www.emblheidelberg.de/predictprotein/submit_meta.html) servers. Secondary structure was predicted on the Predict Protein server (http://www.embl-heidelberg.de/predictprotein/predictprotein.html) and by the local viewers WebLab ViewerPro 3.7 (http://www.accelrys.com/weblab/) and Swiss-PDB Viewer (http://www.ex-pasy.ch/spdbv/). Substitution of matrix residues, phosphorylation, and energy minimization were made in the HyperChem 6.0 package (http://www.hyper.com/products/default.htm). For flexibility estimation, we used the ProtScale tool on ExPASy (http://us.expasy.org/cgi-bin/protscale.pl). We studied the interactions between CD150ct and SH2-domains by a macromolecular docking method using the Hex 2.4 program (http://www.biochem.abdn.ac.uk/hex/). By this program, we generated 128 complexes with various potential energies. The complexes generated were compared with the X-ray structures of SH2-domains in complex with a CD150ct fragment (K276–K286) with phosphorylated and non-phosphorylated Y281 motif (Li *et al.*, 1999; Poy *et al.*, 1999; Morra *et al.*, 2001) and published experimental data (Mikhalap *et al.*, 1999; Latour *et al.*, 2001; Shlapatska *et al.*, 2001; Howie *et al.*, 2002). Affinity in docked complexes (interface area, numbers of hydrogen bonds, and salt bridges) was analyzed by the WebLab ViewerPro 3.7 and Swiss-PDB Viewer. The NetPhos 2.0 server (http://www.cbs.dtu.dk/services/NetPhos/) was used for prediction of phosphorylation sites.

For CD150ct threading, we used the template of gene 5 DNA-binding protein from filamentous bacteriophage fd (M13), PDB:2GN5, and GI:230541. As open forms of SH2D1A, we used PDB:1D4T (SH2D1A with Y281 peptide) and PDB:1D4W (SH2D1A with pY281 peptide) (Poy *et al.*, 1999). SHP-2 (PDB:2SHP, GI:18375642) (Hof *et al.*, 1998) was crystallized in an inactive conformation with a "closed" *N*-terminal SH2-domain, so for docking studies we transplanted an "open" *N*-terminal SH2-domain from PDB:1AYA to the SHP-2 structure. Quality of the model was analyzed on the Eval123 server (http://bioserv.cbs.cnrs.fr/HTML_BIO/valid.html) using Eval123D, Verify3D, ProsaII, and EvTree programs.

3. RESULTS AND DISCUSSION

3.1 Identification of CD150-associated proteins and CD150-initiated pathways

Using different experimental approaches, we identified the molecules that were able to interact with the cell surface receptor CD150 in B lymphocytes. CD150 coprecipitated with the cell surface receptor CD45 (220 kDa), which had two intrinsic tyrosine phosphatase domains. In addition, it associated with the Src-family kinases Lyn and Fgr, SH2-containing protein tyrosine phosphatase SHP-2, SH2-containing inositol phosphatase SHIP, and adaptor protein SH2D1A. We found that in B-cell lines CD150 was differentially associated with tyrosine phosphatase SHP-2 versus the inositol phosphatase SHIP and the adaptor protein SH2D1A. Using mutational analysis, we found that both Y281 and Y237 in TxYxxI/V motif in CD150 cytoplasmic tail are essential for binding of both SHP-2 and SHIP. Apparently, SH2D1A may function as a regulator of alternative interactions of CD150 with SHP-2 or SHIP via a TxYxxV/I motif. Using the experimental system of DT40 transfectants, we found that CD150 ligation activated the ERK pathway and that a CD150-mediated ERK activation occurred independently of SH2D1A. In contrast, another CD150-initiated signaling pathway leading to Akt phosphorylation required SH2D1A expression. Mutations in SH2D1A are found in patients with X-linked lymphoproliferative syndrome (XLP), common variable immunodeficiency (CVID) and familial hemophagocytic lymphohistiocytosis (Sayos *et al.*, 1998; Morra *et al.*, 2001). That is why enlightening the place of SH2D1A in regulation of signal transduction pathways will help to understand pathogenesis of these severe immunodeficiencies and design the effective approaches to the therapy.

3.2 Modeling of CD150ct docking with SH2D1A and SHP-2

To explore possible mechanisms of CD150ct interactions with different SH2-containing proteins and to explain the available experimental data, we applied computational biology methods. The Protein Data Bank (PDB) does not contain protein 3D-structures suitable as templates for CD150ct homology modeling. This, we built our model using a threading approach on FUGUE and SAUSAGE servers. Gene 5 DNA-binding protein, PDB: 2GN5, has a maximum Z-score (distribution of random sequences with the same amino acid composition and sequence length as a query sequence), so it was used as a template for threading. After structure refinement (side chain energy minimization and ligation of phospho groups), CD150ct model and 2GN5 template superimposed with an r.m.s. deviation of only 1.76 Å for C_α atoms. The model of CD150ct has a partially open β-barrel conformation with "Greek-key" topology with a molecular surface of 4197 $Å^2$ and a volume of 9383 $Å^3$ (Figure 1, model 1). According to Structural Classification of Proteins (SCOP), the model of CD150ct, as well as 2GN5 template, belongs to the "all-beta" group of proteins with OB-fold. It contains one antiparallel β-sheet, composed of two strands within a hairpin (positions 283–291). Three tyrosine residues (Y281, Y307, and Y327) have high phosphorylation potential, as well as the threonines at positions 271 and 305 and serine 317. The core ITSM tyrosines Y281 and Y327 in our model are at a distance of only 27.1 Å. Model quality evaluation at Eval123 server and similarity with 1D4T CD150ct–peptide conformation (superimposed with r.m.s.d. of 1.95 Å for C_α atoms) showed the reliability of our CD150ct model. Mapping binding sites for SH2-domains in CD150ct demonstrated that SH2D1A interacted with ITSMs containing Y281/pY281 or pY327 (Li *et al.*, 1999; Poy *et al.*, 1999; Shlapatska *et al.*, 2001).

We performed SH2D1A docking with CD150ct that had different modes of Y281 and Y327 phosphorylation: Y281/pY327–CD150ct (Figure 1, model 2), pY281/Y327–CD150ct (Figure 1, model 3), and pY281/pY327–CD150ct (Figure 1, model 4). SH2D1A docking with CD150ct showed preferential binding to Y281-containing ITSM regardless of CD150ct phosphorylation mode (Figure 1, models 2–4). In all the complexes observed, Y281-containing motif docked to peptide-binding cleft in SH2-domain of SH2D1A (βD and βC strands and αB spiral) in three "prong" fashions similar to NMR structures determined experimentally (Hwang *et al.*, 2002). As a result of SH2D1A docking with Y281/pY327–CD150ct (Figure 1, model 2), we obtained a complex with the CD150ct orientation structurally identical to 1D4T complex (r.m.s.d. = 0.3 Å). All these provide further evidence for the reliability of our CD150ct model.

In all CD150ct–SH2D1A complexes, in addition to Y281-containing ITSM, phosphopeptide-binding cleft of SH2D1A also engaged Y327-motif. The pY327 was hydrogen-bonded to SH2D1A with D48 and Y52 (for model 2) and with Y50 and R55 (for model 4). Even Y327 (model 3) was weakly interacting with the βD-strand of SH2D1A (H bond with Y50).

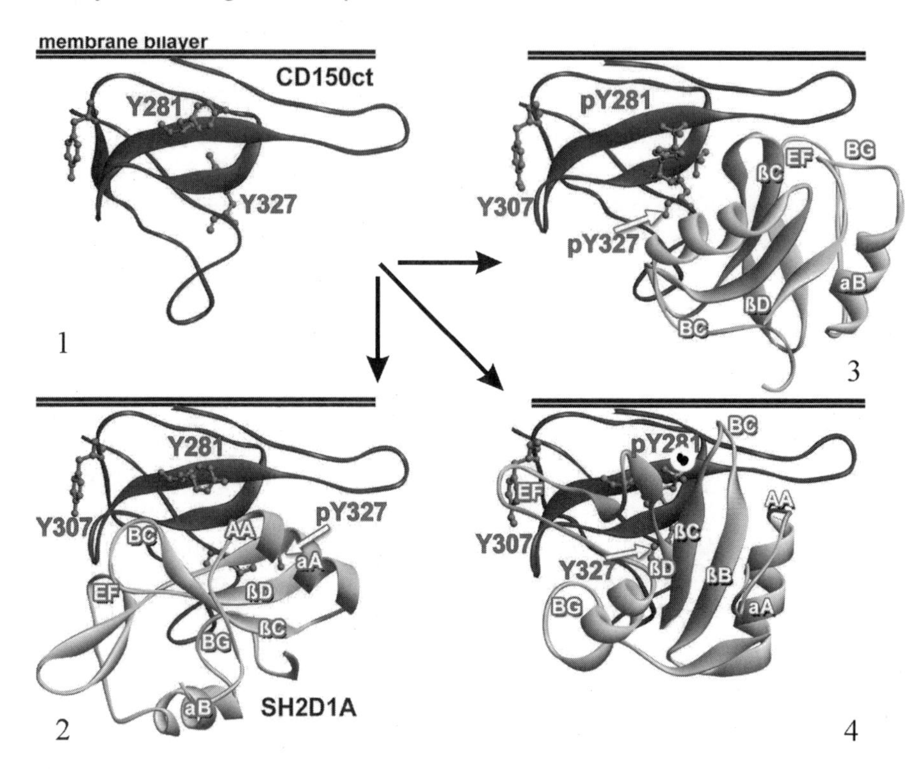

Figure 1. Molecular modeling of CD150ct docking with SH2D1A. SH2D1A interactions with (1) CD150ct model with three modes of tyrosine phosphorylation in ITSMs; (2) Y281/pY327–CD150ct; (3) pY281/Y327–CD150ct; and (4) pY281/pY327-CD150ct (complexes were obtained by Hex 2.4 program; images, generated by WebLab 3.7 viewer).

Our docking studies demonstrated that SH2D1A displayed the highest affinity for pY281/pY327–CD150ct (Figure 1, model 4) with an interface area of 721 Å^2 and predicted 35 H bonds. The complex with pY281/Y327–CD150ct (Figure 1, model 3) formed nine H bonds on an interface area of 388 Å^2; however, the complex with Y281/pY327–CD150ct (Figure 1, model 2) within 612 Å^2 of interface had only four H bonds. Average flexibility calculations for CD150ct model showed that the first five residues proximal to the membrane bilayer had a maximum flexibility. Two more peaks of flexible residues were located between Y281 and Y307 (Figure 2*a*). SH2D1A binding to ITSM with

Y281/pY281 may change the CD150ct conformation that altered Y307 orientation with respect to the cell membrane (Figure 1).

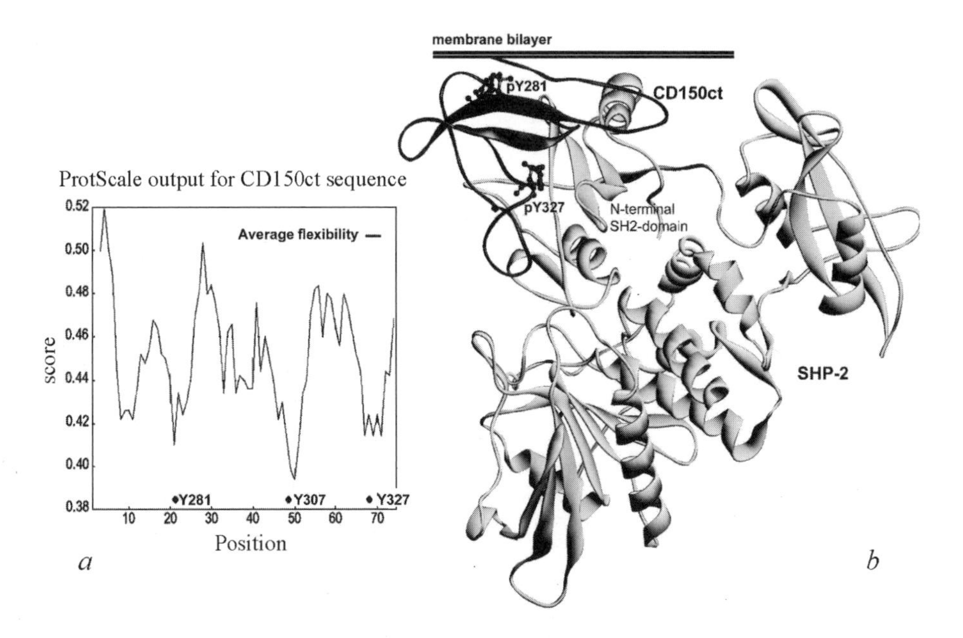

Figure 2. (*a*) ProtScale calculations of CD150ct model average flexibility (three signaling tyrosines are marked) obtained on ExPASY server; (*b*) SHP-2 tyrosine phosphatase docking with pY281/pY327–CD150ct (complexes were obtained by Hex 2.4 program; images, generated by WebLab 3.7 viewer).

It was proposed that SH2D1A competed with tyrosine phosphatase SHP-2 for binding sites in CD150ct (Sayos *et al.*, 1998). To test this hypothesis, we performed docking of SHP-2 with Y281/pY327–CD150ct and pY281/pY327–CD150ct. Despite the homology between the SH2D1A and SHP-2 *N*-terminal SH2-domains, the latter lacks nearly all the "prong" contact residues reported for SH2D1A. Thus, the *N*-terminal SH2-domain of SHP-2 may have markedly reduced the affinity for Y281 compared with the SH2-domain of SH2D1A. Indeed, docking of SHP-2 with CD150ct showed that even in a pY281/pY327–CD150ct, SHP-2 more readily bound to pY327 (80% of complexes) than to pY281. The pY327 positioned in phosphotyrosine-binding pocket of SHP-2 *N*-terminal SH2-domain (Figure 2*b*) in the way similar to pY327 complexed with SH2D1A (Figure 1). The interface area in this complex was only 416 Å^2 with eight H bonds. An assumption that SHP-2 has a higher affinity for pY327 than for pY281 is also supported by experimental data. Mutational analysis of CD150ct demonstrated that in the absence of SH2D1A, the level of SHP-2 that coprecipitated with pY281–CD150ct was sixfold lower than with pY327–CD150ct (Shlapatska *et al*., 2001).

Thus, our modeling studies showed that the flexibility of CD150ct allowed it to change the conformation after SH2D1A binding. Comparison of the interface parameters that determined the affinities of complexes (interface area and H bonds) allowed us to predict the relative affinity of SHP-2 and SH2D1A for CD150ct with different patterns of Y281 and Y327 phosphorylation. SHP-2 preferentially binds pY327; in contrast, SH2D1A has higher affinity for pY281 than for pY327, but also forms H bonds with tyrosine at position 327. Apparently, SHP-2 and SH2D1A utilize different ITSMs for binding to CD150ct, and SHP-2 may compete with SH2D1A only for Y281/pY327 in CD150ct.

4. CONCLUSION

Computer modeling of SH2D1A and SHP-2 docking to CD150ct allowed us to explain experimental data obtained recently by our group and other researchers. Average flexibility calculations showed that the CD150 cytoplasmic tail displayed conformational flexibility. Macromolecular docking studies demonstrated preferential binding of SH2D1A and SHP-2 to different ITSM motifs in CD150 and revealed that CD150ct might change its conformation after docking with SH2D1A. Apparently, SH2D1A may function as a regulator of alternative interactions of SH2-containing molecules with CD150 via a TxYxxV/I motif. However, the initial hypothesis that SH2D1A functions as a natural blocker of interactions between SH2-containing proteins with CD150 seems to hold true only for the SHP-2 phosphatase. For Src-family kinases and SHIP, this adaptor protein facilitates rather than inhibits the interactions. Association of SH2D1A with ITSM motifs was found not only in CD150, but also in CD84, CD229, CD244, and SF2000/NTB-A receptors (Sidorenko and Clark, 2003). Therefore, ITSM motifs may play key roles in the regulation of CD150-mediated signal transduction pathways in normal lymphocytes as well as in X-linked lymphoproliferative disorder, Hodgkin, and non-Hodgkin's lymphomas.

ACKNOWLEDGMENTS

This work was supported by the Howard Hughes Medical Institute (grant No. 76195-548101); INTAS (grant No. 011-2883 to S.P.S.); and CRDF (grants Nos. UB2-531 to S.P.S. and USB-383 to L.M.S.).

PART 3. COMPUTATIONAL EVOLUTIONARY BIOLOGY

STUDY OF THE SPECIFIC CONTEXTUAL FEATURES OF TRANSLATION INITIATION AND TERMINATION SITES IN *SACCHAROMYCES CEREVISIAE*

O.V. VISHNEVSKY[1,2]*, I.V. AVDEEVA[2], A.V. KOCHETOV[1,2]
[1]Institute of Cytology & Genetics, Siberian Branch of the Russian Academy of Sciences, prosp. Lavrentieva 10, Novosibirsk, 630090 Russia, e-mail: oleg@bionet.nsc.ru; [2]Novosibirsk State University, ul. Pirogova 2, Novosibirsk, 630090 Russia
** Corresponding author*

Abstract Investigation of mRNA sequence organization is of importance to reveal the features influencing translation efficiency and specificity. We performed statistical analysis of translation initiation and termination sites of well-studied eukaryotic organism *Saccharomyces cerevisiae*. Yeast mRNAs were analyzed using trinucleotide weight matrices and vocabularies of significant oligonucleotide motifs. A statistically significant difference of nucleotide contexts between high- and low-expressed mRNAs was found. Computer simulation of evolution using genetic algorithm demonstrated that the rate-limiting stage model could explain this phenomenon.

Key words: eukaryotic mRNA, translation initiation, translation termination, weight matrices, oligonucleotide analysis

1. INTRODUCTION

It is known that eukaryotic mRNA sequence organization is an important factor influencing translation efficiency and cytoplasmic stability and, thereby, gene expression rate as a whole. It is known that some mRNA general elements (i.e. 5'-UTR-located stable hairpins, AUG triplets, translation initiation and termination sites, and codon content) determine interaction with the translation machinery. Notably, the context of translation start and stop codons influences efficiencies of translation initiation and termination processes (Kozak, 2002; Matushkin *et al.*, this issue; Pichueva *et al.*, this issue); however, the details of organization of these signals are not clear (Ray *et al.*, 1983; Gallie *et al.*, 1996; Lukaszewicz *et al.*, 2000). The goal of our study was to analyze the contextual features of translation initiation (TIS) and translation termination (TTS) sites of the well-investigated *Saccharomyces cerevisiae* genes using statistical approach and the rate-limiting stage model. We constructed trinucleotide weight matrices describing the contexts of TIS and TTS and found that yeast genes with high expression level displayed correlations between TIS and TTS features. Computer simulation of evolution demonstrated that the rate-limiting stage model could explain this correlation.

2. SEQUENCES USED IN ANALYSIS

We analyzed sequences of the regions encompassing the initiation and termination codons of 6741 mRNAs of *Saccharomyces cerevisiae* from the TransTerm database (Dalphin *et al.*, 1997). Sequences of 30 bp in length, from –21 to +9 with reference to the translation start and of 30 bp from –9 to +21 with reference to the termination codon were examined. For each mRNA, the codon adaptation index (CAI) was retrieved from the same database. This index is known to reflect the correspondence between the frequency distribution of synonymous codons in the mRNA coding part and concentration of the major tRNA fractions in the cell (Sharp and Li, 1987). For *Saccharomyces cerevisiae*, CAI is a promising marker of mRNA expression in the cell. Four samples of 245 sequences each were constructed on the basis of CAI values for the 5'- and 3'-regions of high-expressed (CAI > 0.5) and low-expressed (CAI < 0.055) mRNAs. The region-specific significant oligonucleotide motifs in samples of mRNA sequences were detected using the program ARGO developed previously (Vishnevsky *et al.*, this issue).

3. ANALYSIS OF TIS AND TTS OF HIGH- AND LOW-EXPRESSED *SACCHAROMYCES CEREVISIAE* GENES

3.1 TIS of high- and low-expressed genes differ considerably in the oligonucleotide content

The method searching for degenerate oligonucleotide motifs in a set of sequences includes analysis of oligonucleotide vocabularies of each sequence considered followed by clustering of similar perfect oligonucleotides found in vocabularies of different sequences and subsequent construction of the resulting consensuses that describe the oligonucleotides belonging to one cluster. The search for motives is possible in both four single letter-based code and expanded 15 single letter-based IUPAC code. The oligonucleotide analysis of samples of the 5'-untranslated sequences contrasting in their expression levels demonstrated considerable differences in occurrence rates of oligonucleotide motifs (Table 1). To assess the contrast of occurrence of the motifs, logarithm of the ratio of the occurrence rate of each motif in the sample HIGH to that in the sample LOW is listed in Table 1. For example, the motif MRAUGRSY is 11-fold more frequent (log = 1.05) in the sample of high-expressed mRNAs (46%) than in the sample of low-expressed mRNA (4%; here M denotes A or C nucleotide; R, G or A; S, G or C; and Y, A, G, or C). On the contrary, the motif GNNNAUGN is occurring in 31% of the low-expressed mRNAs and in 4% of high-expressed mRNAs, whereas the motif GNAUGNNN, in 24 and 2%, respectively.

Table 1. Examples of motifs specific of TIS of high-expressed mRNAs (HIGH) and underrepresented in low-expressed mRNAs (LOW) and motifs specific of TIS of low-expressed mRNAs (LOW) and underrepresented in high-expressed mRNAs (HIGH)

Motifs	Presence		
	HIGH	LOW	log(HIGH/LOW)
MR**AUG**KSY	0.46	0.04	1.05
R**AUG**KCYN	0.44	0.04	0.99
AUGKCYRN	0.39	0.04	0.98
UGKCYVVH	0.40	0.04	0.99
HRMA**AUG**G	0.26	0.03	0.85
GNNN**AUG**N	0.03	0.31	–0.93
NN**AUG**VNG	0.02	0.25	–1.02
USN**AUG**NN	0.01	0.19	–1.38
KNCM**UG**NN	0.01	0.17	–1.16
GN**AUG**NNN	0.02	0.23	–0.98

3.2 Classification of mRNAs by TIS oligonucleotide content

The mRNA expression levels (evaluated by CAI) were predicted basing on the context of mRNA TIS and TTS using the program ARGO_VIEWER (Vishnevsky *et al.*, this issue). This program allows the degree of similarity between an unknown sequence and the sequences of training sample to be assessed basing on the occurrence and distribution patterns of specific oligonucleotide motifs detected by the program ARGO. Prediction of expression levels of all the 6741 mRNAs by ARGO_VIEWER demonstrated

(Figure 1) that the content of motifs typical of low-expressed genes is decreased in the majority of yeast mRNAs with CAI > 0.3. Thus, it was shown that the presence of certain motifs in the region of mRNA start codon is incompatible with efficient translation initiation.

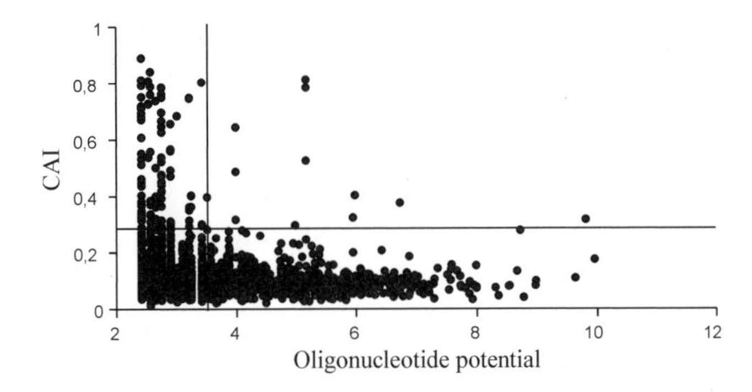

Figure 1. Dependence of CAI on the value of oligonucleotide potential, which is the measure of occurrence and distribution patterns of the motifs specific of low-expressed mRNAs.

3.3 TIS and TTS trinucleotide weight matrices allow yeast genes to be classified according to expression level

The positional context of regulatory regions of mRNAs was estimated using trinucleotide weight matrices. Positional weights were calculated as

$$W(b,k) = \log[P_p(b,k)] - \log[P_n(b,k)],\ b \in A,\ k = 1, \ldots, L, \tag{1}$$

where $P_p(b,k)$ is the frequency of occurrence of trinucleotide at position k of training sequence sample; P_n, frequency of occurrence of this trinucleotide at this position in the sample of negative sequences; A, set of all the trinucleotide values; and L, length of the weight matrix. The value $W(b,k)$ was assumed to be equal to –2 if $W(b,k) < -2$, and 2, if $W(b,k) > 2$. Two types of weight matrices were constructed: (1) $M_{rand\_3'}$- and $M_{rand\_5'}$-, for which the sample of randomly generated sequences with neutral mononucleotide composition was used as negative, and (2) $M_{contr\_3'}$- and $M_{contr\_5'}$-, for which samples of sequences contrasting in their expression levels were used as negative.

The score of an unknown sequence S of the length L during recognition by a weight matrix W was calculated as a sum of the weights of the corresponding positions:

$$S_L = \Sigma_{k=1, \ldots, L} W(b_k, k). \qquad (2)$$

Trinucleotide weight matrices were constructed (Table 2) for description of the translation initiation ($M_{rand\_5'-}$) and termination ($M_{rand\_3'-}$) sites in high- and low-expressed mRNA fractions in comparison with random sequences.

Table 2. Positional nucleotide weights for the regions of (a) translation initiation context $M_{rand\_5'-}$ and (b) termination codon context $M_{rand\_3'-}$ in high- and low-expressed mRNA fractions (positions whereat the corresponding trinucleotide is never met are boldfaced)

	Position of start	ATG (high)	ATG (low)		Position of stop	TAA (high)	TAG (high)	TGA (high)	TAA (low)	TAG (low)	TGA (low)
a)	–21	**–2**	**–2**	b)	–8	**–2**	**–2**	0.32	0.04	0.13	0.16
	–20	**–2**	–0.38		–7	0.28	–0.62	0.11	–0.1	–0.08	–0.13
	–19	**–2**	0.1		–6	**–2**	**–2**	**–2**	**–2**	**–2**	**–2**
	–18	**–2**	**–2**		–5	–0.81	**–2**	0.15	0.09	0.04	0.04
	–17	**–2**	–0.22		–4	–0.45	–0.37	0.31	–0.05	–0.2	0.04
	–16	**–2**	–0.7		–3	**–2**	**–2**	**–2**	**–2**	**–2**	**–2**
	–15	**–2**	–0.2		–2	**–2**	**–2**	**–2**	**–2**	**–2**	**–2**
	–14	**–2**	–0.38		–1	**–2**	**–2**	**–2**	**–2**	**–2**	**–2**
	–13	–0.53	0.1		0	1.23	0.83	0.47	1.1	1.13	1.24
	–12	**–2**	**–2**		1	**–2**	**–2**	**–2**	**–2**	**–2**	**–2**
	–11	**–2**	0.13		2	**–2**	**–2**	**–2**	**–2**	**–2**	**–2**
a)	–10	–0.5	–0.12	b)	3	–1.06	**–2**	–0.37	0	–0.43	0.21
	–9	**–2**	–0.69		4	–0.1	0.02	0.12	0.14	**–2**	0.04
	–8	**–2**	–0.1		5	–0.07	0.01	0.01	0.01	**–2**	–0.12
	–7	**–2**	–0.32		6	0.11	–0.1	–0.06	0.1	–0.04	–0.17
	–6	**–2**	–0.71		7	–0.59	**–2**	0.37	0.12	–0.71	0.06
	–5	**–2**	–0.68		8	–0.14	–0.38	–0.11	0.19	–0.13	–0.41
	–4	**–2**	–0.12		9	0.46	0.15	0.63	–0.31	0.11	–0.17
	–3	**–2**	**–2**		10	–0.47	**–2**	0.07	–0.27	–0.21	–0.38
	–2	**–2**	**–2**		11	–0.1	–0.24	0.09	0.07	–0.38	–0.67
	–1	**–2**	**–2**		12	0.01	–0.22	–0.01	0.03	–0.09	0.13
	0	1.87	1.66		13	–0.24	–0.67	0.03	0.04	–0.69	–0.71
	1	**–2**	**–2**		14	–0.06	0.1	0.02	–0.32	0.03	–0.14
	2	**–2**	**–2**		15	0.21	–0.67	0.35	0.02	–0.09	–0.01
	3	**–2**	0.14		16	–0.32	–0.7	–0.34	–0.13	0.25	–0.23
	4	**–2**	0.08		17	0.01	**–2**	–0.02	–0.2	–0.04	–0.39
	5	**–2**	–0.39		18	–0.02	–0.16	–0.13	–0.88	–0.08	0.14
	6	–0.04	0.26		19	–0.37	–0.07	0.02	–0.25	–0.08	**–2**

It appeared that unlike low-expressed mRNAs, 5'-regions of high-expressed mRNAs displayed a complete absence of certain trinucleotides. In particular, the presence of ATG is prohibited in virtually the entire TIS of high-expressed mRNAs. On the contrary, this prohibition does not work for low-expressed mRNAs. For example, the AUG codon is never met at position –11 with reference to the start codon of high-expressed mRNAs ($W_{(AUG,-11)} = -2$), whereas in low-expressed mRNAs, the frequency of this codon at this position slightly exceeds the random value ($W_{(AUG,\ -11)} = 0.1$). It coincides well with the commonly known fact that the presence of multiple ATGs near the transcription start can bring about false transcription starts, which is extremely unfavorable for high-expressed mRNAs (Kochetov *et al.*, 1998). Some differences in positional weights also occur in the context of the termination codon. For example, the trinucleotides TAA and TAG are never met at the –8 position with reference to the termination codons of high-expressed mRNAs ($W_{(TAA,-8)} = -2$, $W_{(TAG,-8)} = -2$), whereas this was not observed in low-expressed mRNAs ($W_{(TAA,-8)} = 0.04$, $W_{(TAG,-8)} = 0.13$). To detect all the positional differences between 5'- and 3'-untranslated mRNA regions with contrasting mRNA expression levels, we constructed contrasting weight matrices $M_{contr\_5'}$- and $M_{contr\_3'}$-. Their analysis shows significant positional differences in the regulatory regions of high- and low-expressed mRNAs. The dependences of yeast mRNA CAI on the context of (a) TIS and (b) TTS are shown in Figure 2.

Note that no relation between these values is observed when the total set of mRNA sequences is considered. However, the fraction of high-expressed mRNAs (CAI > 0.3) shows significant correlations between the context of 5'- and 3'-regions and CAI. We specified this threshold (CAI > 0.3) basing on the previous analysis (Figure 1), which demonstrated that oligonucleotide composition of the 5'-regions of yeast mRNAs with CAI > 0.3 behaved in a similar manner in comparison with all other sequences. It means that the value of CAI ~ 0.3 separates the high-expressed yeast mRNA fraction from mRNAs of all the other fractions. Thus, in high-expressed mRNAs, on the one hand, specific contextual features of the translation start and mRNA coding sequence are correlated and, on the other, specific contextual features of the termination codon and mRNA coding sequence display the correlation as well.

3.4 Contexts of TIS and TTS in high-expressed mRNAs correlate

Of special interest is the correlation between the context of initiation and termination codons. Figure 3 (filled circles) shows that this correlation is significant. Thus, the evolution of yeast genomes established correlations between the contexts of the initiation codon, coding region, and termination codon for high-expressed mRNAs.

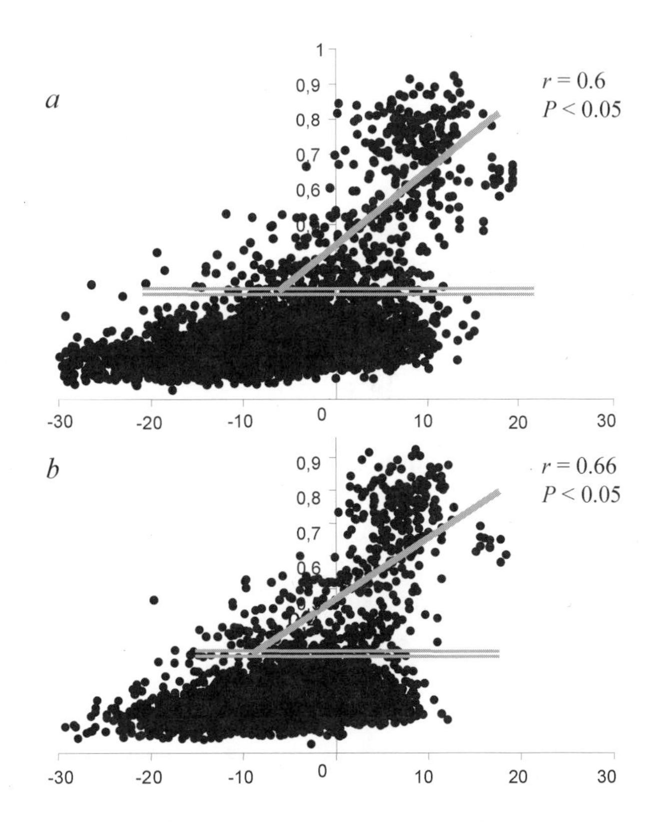

Figure 2. Dependence of CAI (*Y* axis) on the context of (*a*) AUG codon and (*b*) the termination codon (*X* axis). High-expressed mRNAs (CAI > 0.3) are separated from the other mRNA fractions with a double horizontal line. Linear regression dependencies were constructed for these mRNAs.

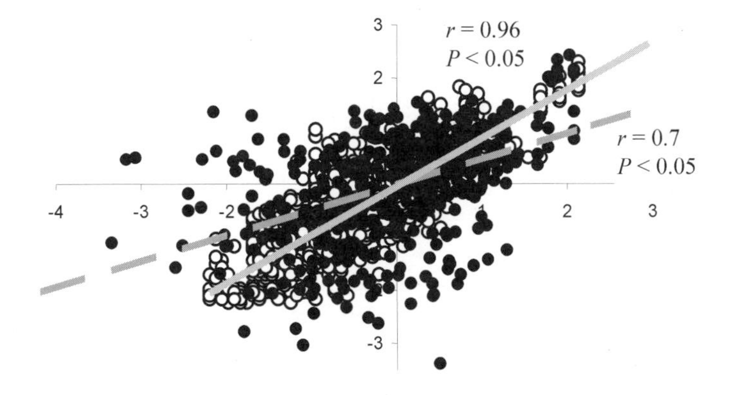

Figure 3. Filled circles, correlations between the contexts of AUG codon (*X* axis) and termination codon (*Y* axis) for the yeast mRNA fraction with CAI > 0.3; open circles, correlations between AUG codon and termination codon for mRNAs with CAI > 0.3 simulated using genetic algorithm; dash line, a regression line describing the behavior of yeast mRNAs; and solid line, a regression line describing the behavior of the computer-simulated mRNAs.

4. DISCUSSION

The analysis performed demonstrates that TIS and TTS of high and low expression genes of *Saccharomyces cerevisiae* differ in numerous contextual specific features.

4.1 Correlation between the TIS and TTS contexts in high-expressed mRNAs is similar to the correlation obtained for computer-simulated mRNAs

We applied genetic algorithm as a simulation method to clarify the mechanisms underlying emergence of these correlations. The evolution of a population of mRNA sequences, including the 5'-region, coding region, and 3'-region, was considered. It was governed by (1) recombinations exchanging fragments of 5'-, coding, and 3'-regions between mRNA molecules; (2) point mutations; and (3) selection directed to increase in translation rate F according to the rate-limiting stage model.

The rate of mRNA translation F was determined as

$$F = \min \begin{cases} \text{Score(5' - region)} \\ \text{CAI(coding\_region)} \\ \text{Score(3' - region)} \end{cases} . \tag{3}$$

Here, Score(5'-region) and Score(3'-region) are scores of the 5'- and 3'-regions calculated from the corresponding weight matrices and CAI(coding_region) is the aforementioned CAI.

Translation of mRNA proceeds through three consequent processes: initiation, elongation, and termination. Obviously, the greater is each of the indices Score(5'-region), Score(3'-region), or CAI(coding_region), the higher is the rate of each of the three processes. However, according to the rate-limiting stage model, the overall efficiency of mRNA translation is limited by the rate of the slowest process, that is, the least efficient stage.

The simulation shows (Figure 3, open circles) that when $CAI > 0.3$, the behavior of natural mRNAs reflects the actual correlation between the specific contextual features of the 5'- and 3'-untranslated regions of high-expressed mRNAs. Posttranscriptional regulation, which determines the translation rate of a particular mRNA, involves translation initiation, elongation, and termination. We have described it according to a model of an unbranched consequent molecular process involving three stages, which determine the yield of the final product, protein. The model of rate-limiting stage is a good approximation to such linear processes (Figure 4). Its application to biological systems and processes was most comprehensively described by Poletaev (1973) and Ratner (1990).

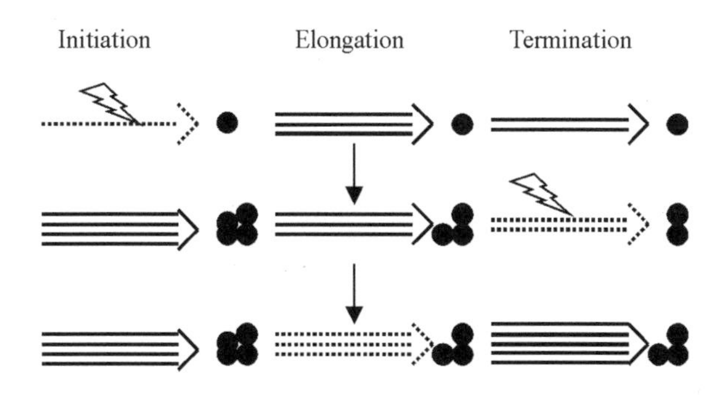

Figure 4. The model of rate-limiting stage. The limiting link of each stage is marked with dashes; lightning, the mutation causing an increase in efficiency of the particular link and making it nonlimiting.

Limiting links are those links in a chain of reactions that determine the yield of the final product of the overall chain. Thus, the change in the reaction rate in the limiting link can alter significantly the yield of the final product, whereas changes in nonlimiting links do not bring about notable yield changes. Hence, control of the limiting stage of a system is the most efficient method for controlling the overall system (Ratner, 1990). As soon as a mutation makes a certain stage nonlimiting, the regulation is determined by the next limiting link. Description of evolution of mRNA molecules according to this model means that the overall efficiency of the translation system can be increased only by mutations removing such limitations. Other (nonlimiting) elements of mRNA context organization undergo neutral evolution independently of one another.

Our results show that the evolution of high-expressed mRNAs optimized three main mRNA regions affecting translation rate: the 5'-untranslated, coding, and 3'-untranslated regions. This explains the correlation of specific contextual features of these regions in high-expressed mRNAs (CAI > 0.3), which have experienced the greatest selection pressure directed to an increase in the translation rate. We observed this by the example of yeast (Figure 3, filled circles). This is the sample of high-expressed mRNAs with CAI > 0.3 that exhibits a significant correlation between the scores of the 5'- and 3'-untranslated mRNA regions (Figure 3). A low translation level can be determined by the presence of a limiting link in any of the three regions. In this case, mRNAs with good context properties of the translation start and coding region may have a low overall translation rate because of the presence of a limiting link in the termination codon. Similarly, the presence of a limiting link at the level of a coding codon may determine a low translation rate even with good contexts of the initiation and termination codons and so on. Obviously, in the case of low-expressed mRNAs, a correlation of contextual properties of the three regions is unlikely. Some differences between the actual and theoretical correlations can be presumably explained by an incomplete consideration of the fine specific features in the dependence of the overall translation rate on the ratios of contributions of the three components according to the limiting-stage model in functional (3). Note that this approach can also be used for solving the inverse problem: evaluation of the selection pressure on the evolution of the contexts of 5'-, coding, and 3'-untranslated mRNA regions.

ACKNOWLEDGMENTS

The authors are grateful to V.A. Likhoshvai for the data supplied and fruitful discussions. The study was supported in part by the Russian Foundation for Basic Research (grants Nos. 01-07-90376-в, 02-07-90355, 03-07-96833-p2003, 03-07-96833, 03-04-48829, 03-07-06082-мас); grant PCB RAS (No. 10.4); the Siberian Branch of the Russian Academy of Sciences (integration project No. 119); Russian Ministry of Industry, Science, and Technologies (grants Nos. 43.073.1.1.1501 and 43.106.11.0011); NATO (grants Nos. LST.CLG.979816 and PDD(CP)–(LST.CLG 979815)); the U.S. Civilian Research & Development Foundation for the Independent States of the Former Soviet Union (CRDF) No. NO-008-X1.

CONTRIBUTION OF COORDINATED SUBSTITUTIONS TO THE CONSTANCY OF PHYSICOCHEMICAL PROPERTIES OF ATP-BINDING LOOP IN PROTEIN KINASES

D.A. AFONNIKOV
Institute of Cytology & Genetics, Siberian Branch of the Russian Academy of Sciences, prosp. Lavrentieva 10, Novosibirsk, 630090 Russia, e-mail: ada@bionet.nsc.ru; Novosibirsk State University, ul. Pirogova 2, Novosibirsk, 630090 Russia

Abstract It is known that the physicochemical properties of a protein that determine the specific folding of its polypeptide chain and functional features remain stable in the course of evolution. Search for such conserved characteristics by analysis of homologous sequences may aid understanding the function, structure, and evolution of proteins. Coordinated substitutions of amino acid residues are one of the plausible mechanisms maintaining these characteristics. The contribution of coordinated amino acid substitutions to the constancy of several integral physicochemical properties of the ATP-binding loop in protein kinases was studied. It was shown that coordinated substitutions contributed to the conservation of several integral characteristics of the loop related to the charge, hydrophobicity, and β-structure.

Key words: amino acid sequences, coordinated substitutions, physicochemical properties of amino acids, protein kinases

1. INTRODUCTION

Search for the most conserved features of proteins is one of the main tasks in the analysis of their homologous sequences. At the level of primary structure, this appears as conserved amino acid patterns. Moreover, proteins can be conservative with regard to their physicochemical properties depending on the amino acid residues at several positions of the protein (integral characteristics). A total volume of the hydrophobic core (Lim and

Ptitsyn, 1970; Gerstein *et al.*, 1994), total volume of residues forming an active site (Eisenhaber *et al.*, 1998), net charge, etc., are examples of such integral characteristics. The variance of characteristic is taken as a measure of its conservation. One of the plausible mechanisms for maintaining integral characteristics is coordinated substitutions of amino acid residues. They are assumed to result from interactions between amino acid residues in a protein. The program package CRASP developed by Afonnikov *et al.* (2001) allows for searching for coordinated amino acid substitutions in families of homologous sequences and evaluation of their contribution to the conservation/variation of integral physicochemical characteristics of proteins.

In this study, the effect of coordinated amino acid substitutions on the constancy of some integral characteristics depending on the physicochemical properties of amino acids in the ATP-binding loop of protein kinases is analyzed. Eukaryotic protein kinases form a large superfamily of catalytic proteins (Manning *et al.*, 2002). These proteins have similar catalytic domains, formed of 250–300 amino acid residues, responsible for binding and correct positioning of the phosphate group donor, ATP complexed to bivalent cation; binding and correct positioning of the protein (peptide) substrate; and transfer of γ-phosphate from ATP to hydroxyl group of the residue (Ser, Thr, or Tyr) of the protein substrate (Hanks and Quinn, 1991).

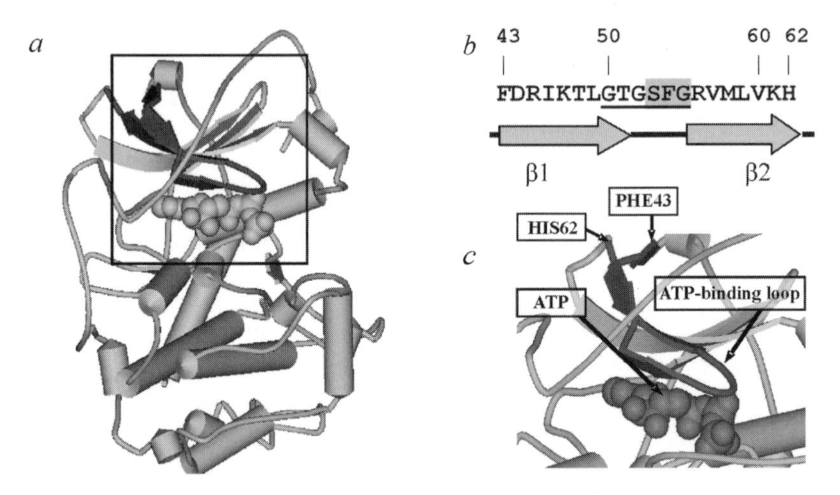

Figure 1. Structural properties of ATP-binding loop: (*a*) diagram of the spatial structure of the protein kinase catalytic domain (PDB ID 1atp:e; Zheng *et al.*, 1993); (*b*) amino acid sequence and secondary structure of the ATP-binding loop: the residues forming glycine motif are underlined, residues whose main chains form hydrogen bonds with ATP are shadowed, and β-sheets are indicated below; (*c*) spatial structure of the ATP loop (indicated dark-gray).

Structure of the catalytic domain is known (Taylor *et al.*, 1992; Zheng *et al.*, 1993; Figure 1). It consists of two large lobes: the *N*-terminal lobe (residues 1–113) binds ATP (GTP), while the *C*-terminal lobe (residues 138–297), the protein substrate (hereinafter, the numbers of amino acid residues correspond to amino acids and positions as given in Zheng *et al.*, 1993). The ATP-binding loop is an important region within the catalytic domain. It is localized to the *N*-terminus of the catalytic domain (positions 43–64), consists of two β-strands (numbered 1 and 2) and a turn between them, containing three glycine residues (Figure 1). Although the glycine motif G50-X-G52-X-X-G55 is highly conservative, substitutions there are not forbidden. However, specific features of the structure and function of this region may influence the pattern of amino acid substitutions. We studied evolutionary conservation of the total values of the 20 various physicochemical amino acid properties reflecting their structural and functional interactions in ATP-binding loop. It was found that the contribution of coordinated substitutions was conservative for certain physicochemical properties.

2. METHODS AND ALGORITHMS

2.1 Integral physicochemical characteristics

An integral characteristic F_j, corresponding to the *j*th physicochemical property of amino acids in a sequence of the length *L* is considered to be the total value of this property at the positions of the sequence

$$F_j = \sum_{i=1}^{20} f_{ij} \,, \quad (1)$$

where f_{ij} is the value of the property *j* for amino acid at position *i*.

Variance of F_j in the sample of homologous proteins was used as a measure of its variability. To estimate characteristic variances, the weighting according to Felsenstein (1985) was used to take into account the evolutionary relations of sequences.

2.2 Criterion for conservation of an integral characteristic

The contribution of amino acid substitutions to the conservation of integral characteristics is considered. For this purpose, the criterion proposed in the earlier study (Afonnikov *et al.*, 2001) was used.

The variance of an integral characteristics $D(f_i)$ can be expressed as a sum of two constituents: D_{var}, determined by the variability of protein positions, and D_{cov}, determined by coordinated amino acid substitutions:

$$D(F_j) = D_{var} + D_{cov}.$$

Note that D_{var} is always equal to or greater than zero, whereas D_{cov} may be positive, negative, or null. The last case can be used as the null hypothesis for verifying the significance of the contribution of coordinated substitutions to $D(F_\alpha)$. In this case (for all $r_{ij} = 0$), the expected variance of the physicochemical characteristic $D_{exp}(F_\alpha)$ is equal to D_{var}:

$$D_{exp}(F_j) = D_{var} = \sum_{i=1}^{L} D(f_{ij}).$$

Coordinated substitutions contribute to the stability of the integral index F_j if

$$D(F_j) < D_{exp}(F_j). \tag{2}$$

For validation of inequality (2), we may use a ratio of dispersions $\lambda = D_{exp}(F_j)/D(F_j)$. Thus, λ may serve as an index characterizing the contribution of coordinated substitutions to the variance of integral characteristic F_j: this contribution decreases the variance (increases the degree of conservation) at $\lambda > 1$ and increases the variance (increases the variation) at $\lambda < 1$. If $\lambda \approx 1$, the contribution of coordinated substitutions is insignificant. To test the significance of deviation of the parameter λ from unity, note that under the null hypothesis of equal dispersions, λ obeys the F distribution with $L(N - 1)$ and $N - 1$ degrees of freedom. Here, N is the number of sequences and L, number of positions (Selvin, 1998).

Additionally, a Monte Carlo technique is applied to estimate the statistical significance of the observed deviation of the parameter λ from unity. We generate normally distributed independent numbers with the mean and variance equal to their estimates at each position considered and repeat this procedure N times.

Then, we estimate the dispersion $D_{rand}(F_j)$ for such random samples (its value is expected close to unity). We repeat this procedure $M = 100\,000$ times and count samples with $\lambda_{rand} > \lambda$; the ratio of this number to M, p_{MC}, is an estimate of $p(\lambda_{rand} > \lambda)$ under the null hypothesis. These values were also compared to tabulated probabilities p_{tab}.

2.3 Sequence sample

The sample of protein kinase sequences from Hanks and Quinn (1991; http://www.sdsc.edu/kinases/pk_home.html) was analyzed. After discharge of identical sequences, its size N was 388 sequences. We considered positions 43–62, because some sequences in the alignment had deletions at loop positions 63–64. The positions analyzed correspond to alignment positions 1–20. The number of degrees of freedom in the equation (2) was $L(N-1)=7740$ and $N-1=387$.

3. RESULTS AND DISCUSSION

Overall, 19 characteristics from the AAIndex database (Tomii and Kanehisa, 1996) and one characteristic unrelated to any properties (the ordinal numbers of amino acids) were analyzed. The results, listed in Table 1, demonstrate that the contributions of coordinated substitutions to the conservation of various characteristics of the ATP-binding site differ considerably. The properties of amino acids are shown in Table 1 in a decreasing order of the parameter p_{tab}. A small value of this parameter shows that the coordinated substitutions add to the conservation of integral characteristics. A larger value indicates that the coordinated substitutions increase the variation.

The least values of p_{tab} and p_{MC} are characteristic of amino acid properties determining their hydrophobicity (Table 1, lines 1–6, 9), charge (isoelectric point, line 7), and occurrence in the β-structure (line 8). Another specific feature is a high value (exceeding 1.2) of the parameter λ. The probability to observe such values of λ for all the characteristics listed amounted to less than 1% for both Monte Carlo estimates and estimates according to the F distribution. Hence, coordinated substitutions support the conservation of these amino acid characteristics in the ATP-binding loop. On the contrary, coordinated substitutions do not contribute significantly to the variances of another group of characteristics (lines 10–19). The values of parameter λ for this group of characteristics fall in the range of 1.14–0.92; of parameters p_{tab} and p_{MC}, in the range of 0.028–0.99. Note that this group includes the ordinal number of amino acids, which has no physical sense (line 11). Analysis of the helix formation index (line 20) shows that coordinated substitutions significantly increase the variance of this characteristic (p_{tab} and $p_{MC} > 0.99$).

Let us consider compliance of the results we obtained with the known models of structure–function organization of the protein kinase catalytic domain. Note first that the conservancy of integral characteristics in proteins reflects their selective importance in the course of evolution. The selective importance may result from the functional or structural role played by a characteristic in the protein. Thus, the ranking of integral characteristics with reference to the degree of conservation allows for estimation of their structural or functional importance, thereby providing important information of the mechanism of protein function (Gerstein *et al.*, 1994). Coordinated substitutions may contribute considerably to maintenance of the conservation and reflect the selection pressure directed to preserve a characteristic at a constant level. Total values of various physicochemical characteristics studied here represent the integral interactions of various types between residues constituting the ATP-binding loop. The results demonstrated that, depending on particular physicochemical characteristic, the contribution of coordinated substitutions might decrease the variation of the characteristic, be insignificant, or increase the variation. This allows various types of interactions to be ranked with reference to their functional importance as well as potential mechanisms underlying operation of the residues in ATP-binding loop to be hypothesized.

Note that the contribution of coordinated substitutions to the characteristics related to hydrophobicity of residues appeared the most conservative. These characteristics include hydropathy index (line 1 in Table 1), HPLC parameter (line 2), solvation free energy (line 3), free energy of transfer to surface (line 4), hydrophobicity according to Eisenberg (line 5), hydrophobicity according to Hopp and Woods (line 6), and average accessible surface area (line 9). Analysis of interactions between the ATP molecule and ATP-binding loop demonstrates that a part of the residues in the loop are, first, involved in formation of the hydrophobic pocket housing the adenine ring of ATP. These residues are mainly localized to the loop's ends (Taylor *et al.*, 1992; Zheng *et al.*, 1993). Second, side groups of the residues localized to the turn region form hydrophobic contacts with the ATP molecule. Thus, hydrophobic interactions are actually of functional importance in formation of the ATP-catalytic domain complex.

The total value of isoelectric points of the residues forming the ATP-binding loop is also a conservative characteristic. This characteristic reflects the interactions between the charges of these residues. The importance of their contribution may stem from the fact that the ATP molecule has three phosphate groups, carrying a negative charge. The total charge of the loop's residues may compensate in part for this negative charge and, consequently, be preserved at a constant level in the course of evolution. Constancy of the characteristic reflecting the probability to form β-structure may result from the necessity to maintain a β-structure conformation of the residues of the residues in the loop.

Table 1. Analysis of the contribution of coordinated substitutions to variances of 20 integral characteristics calculated according to (1) for 20 physicochemical properties of amino acids: Names of the properties in the AAIndex database (Tomii and Kanehisa, 1996; full references can be found in AAIndex database); ratio λ of the characteristic variance to the expected value D_{exsp} in the absence of coordinated substitutions; proportion p_{MC} of randomized samples with $\lambda < \lambda_{rand}$; and estimate of this value according to the *F* distribution of p_{tab}

	Physicochemical property	λ	p_{MC}	p_{tab}
1	Hydropathy (Kyte and Doolittle, 1982)	1.53	0.00000	0.000000
2	HPLC parameter (Parker *et al.*, 1986)	1.51	0.00000	0.000000
3	Solvation free energy (Eisenberg-McLachlan, 1986)	1.47	0.00000	0.000000
4	Free energy of transfer to surface (Bull and Breese, 1974)	1.45	0.00000	0.000001
5	Hydrophobicity (Eisenberg *et al.*, 1984)	1.45	0.00000	0.000001
6	Hydrophilicity (Hopp and Woods, 1981)	1.40	0.00000	0.000008
7	Isoelectric point (Zimmerman *et al.*, 1968)	1.34	0.00003	0.000078
8	Normalized frequency of beta-sheet weights (Levitt, 1978)	1.28	0.00033	0.000670
9	Average accessible surface area (Janin *et al.*, 1978)	1.25	0.00089	0.001822
10	Average flexibility indices (Bhaskaran and Ponnuswamy, 1988).	1.14	0.02868	0.042569
11	Amino acid ordinal number	1.09	0.10091	0.128731
12	Normalized frequency for reverse turn with weights (Levitt, 1978)	1.07	0.17672	0.187593
13	Polarity (Zimmerman *et al.*, 1968)	1.07	0.14008	0.187593
14	Free energy in beta-strand region (Munoz and Serrano *et al.*, 1994)	0.95	0.75417	0.765414
15	Refractivity (Jones, 1975)	0.93	0.85991	0.878659
16	Normalized frequency of alpha helix with weights (Levitt, 1978)	0.92	0.87097	0.878659
17	Free energy in alpha-helical region (Munoz and Serrano *et al.*, 1994)	0.86	0.98904	0.982859
18	Volume (Chothia, 1975)	0.86	0.98756	0.982859
19	Residue volume (Bigelow, 1967)	0.86	0.98403	0.982859
20	Helix formation parameters (delta G; O'Neil and DeGrado, 1990)	0.77	0.99997	0.999898

Note that hydrophobicity and charge may be considered the interactions that are weakly dependent on the precise positioning of the side groups, that is, the nonspecific interactions. On the contrary, the substitutions coordinated with reference to the side group volumes reflect close interactions of residues (steric interactions). The contribution of coordinated substitutions is not significant in the case of the total volume of the side groups (lines 18 and 19, Table 1). Presumably, the precise coincidence of the sizes of side groups of the residues and ATP molecule is not crucial for interaction of the loop's residues and ATP. Apparently, the ATP–loop interactions do not follow a key–lock mechanism; they rather interact in a nonspecific manner (while other residues of the protein provide for specific binding). The fact that main chain atoms of the residues L49, G52, S53, G55, and F54, not of the side groups, are involved in hydrogen bonds, requiring a precise positioning of the donor and acceptor, also favors this hypothesis. Formation of these hydrogen bonds may depend weakly on the type of the substituted residue and positioning of its side group.

Thus, analysis of the contribution of coordinated substitutions to conservation of various physicochemical characteristics demonstrates that nonspecific remote interactions (hydrophobic and electrostatic) of the residues constituting the ATP-binding loop make a major contribution to the formation of catalytic domain–ATP complex. In addition, evolutionary maintenance of a β-structure of these residues also plays an important role. This complies with the currently existing model describing formation of the complex between the catalytic domain and ATP, based on X-ray structure data (Taylor *et al.*, 1992; Zheng *et al.*, 1993).

The results obtained in this work demonstrate that the degree of conservation of various integral physicochemical characteristics of proteins can be assessed by analyzing the contributions of coordinated substitutions of amino acid residues to the constancy of these characteristics. This allows the importance of contribution of particular interactions to protein stability or function to be determined. Such analysis is capable of revealing new information on the structure–function organization of protein macromolecules and the mechanisms underlying their operation.

ACKNOWLEDGMENTS

The work was supported in part by the Russian Foundation for Basic Research (grants Nos. 01-07-90376-в, 01-07-90084, 03-07-96833-p2003, 03-07-96833); Russian Ministry of Industry, Science, and Technologies (grant No. 43.073.1.1.1501, subcontract No. 28/2003); grant PCB RAS (No. 10.4); the Siberian Branch of the Russian Academy of Sciences (integration projects Nos. 119 and 65); the U.S. Civilian Research & Development Foundation for the Independent States of the Former Soviet Union (CRDF) No. Y1-B-08-20; US National Institutes of Health (grant No. 2 R01-HG-01539-04A2); US Department of Energy (grant No. 535228 CFDA 81.049).

SPORADIC EMERGENCE OF LATENT PHENOTYPE DURING EVOLUTION

V.A. LIKHOSHVAI[1,2,3]*, Yu.G. MATUSHKIN[1,2]
[1]*Institute of Cytology & Genetics, Siberian Branch of the Russian Academy of Sciences, prosp. Lavrentieva 10, Novosibirsk, 630090 Russia, e-mail: likho@bionet.nsc.ru;* [2]*Novosibirsk State University, ul. Pirogova 2, Novosibirsk, 630090 Russia;* [3]*Ugra Research Institute of Information Technologies, ul. Mira 151, Khanty-Mansyisk, 628011 Russia*
* *Corresponding author*

Abstract Development of the mathematical theory of gene network operation is a topical problem of informational biology. A simplest self-reproducing system (SRS), comprising only two genes, is described in the work. It is demonstrated that in the case of resource deficiency, SRS acquires the ability to exist in two states (two phenotypes) due to random mutations influencing the rates of biochemical processes. However, the operation mode of SRS genome remains unchanged.

Key words: mathematical model, computer model, regulation, translation, evolution, epigenetics

1. INTRODUCTION

In 1982–1983, it was demonstrated that certain RNA species existed in nature that, similar to proteins, displayed highly specific catalytic activities (Kruger *et al.*, 1982; Guerrier-Takada *et al.*, 1983). These RNA catalysts were named ribozymes. Now, ribosomes are also regarded as a ribozyme, as all the relevant experimental data accumulated so far suggest that the synthesis of polypeptide chain in the ribosome is catalyzed by ribosomal RNA, not ribosomal proteins. The major RNA functions known are genetic replication (in certain viruses), coding for polypeptide in linear nucleotide sequences, highly specific spatial interactions with other macromolecules, and specific catalysis of chemical reactions by ribozymes. Thus, RNA is capable of performing functions of both polymers that are basically important for the life—DNA and proteins. Accumulation of the knowledge on genetic code, nucleic acids, and protein biosynthesis gave birth to a principally new idea that not proteins but RNAs were the original building blocks of life (Woese, 1967; Orgel, 1968). Discovery of the catalytically active RNA species gave a powerful impulse to further development of the idea of RNA antecedence in the origin of life, resulting in the concept of the all-sufficient world of RNA prior to the modern world of ours (Gilbert, 1986; Joyce and Orgel, 1993). Emergence of proteins in the RNA world allowed for increase in the specificity of catalysts and, correspondingly, rates of biochemical reactions, translation included. The self-reproduction rate of coupled RNA–protein systems determined their fitness under conditions of resource deficiency. Optimization of reaction constants of the translation process leads to an increase in self-reproduction rate, conferring selection advantage on the system. In this work, a mathematical model of evolution of a single cycle of the simplest RNA–protein self-reproducing system is analyzed.

2. MODEL

2.1 Single development cycle

Let us consider that the simplest self-reproducing system (SRS) occupies a finite constant volume, containing SRS soluble part (mRNA and proteins) and the genome, determining the program of single development cycle.

The genome of SRS contains two RNA genes; the first encodes the protein P_r, subunit of a pooled ribosome; the second, the pooled structural protein P. As for the soluble SRS part, it comprises four components: pooled ribosome R, assembled of several P_r molecules; the mRNA F_r, forming the templates for translating proteins P_r; and mRNA F_p, templates for translating the pooled protein P. During a single SRS cycle, mRNAs and proteins are produced in amounts sufficient for assembling a new SRS copy. On completion of the single cycle, two SRSs appear instead of one, both containing all the necessary soluble components and full-fledged copies of the genome. The model of SRS single cycle describes in an explicit form only the translation processes from mRNA (via ribosomes) inherited from the maternal SRS of the previous development cycle. We do not consider synthesis of mRNAs and involvement of newly synthesized ribosomes and mRNAs in the synthesis. Processes of protein synthesis are described in the general case with the following reactions:

$$R^{(i)} + F_r^{(i)} \xrightarrow{b_r z^{(i)}} F_r^{(i)*} \qquad (1.1)$$

$$F_r^{(i)*} \xrightarrow{k_e z^{(i)} / n_s} F_r^{(i)} + R_r^{(i)} \qquad (1.2) \qquad\qquad (1)$$

$$R_r^{(i)} \xrightarrow{k_e z^{(i)} / (n_r - n_s)} R^{(i)} + P_r^{(i)}, \qquad (1.3)$$

$$R^{(i)} + F_p^{(i)} \xrightarrow{b_p z^{(i)}} F_p^{(i)*} \quad (2.1)$$

$$F_p^{(i)*} \xrightarrow{k_e z^{(i)} / n_s} F_p^{(i)} + R_p^{(i)} \quad (2.2) \qquad (2)$$

$$R_p^{(i)} \xrightarrow{k_e z^{(i)} / (n_p - n_s)} R^{(i)} + P^{(i)}. \quad (2.3)$$

Here, the superscript (i) denotes the current single development cycle of the SRS, which we will call maternal. The pooled resource $z^{(i)}$ is spent for reproduction. This resource is spent at the stages of both initiation and elongation; therefore, equations contain constants of translation initiation rate and elongation rate. The resource is inputted from the environment (see below the calculation of the pooled resource). Assemblage of ribosomes from subunits is not described; however, it is taken into account in the system of equations. $F_r^{(i)*}$, $F_p^{(i)*}$, $R_r^{(i)}$, and $R_p^{(i)}$ are intermediate complexes. Reactions (1.1) and (2.1) describe translation initiation; reactions (1.2) and (2.2), translation of the initial gene region, during which translation initiation site remains shielded by ribosome due to its considerable size, making the next event of translation initiation impossible. Reactions (1.3) and (2.3) describe the stage of translation of the rest gene, when the translation initiation site is free from ribosome, allowing for the next initiation event. The designations of parameters and biological sense of variables are listed in Table 1. We assume that the processes proceed under conditions of ideal mixing; this allows us to write the following system of ordinary differential equations from reactions (1.1)–(2.3):

$$\begin{cases} dF_r^{(i)*} / dt = b_r z^{(i)} F_r^{(i)} R^{(i)} - k_e z^{(i)} F_r^{(i)*} / n_s \\ dF_r^{(i)} / dt = -b_r z^{(i)} F_r^{(i)} R^{(i)} + k_e z^{(i)} F_r^{(i)*} / n_s \\ dR_r^{(i)} / dt = -k_e z^{(i)} R_r^{(i)} / (n_r - n_s) + k_e z^{(i)} F_r^{(i)*} / n_s \\ dF_p^{(i)*} / dt = b_p z^{(i)} F_p^{(i)} R^{(i)} - k_e z^{(i)} F_p^{(i)*} / n_s \\ dF_p^{(i)} / dt = -b_p z^{(i)} F_p^{(i)} R^{(i)} + k_e z^{(i)} F_p^{(i)*} / n_s \\ dR_p^{(i)} / dt = -k_e z^{(i)} R_p^{(i)} / (n_r - n_s) + k_e z^{(i)} F_p^{(i)*} / n_s \\ dR^{(i)} / dt = -(b_r F_r^{(i)} + b_p F_p^{(i)}) z^{(i)} R^{(i)} + \\ + k_e z^{(i)} (R_r^{(i)} / (n_r - n_s) + R_p^{(i)} / (n_p - n_s)) \\ dP_r / dt = k_e z^{(i)} R_r^{(i)} / (n_r - n_s) \\ dP / dt = k_e z^{(i)} R_p^{(i)} / (n_p - n_s). \end{cases} \qquad (3)$$

Let the SRS of the previous generation donate the following amounts of components to maternal SRS at the initial time point: $R_0^{(i)}$ pooled ribosomes, $F_{r0}^{(i)}$ mRNA encoding proteins of the pooled ribosome, and $F_{p0}^{(i)}$ mRNA encoding the pooled protein. Then, solving system (3) with the initial values $R_0^{(i)}$, $F_{r0}^{(i)}$, and $F_{p0}^{(i)}$, it is possible to calculate the number of synthesized molecules of the pooled protein $P^{(i)}$ at any moment t elapsed from the SRS single cycle commencement. Let us consider that the duration of an SRS single cycle equals the time required for synthesizing the threshold number π of pooled protein molecules. This is the moment when two identical daughter SRSs emerge instead of the maternal SRS. Let us assume that the amount of initial mRNAs donated to the daughter SRSs is identical to the amount contained in the maternal SRS, i.e. $F_{r0}^{(i+1)} = F_{r0}^{(i)}$ is the amount of mRNA encoding the proteins of pooled ribosome and $F_{p0}^{(i+1)} = F_{p0}^{(i)}$, mRNA encoding the pooled protein. Let us consider that the total number of ribosomes, consisted of the maternal SRS ribosomes and ribosomes synthesized *de novo,* are divided between daughter SRSs in equal shares: $R_0^{(i+1)} = (R_0^{(i)} + P_r^{(i)}/P_{\text{rib}})/2$, where P_{rib} is the number of subunits in one ribosome.

2.2 Consumption of pooled resource

The model of SRS single cycle does not consider explicitly the process of mRNA synthesis. However, we assume that the resource is spent for syntheses of all the system's components.

To derive the balance equation, we consider that the internal variables of SRS single cycle are in equilibrium with both one another and the environment. Then, the balance equation has the following form (the parameters are described in Table 1).

Let us designate

$$v_r^{(i)} = F_{r0} / (n_s / k_e + 1/(b_r R^{(i)})), \quad v_p^{(i)} = F_{p0} / (n_s / k_e + 1/(b_p R^{(i)})).$$

Then,

$$z^{(i)} = s_z / (d_z + \varepsilon_n (m_r F_{r0}^{(i)} + m_p F_{p0}^{(i)}) v_p^{(i)} / \pi + \varepsilon_a (v_r^{(i)} + v_p^{(i)}), \tag{4}$$

where $R^{(i)}$ is found as the single positive solution of the ribosome balance equation

$$R^{(i)} + (n_r / k_e) v_r^{(i)} + (n_p / k_e) v_p^{(i)} = R_0^{(i)}. \tag{5}$$

Duration of the *i*th SRS single cycle is calculated using the equation

$$T^{(i)} = \pi/(v_p^{(i)} z^{(i)}). \tag{6}$$

The following equation is used to calculate the amount of proteins of the pooled ribosomes produced in the maternal SRS during the time $T^{(i)}$:

$$P_r^{(i)} = v_r^{(i)} z^{(i)} T^{(i)}. \tag{7}$$

Thus, knowing the values of constants for maternal SRS as well as the initial concentrations of ribosomes and mRNAs, it is possible to calculate the parameters of SRS homeostasis and duration of the SRS single cycle using equations (4)–(7) and use these data to determine the concentrations of mRNAs and ribosomes of its descendants:

$$R_0^{(i+1)} = (R_0^{(i)} + P_r^{(i)}/P_{\text{rib}})/2,\ F_{p0}^{(i+1)} = F_{p0}^{(i)},\ F_{r0}^{(i+1)} = F_{r0}^{(i)}. \tag{8}$$

Evidently, applying consecutively equations (4)–(8) for an increasing number of generation *i*, it is possible to calculate the single cycle parameters of any descendant of the initial SRS.

All the parameters of the model and most important characteristics of the single cycle introduced in equations (1)–(8) are described in Table 1.

Table 1. Parameters of the SRS computer model

No.	Designation	Unit	Initial state[2]	State after evolution[3]		Comments
				Phenotype 1	Phenotype 2	
1	F_{r0}^{1}	Ps.	50	376		Total amount of the mRNA encoding ribosomal proteins
2	F_{p0}^{1}	Ps.	50	729		Total amount of the mRNA encoding structural protein
3	b_r^{1}	Ps.$^2 \times$ sec^{-1}	10^{-6}	0.0163		Translation initiation rate constant of F_r mRNA
4	b_p^{1}	Ps.$^2 \times$ sec^{-1}	10^{-6}	0.0611		Translation initiation rate constant of F_p mRNA
5	T^{4}	Sec	554 767	4886	5376	Duration of SRS single cycle
6	Z^{4}	Ps.	1.97	0.398	0.200	Equilibrium amount of pooled resource

No.	Designa-tion	Unit	Initial state[2]	State after evolution[3]		Comments
				Phenotype 1	Phenotype 2	
						arbitrary units established during SRS development
7	R_0[4.5]	Ps.	37 018	8330	16 887	Total concentration of ribosomes
8	P_{rib}[6.9]	Ps.		54		Amount of proteins in ribosomes
9	π[6.9]	Ps.		2×10^6		Threshold value with respect to structural protein determining the duration of SRS single cycle
10	n_r[6.9]	A.a.		100		Length of pooled ribosomal protein in amino acids
11	n_p[6.9]	A.a.		300		Length of pooled structural protein in amino acids
12	m_r[6.9]	Nt.		400		Length of F_r mRNA in nucleotides
13	m_p[6.9]	Nt.		1000		Length of F_p mRNA in nucleotides
14	n_s[6.9]	Nt.		45		Steric size of ribosome in nucleotides
15	k_e[6.9]	$Ps.^{1} \times sec^{-1}$		40		Averaged elongation rate constant
16	ε_n[7.9]	Arb. unit of resource		3×10^{-4}		Relative energy value of one nucleotide
17	ε_a[7.9]	Arb. unit of resource		10^{-4}		Relative energy value of one amino acid
18	s_z[8.9]	Sec^{-1}		20		Pooled resource input rate constant
19	d_z[8.9]	Sec^{-1}		10		Pooled resource degradation rate constant

[1] Parameters that can be changed by mutations; [2] the only stable phenotype of the initial SRS; [3] two stable SRS phenotypes; [4] homeostasis parameters (phenotypic traits) calculated using equations (4)–(8); [5] fluctuation-prone concentration; [6] SRS structural parameters; [7] parameters of SRS interactions with the environment; [8] parameters of the environment; and [9] specified and remain constant at all the stages of model's operation.

2.3 Simulation of mutations and evolutionary selection

In the absence of mutations and fluctuations, the daughter SRS inherits the parameters indicated in lines 1–4 and 8–19 of Table 1 without any changes. In general case, the number of ribosomes, determined by equation (8) is not equal to that of the previous generation, i.e. $R_0^{(i+1)} \neq R_0^{(i)}$.

Let us assume that random mutations at the genomic level accompany generation of daughter SRSs. Let us consider that the mutations has no effect on the structure–function organization of the single cycle gene network, i.e. mutation effects on the reproduction processes consist in changes in rates of processes within the genetic program specified. Applied to the model, this means direct random changes in model's constants without alterations in the form of equations. Overall, the model contains 16 parameters. However, all the parameters of SRS single cycle homeostasis (characteristic of SRS homeostasis are equilibrium values of the variables of model (3)) may be expressed via $R^{(i)}$, which is found as a positive root of equation (5), and the four pooled parameters s_1F_p, s_2F_r, c_1/b_r, and c_2/b_p, where s_1, s_2, c_1, and c_2 depend only on the parameters listed in Table 1 lines 8–15. This allows us to limit ourselves to considering changes in only four parameters: the concentrations of mRNAs, F_r and F_p, and the translation initiation rates, b_r and b_p. Without limiting the generality, the parameters present in s_1, s_2, c_1, and c_2, may be considered constant. Let us also admit the hypothesis that both the environment and SRS relations with the environment are invariant. Consequently, the parameters ε_n, ε_a, s_z, and d_z are assumed constant during the entire evolution.

Since the mutations in population are fixed not in an instant manner, but upon a certain number of generations, we will evaluate the evolutionary value of a mutation according to the reproduction rate of a remote descendant of the mutant SRS. Let us use the limit variants of equations (1)–(8), obtained if i is approaching infinity, to calculate the homeostasis characteristics of SRS remote descendants.

An elementary evolutionary event is simulated in the following way. Let us have an SRS with a specified set of homeostasis parameters. The value of one of the parameters F_r, F_p, b_r, and b_p selected randomly (the probability of selection of a parameter is assumed equaling 1/4) is randomly altered within the specified ranges of alterations (a "mutation" is introduced). A certain number M_u of the "mutant" individuals, which is also a prespecified parameter of the model of evolution, is generated. Then, characteristics of the single cycle of a remote descendant generated by the initial and mutant SRSs are calculated assuming that other mutations and fluctuations are absent.

It is considered that the individual with a minimal duration of the single cycle of its remote descendant "survives", whereas the rest "die". This may be either a mutant or the initial SRS. The parameters of this survived individual are considered the norm, and mutations are generated again to select the best individual. Evolution is calculated as a sequence of elementary evolutionary events. Consequently, we obtain an SRS movement trajectory in the space of the altered parameters F_r, F_p, b_r, and b_p.

2.4 Fluctuations

A random deviation in concentrations of soluble fractions from the average values during division of the maternal system are considered as fluctuations. Since the concentrations of mRNAs are parameters of the model, the fluctuations are considered only for ribosomes.

3. RESULTS AND DISCUSSION

Study of the SRS operation mode in the absence of mutations and fluctuations demonstrates that each series of generations converges to a certain limit. Proceeding to the limit of i in equations (4)–(8), we obtain the following cubic equation:

$$\begin{aligned}
&R^3 - a_2R^2 + a_1R - a_0 = 0,\\
&\text{where } a_0 = (\pi F_{r0}/F_{p0}/P_{\text{rib}}) \times (k_e/n_s/b_r),\\
&a_1 = k_e^2/n_s^2/b_r/b_p + F_{r0}n_rk_e/n_s^2/b_p + F_{p0}n_pk_e/n_s^2/b_r -\\
&- (\pi F_{r0}/F_{p0}/P_{\text{rib}}) \times (k_e/n_s/b_p),\\
&\text{and}\\
&a_2 = (\pi F_{r0}/F_{p0}/P_{\text{rib}}) - (k_e/n_s/b_r + k_e/n_s/b_p + F_{r0}n_r/n_s + F_{p0}n_p/n_s).
\end{aligned} \tag{9}$$

The positive roots of equation (9) have the sense of an equilibrium concentration of free ribosomes. There is no difficulty to find out that at least one positive root always exists at any positive values of the parameters n_s, n_r, n_p, k_e, P_{rib}, π, F_{r0}, F_{p0}, b_r, and b_p. However, if values of the parameters are selected to meet the below conditions

$$\begin{aligned}
&a_1 > 0, \quad a_2 > 0, \quad a_2^2 \geq 3a_1,\\
&(a_2 + \sqrt{a_2^2 - 3a_1})^2(a_2 - 2\sqrt{a_2^2 - 3a_1}) \leq 27a_0 \leq\\
&\leq (a_2 - \sqrt{a_2^2 - 3a_1})^2(a_2 + 2\sqrt{a_2^2 - 3a_1}),
\end{aligned} \tag{10}$$

equation (9) has three positive roots taking into account their repetition factor. Taking the value of any of the positive roots as the concentration of free ribosomes and using equations (4)–(6) without the upper indices, it is possible to calculate unambiguously the following parameters: R_0, the equilibrium concentration of pooled ribosomes; Z, concentration of pooled resource; and T, duration of SRS single cycle. These values are equilibrium in a series of generations in the sense that if they are ascribed to the initial SRS, all its descendants will display the same characteristics. Consequently, a certain SRS corresponds to any positive root of equation (9); let us name this SRS a limit SRS corresponding to the particular root.

3.1 Variants of SRS development

If equation (9) has the only positive root with a repetition factor of unity, than all the generation series will converge to the limit SRS corresponding to this root. Let us have three different positive roots. If the number of free ribosomes $R_0^{(0)}$ in the initial SRS is smaller than the concentration of free ribosomes in the limit SRS corresponding to the medium root value of equation (9), than the series of generations of this SRS converges to the limit SRS corresponding to the smaller root. Otherwise, all the series of SRS generations converge to the limit SRS determined by the larger root of equation (9). None of the series converge to the SRS corresponding to the medium root, except for that constructed using the limit SRS itself. In this sense, the SRSs corresponding to the smaller and larger roots are stable, whereas the SRS determined by the medium root is unstable.

3.2 Effect of fluctuations

Let us assume occurrence of random fluctuations in distribution of ribosomes between the daughter SRSs. Evidently, the fluctuations will randomize the kinetics of single cycle. Let, first, equation (9) have one positive root with a repetition factor equaling unity. Then, as described above, development of any system reduces to the single limit SRS. This is also true in the case of development of an SRS whose internal content changed due to a fluctuation. Consequently, in this case, the fluctuations occurring during development of the system are leveled and incapable of changing the qualitative behavior of the system. Let us name "phenotype" the observable properties of SRS model, i.e. dynamic characteristics of SRS single cell. Correspondingly, the two genes considered in the model will represent the genotype. Then, we can infer that the phenotype is unambiguously determined by genotype in the domain of existence of the single positive root, with a unity repetition factor, of equation (9).

Fluctuations introduce random changes into the phenotype, which fail to influence the stability of SRS development in time.

The situation is quite different, if equation (9) has three positive unequal roots. Let us assume that the initial SRS is in the domain of attraction of the limit SRS corresponding to the smaller root. Let us also assume that during division of the maternal SRS, the probability of fluctuations transiting the daughter SRS into the domain of attraction of the larger root is non-zero. Then, during development of the initial SRS, a descendant falling into the domain of attraction of the limit SRS corresponding to the larger root will appear with a unity probability and *vice versa*. Consequently, the unambiguity of phenotype reproduction in this domain is lost. Development also ceases being stable, as the system will be attracted alternately to one or the other limit SRS during the development.

Thus, the space of admitted values of the SRS parameters falls into two qualitatively distinct regions (Figure 1): (1) the region where the phenotype is determined unambiguously by the genotype and SRS development is stable (region D_1) and (2) the region where two development variants of the self-reproduction genetic program exist and the development is unstable (region D_2). Behavior of SRS in the region D_1 is completely predictable, whereas in the region D_2, it is impossible to foretell the phenotype of a descendant of the initial SRS, because not only the genetic program determines the phenotype, but also epigenetic factors prone to fluctuations. Thus, the SRS behavior differs in complexity in various parts of the parametric space, being more complex in the region D_2 compared to the region D_1.

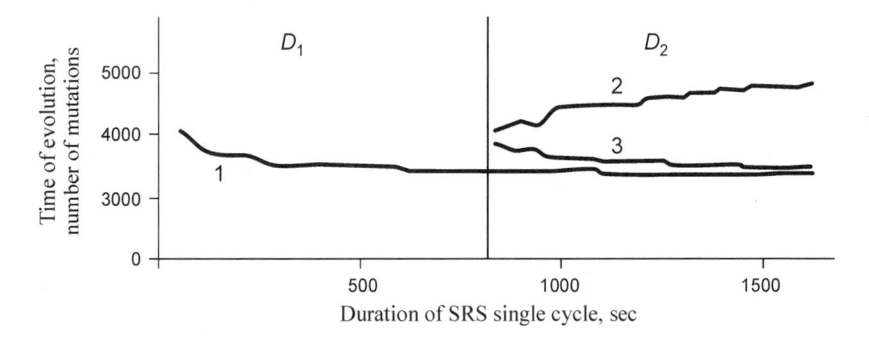

Figure 1. Changes in SRS single cycle duration during evolution (calculation according to the model): (1) optimal phenotype; (2) latent phenotype; and (3) separatrix; the vertical line is the boundary between D_1 and D_2 regions.

We studied the patterns of SRS development trajectories in the parametric space taking into account mutations and selection. In this work, we considered the simplest script of SRS evolution: selection of the most fitted system among individual mutants (for details, see section 2.3 *Simulation of mutations and evolutionary selection*). We analyzed the trajectories calculated numerically. Overall, nine batches of evolutionary trajectories were calculated. Each batch corresponded to certain value of the parameter s_z, the rate of pooled resource input, and certain number of mutations M_u generated and comprised 300 trajectories. For each initial SRS, the parameters F_r, F_p, b_r, and b_p were randomly selected from a four-dimensional parallelepiped ($1 \leq F_r \leq 10^4$, $1 \leq F_p \leq 10^4$, $10^{-6} \leq b_r \leq 10^{-4}$, and $10^{-6} \leq b_p \leq 10^{-4}$); the duration of SRS single cycle $T^{(0)}$ was selected from the range $T^{(0)} \in$ [7000 sec, 8000 sec]. The used conditions for selecting the parameters F_r, F_p, b_r, and b_p are not mandatory (Table 1 shows the example of SRS initial status, which is distinguished from described above but evolutionary drifted to region D_2). They are introduced for the sake of convenience, as all the initial SRSs in this case display approximately similar fitness to the environmental conditions. All the evolutionary trajectories initially started from the region of unambiguity D_1. Each trajectory was calculated until it entered the region D_2 or reached the upper evolutionary time limit, measured in elementary evolutionary events. We took the maximal number of evolutionary events equaling 10^6. Calculation results demonstrate that: 1. Under conditions of limited resource input rate, all the evolutionary trajectories transit into the region D_2 during the evolutionary time span specified. 2. An increase in the resource input rate brings about the trajectories that yet remain in the region D_1 when the span of 10^6 evolutionary events, specified in the numerical experiment, already elapsed. 3. On reaching a certain resource input rate, all the trajectories at all the values of parameters M_u remain in the region D_1 during the entire period of evolution considered.

The overall data suggest the following general inference—*the region* D_2 *is the domain of attraction* in the evolutionary scheme chosen under conditions of limited resource. Lifting the restriction on resource retains all the trajectories in the region D_1.

Thus, numerical analysis demonstrates that the SRS evolves from the region D_1 to the region D_2 under certain conditions. Resulting from this transition is *complication of the system's dynamic behavior.* Noteworthy that dynamic complication of the system occurs due to random changes in its operation parameters and selection of the most fitted individuals under condition of the limited resource. Moreover, the random changes in constants, resulting in attraction of SRS into the domain of two potential phenotypes in the case of limited resource, may emerge not only through mutations, but also due to changes in environment (for example, alteration in temperature).

4. CONCLUSION

These results, obtained by both analytical study (a "compressed model") and computer-assisted portrait simulation (an "expanded model"), demonstrate that SRS may exist in different states (display different phenotypes) having one and the same (simplest) genotype. Moreover, the possibility of two states is not encoded in the genome—it appears during selection for increase in reproduction rate as a result of deficiency in the resource.

Note that it is not necessary that the model's variables have the sense of mRNA, ribosomes, etc., to obtain the result observed. It is possible to consider DNA instead of mRNA and mRNA instead of proteins without changing the formal structure of the model, i.e. obtaining the similar results of modeling. The assumption on constant SRS size and exclusion of mRNAs and proteins synthesized de novo also do not represent the phenomenon-forming factors. Two issues are crucial in the model: (1) a two-stage description of protein synthesis in the model, i.e. separate consideration of the early stage, when the translation initiation site is unavailable for the next initiation event, and the late stage, when this site is free for initiation of the next event in the protein synthesis (from the mathematical standpoint, this provides the necessary degree of nonlinearity), and (2) separate consideration of two moieties within the SRS—the first moiety, structural protein, is the main in the sense that it determines the duration of a single cycle, whereas the rest components are subordinate. The sole role of ribosomes is to synthesize structural proteins. However, SRS should be reproduced in an infinite succession of generations for the system to survive; therefore, ribosomes should also be reproduced in parallel with the structural proteins, and ribosomes themselves are responsible for their reproduction. However, ribosomes are reproduced only until the amount of structural protein necessary for the daughter copy of SRS is synthesized. In this sense, the ribosomes are synthesized according to a residual principle, and distributed between the daughter cells is not a particular number of ribosomes, but just the actual amount that has been synthesized during a single cycle. Thus, two basic principles, inherent in all the living systems, form the background for emergence of the ambiguity region in reproduction of the phenotype from the genotype: template-based synthesis of mRNAs/proteins and the ability to self-reproduce. A supplementary requirement is reproduction of a part of the system as the main, structural moiety, whereas the rest part is reproduced according to a residual principle. The ambiguity of phenotype reproduction is not a function of the genome and cannot be derived from it. Here the epigenetic factor—i.e. all the constituents of the system, including mRNAs, proteins, low-molecular-weight compounds, etc.—plays the most important role. Formation of the phenotypic traits within a species is determined by the genotype (genetic program or gene network) and influence of the environment. In this process, the ontogenetic program encoded in the genotype may imply realization of alternative (to a certain degree) phenotypes depending on the environmental conditions. Thus, in addition to minor phenotypic differences (polymorphism within the norm of response), larger distinctions determined by alternative ontogenetic pathways may emerge (Tsurimoto and Matsubara, 1983).

Such alternative pathways are exemplified by the lysogenic and lytic variants of λ phage ontogenesis. This triggering pattern in switching the system is a systemic property in the sense that no subsystem of λ phage alone possesses this property. We have demonstrated here that in addition to the two major causes of the distinction in phenotypes described above, the phenotype may be determined by epigenetic factors; however, the gene network of an organism lacks specialized subnetworks (genetic subprograms) that would be responsible for realization of these morphological variants. We believe that the emergence of the second (latent) phenotype may also realize in the case of organisms more intricately organized compared with the simplest haploid SRS considered in this work. The second phenotype is realized in a random manner; however, if the environmental conditions render it selectively advantageous compared with the first (basic) phenotype, the second phenotype may be later fixed by genetic mechanisms independent of the parameters. In this case, the new organism with the new genotype may enter during its evolution under a tense natural selection the region where the second phenotype would manifest itself.

From the genetic standpoint, we are performing selection with reference to the rate of reproduction, which, upon evolution completion, results in emergence of the new trait (latent phenotype), however, manifesting in a sporadic manner. Similar situation was described by Academician D.K. Belyaev in his works (Belyaev, 1987 and others) on destabilizing selection. In these experiments, selection in a population of silver-black foxes involved the trait of friendliness to humans, resulting with time in emergence and random manifestation of new purely morphological traits (skewbalds, dapples, lop ears, etc.).

ACKNOWLEDGMENTS

Authors are grateful to N.A. Kolchanov for fruitful discussion of the results obtained. The work was supported in part by the Siberian Branch of the Russian Foundation for Basic Research (grants Nos. 02-04-48802, 02-07-90359, 01-07-90376, 03-01-00328, 03-07-96000, 03-04-48829); grant PCB RAS (No. 10.4); the Siberian Branch of the Russian Academy of Sciences (integration projects Nos. 148 and 119); Basic Research Programm of RAS (grant No. 12.4).

SIMILARITY ANALYSIS OF INVERSION BANDING SEQUENCES IN CHROMOSOMES OF *CHIRONOMUS* SPECIES (BREAKPOINT PHYLOGENY)

I.I. KIKNADZE[1,3], L.I. GUNDERINA[1]*, A.G. ISTOMINA[1], V.D. GUSEV[2], L.A. NEMYTIKOVA[2]

[1]*Institute of Cytology & Genetics, Siberian Branch of the Russian Academy of Sciences, prosp. Lavrentieva 10, Novosibirsk, 630090 Russia, e-mail: gund@bionet.nsc.ru;* [2]*Sobolev Institute of Mathematics, Siberian Branch of the Russian Academy of Sciences, prosp. Koptyuga 4, Novosibirsk, 630090 Russia, e-mail: gusev@math.nsc.ru;* [3]*Novosibirsk State University, ul. Pirogova 2, Novosibirsk, 630090 Russia*

* *Corresponding author*

Abstract Comparison of polytene chromosome banding sequences and estimation of the degree of chromosomal divergence according to the number of intrachromosomal breakpoints were applied to reconstruct the chromosomal phylogeny in the genus *Chironomus.* A measure for detecting differences in the banding sequences based on the concept of complexity decomposition (Gusev *et al.*, 2001) was developed. For the first time, a phylogenetic tree based on analysis of the banding sequences in five chromosome arms (A, C, D, E, and F), comprising about 70% bands of the genome, was constructed. The ArmsACDEF tree displayed distinct clusters of banding sequences complying with the species taxonomy (affiliation with groups of sibling species and individual cytocomplexes). These clusters comprised 79% of the sequences studied; the rest 21% appeared unclustered and belonged to the species outside any sibling group. Comparison of the phylogenetic trees for banding sequences based on different set of arms demonstrated a relative independence of banding sequence evolution in individual arms. Distinctness and stability of clusters in the phylogenetic tree increased with the number of arms involved in the analysis. The data obtained demonstrate an essential influence of changes in gene arrangement on species divergence.

Key words: chromosomal phylogeny, banding sequences, break point, Chironomidae

1. INTRODUCTION

A large number of cytological and molecular genetic maps have been built for various microorganisms, animals, and plants, thereby intensely stimulating development of comparative genomics. Evolutionary patterns at the chromosomal and nucleotide levels differ considerably. Chromosomes are rearranged mainly due to inversions, not by deletions, insertions, and duplications, as typical of DNA. Consequently, computer analysis of chromosome evolution require specialized methods based, as a rule, on a pairwise comparison of gene arrangements in syntene linkage groups followed by phylogenetic tree construction basing on pairwise similarity matrices (Nadeau and Taylor, 1984; Sankoff and Nadeau, 1996; Zakharov *et al.*, 1997). However, use of breakpoint phylogeny methods is occasionally limited, because specific features of genomic organization of certain organisms and the markers used require considerable modifications of the software for calculation of pairwise similarity matrices. Such modifications also appeared necessary when studying the evolution of polytene chromosome banding sequences of the species belonging to the genus *Chironomus* (numerous inversion sequences have been found in chromosomes of these species from various Holarctic zoogeographical divisions). The chironomids or midges, belonging to the genus *Chironomus* (Chironomidae, Diptera), have giant polytene chromosomes, displaying strict patterns of bands with varying thicknesses and morphology. The band is a structural unit of polytene chromosome, comprising one to several genes. The karyotype of chironomids contains up to 2500 bands. The banding pattern is specific of each chromosome and species-specific. Large number and specific morphology of bands make them reliable markers of chromosome rearrangements. *Chironomus* chromosomes are most frequently rearranged due to breaks at certain points followed by inversion, a turn of individual chromosome region by 180°. Inversions change visually the observed banding patterns of polytene chromosomes and usually involve two or more sequential bands (Figure 1).

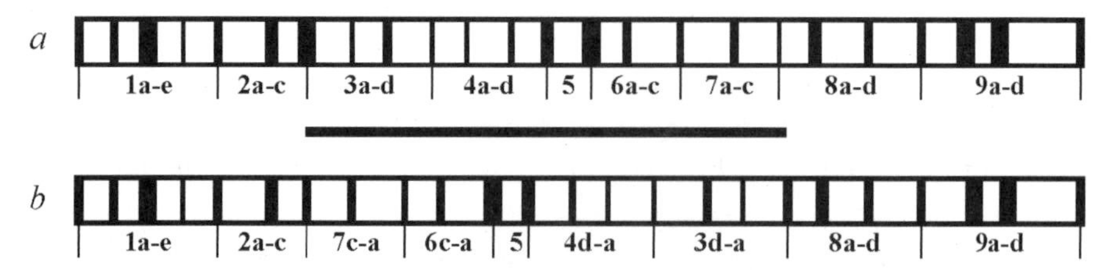

Figure 1. Scheme of an inversion rearrangement of banding sequence in chromosome: (*a*) initial banding sequence; (*b*) banding sequence carrying inversion in the region 3–7 (chromosome is divided into nine regions designated by digits; bands, by letters; and inversion sequence is indicated by horizontal line).

In this work, a measure of similarity between polytene chromosome banding sequences of 63 *Chironomus* species was for the first time estimated by a somewhat unconventional tool for comparing pairs and groups of sequences (Gusev *et al.*, 2001). This measure was used to build phylogenetic trees demonstrating a progress of chromosomal phylogeny in the genus *Chironomus*. The phylogenetic trees based on banding sequences of the most variable arms C and D and on five chromosome arms of the karyotype (A, C, D, E, and F) differ essentially, suggesting that chromosome arms have evolved independently. The phylogenetic trees display distinct clusters of banding sequences complying with the species taxonomy (affiliation with groups of sibling species and individual cytocomplexes). Distinctness and stability of clusters in the phylogenetic tree increased with the number of arms involved in the analysis. The data obtained demonstrate an essential influence of changes in gene arrangement on species divergence.

2. METHODS, ALGORITHMS, AND MATERIAL

1. To study the chromosomal breakpoint phylogeny of *Chironomus* species, specialized software allowing for comparative analysis of polytene chromosome banding sequences by estimation of pairwise similarities of sequences with different band arrangements resulting from simple or complex inversions was developed. Specificity of the initial data stems from (i) the equal lengths of all the sequences in a fixed set of chromosome arms and (ii) the absence of repeated elements in each sequence. Both the lengths of sequences processed and the size of alphabet grow with the increase in the number of simultaneously processed arms. In particular, analysis of five arms required an alphabet of 617 symbols. As is noted above, the evolution of sequences in the situation considered involves breaking of chromosomes in two points followed by inversion of the band arrangement between these points.

2. The developed database (electronic library) "Chironomidae: species, populations, genetic variability" (Golygina *et al.*, 2001) available at http://www-sbras.nsc.ru/win/elbib/atlas/chironomids/, allowed us to create a unique inventory of the banding sequences for the genus *Chironomus*, one of the largest in the family Chironomidae, and formed the background for a comprehensive study of *Chironomus* chromosomal phylogeny. For this purpose, 63 *Chironomus* species with known banding sequences of five chromosome arms (A, C, D, E, and F) were chosen from the electronic library. Note that the *Chironomus* genome comprises seven chromosome arms; however, banding sequences of the arms B and G are yet unmappable by the conventional method of Keyl (1962), as the chromosomal rearrangements having accumulated evolutionary are too intricate. Seven chromosome arms may form various

combinations in the four chromosomes of haploid genome of *Chironomus* species; this unites the species into several cytocomplexes. Species of the genus *Chironomus* and their classification according to cytocomplexes and sibling species groups are as follows:

I. ***Thummi* complex (thu)**, arm combination AB CD EF G:

1. Gr. *plumosus* (plu): *C. agilis*, *C. agilis* 2, *C. bonus*, *C. borokensis*, *C. entis*, *C.* sp. *J*, *C. muratensis*, *C. nudiventris*, *C. plumosus*, *C. suwai*;
2. Gr. *aberratus* (abe): *C. aberratus*, *C. beljaninae*, *C. fraternus*, *C. jonmartini*, *C. sororius*;
3. Gr. *decorus* (dec): *C. blaylocki*, *C.* sp. B1, *C. utahensis*;
4. Gr. *obtusidens* (obt): *C. acutiventris*, *C. arcustylus*, *C. heterodentatus*, *C. obtusidens*, *C. sokolovae*;
5. Gr. *riihimakiensis* (rii): *C. nigrifrons* (sp. Tu1), *C. riihimakiensis*, *C. tuvanicus*, *C.* sp. Al1, *C.* sp. Le 1;
6. Gr. *piger* (pig): *C. piger*, *C. riparius* (thummi);
7. Gr. *staegeri* (sta): *C. crassicaudatus*, *C. staegeri*;
8. Gr. *pilicornis* (pil): *C. pilicornis*, *C. heteropilicornis*;
9. Gr. *tenuistylus* (tnu): *C. longistylus*, *C. tenuistylus*;

Species outside sibling groups: *C. albimaculatus*, *C. annularius*, *C. anthracinus*, *C. atrella*, *C. behningi*, *C. cucini*, *C. esai*, *C.* sp. Is, *C. nuditarsis*, *C. trabicola*;

II. ***Pseudothummi* complex (pst)**, arm combination AE CD BF G:
C. acidophilus, *C. alluaudi*, *C. aprilinus*, *C. dorsalis*, *C. luridus*, *C. melanescens*, *C. pankratovae*, *C. pseudothummi*, *C. yoshimatsui*;

III. ***Maturus* complex (mat),** arm combination AF BE CD G:
C. fundatus, *C. maturus*, *C. whitseli*;

IV. ***Camptochironomus* complex (cam)**, arm combination AB CF ED G:
C. dilutus, *C. novosibiricus*, *C. pallidivittatus*, *C. setivalva*, *C. tentans.*

Overall, 315 banding sequences of the 63 species chosen were analyzed, as we took not all the available sequences of the arms in question, but only those met in all the populations studied with a dominating frequency. Thus, the chromosomal sequences analyzed at this stage represented species-specific and basic pattern sequences.

3. Phylogenetic tree was constructed through calculation of the matrix of pairwise distances. A strict calculation of inversion distances is an intractable problem; hence, various approximations are actually used, mostly based on calculation of the number of breakpoints or number of conservative fragments in the compared sequences S_1 and S_2. A breakpoint exists in S_1, if two elements are adjacent one another in S_1, but not in S_2 (similarly, for S_2). Evidently, the number of breakpoints in S_1 is equal to the number of breakpoints in S_2.

The measure we are calculating is tightly bound to the number of breakpoints and is based on the concept of “complexity decomposition” (Gusev *et al.*, 2001). This comparison exploits the idea of representing one of the sequences compared, for example, S_1, as a minimal set (concatenation) of the fragments from the other sequence S_2, so that the S_2 fragments may be used in both direct and reverse orientations. S_1 is covered with fragments from S_2 from left to right, as illustrated in Figure 2; the coverage is further designated as $H(S_1/S_2)$. Selection of the maximal (inextensible) fragment from S_2 at each step guarantees the minimal coverage. Let us designate the number of components in decomposition of S_1 with respect to S_2, that is, the number of fragments in $H(S_1/S_2)$, as $C(S_1/S_2)$. Similarly, decomposing S_2 with respect to S_1, we obtain $C(S_2/S_1)$. The smaller are the values $C(S_1/S_2)$ and $C(S_2/S_1)$, the closer are the texts analyzed. In the situation considered, a double decomposition is unnecessary, because equal lengths of the sequences and the absence of repeated elements give us $C(S_1/S_2) = C(S_2/S_1)$. In a general case, $C(S_1/S_2) \neq C(S_2/S_1)$, as the number of fragments in S_1 with respect to S_2 and in S_2 with respect to S_1 depends on the text lengths and compositions. In this case, the value of $C(S_1/S_2) = 1/2(C(S_1/S_2) + C(S_2/S_1))$ reflects similarity between the texts. The running time of calculation of $C(S_1/S_2)$ depends linearly on the sum of S_1 and S_2 lengths.

S_1:	a	b	c	d	e	f	g	h	i	j	k	l
S_2:	a	f	e	d	h	g	b	c	i	k	j	l
$H(S_1/S_2) =$	(a)	(b	c)	[d	e	f]	[g	h]	(i)	[j	k]	(l)
$H(S_2/S_1) =$	(a)	[f	e	d]	[h	g]	(b	c)	(i)	[k	j]	(l)

$$C(S_1/S_2) = C(S_2/S_1) = 7 = C(S_1, S_2)$$

Figure 2. An example of decomposition of the sequence S_1 with respect to S_2 and the sequence S_2 with respect to S_1.

4. An example of decomposition of the sequence S_1 with respect to S_2 and the sequence S_2 with respect to S_1 is shown in Figure 2: S_2 is obtained from S_1 by three inversions (first, **bcdef** is inverted, forming an intermediate sequence of **afedcbghijkl**; then, **cbgh**, and finally **jk**). In these decompositions, the components obtained by a direct copying are in parenthesis; by a symmetric, in square brackets. For example, the component (**bc**) in $H(S_1/S_2)$ is a direct inextensible copy of the analogous fragment from S_2, whereas the component [**def**] is copied from the fragment fed from S_2 by reading it in a reverse direction, that is, from right to left.

The package MEGA (Kumar *et al.*, 2001) was used to construct phylogenetic trees basing on the matrix of pairwise distances. Neighbor-joining (NJ) method (Saitou and Nei, 1987), based on a stepwise minimization of the sum of branch lengths of the tree reconstructed, was used for this purpose.

3. RESULTS AND DISCUSSION

1. The phylogenetic tree constructed using the set of banding sequences from five chromosome arms (ArmsACDEF tree) is shown in Figure 3. Topography of this tree displays numerous clusters, formed in part by sequences belonging to individual cytocomplexes—*pseudothummi* (pst), *camptochironomus* (cam), and *maturus* (mat). However, the largest cytocomplex *thummi* fails to form a single cluster—it comprises clusters of sequences belonging to various groups of sibling species (related species lacking morphological distinctions) typical of this cytocomplex, namely, gr. *plumosus* (plu), gr. *tenuistylus* (tnu), gr. *riihimaekiensis* (rii), gr. *aberratus* (abe), gr. *decorus* + gr. *obtusidens* (dec + obt), and gr. *pilicornis* (pil). In addition to the sequences within clusters of sibling species, certain sequences from the cytocomplex *thummi* lie separately or in pairs; they do not belong to sibling species. Evolution of these sequences is of special interest for further analysis of phylogenetic relationships both within the cytocomplex *thummi* and between cytocomplexes, because some of the sequences in question display certain relations with the cytocomplexes mat, cam, and pst.

When examining clusters formed by individual cytocomplexes, note several specific aspects. For example, the species *C. piger* and *C. riparius*, belonging to the cytocomplex *thummi* by the arm combination, cluster with pst. However, it is long since it was found that these species belong to *pseudothummi* according to the morphological, ecological, and molecular characteristics as well as banding pattern. (These species are noted by asterisk in Figure 3).

Interestingly, the cam cluster contains four species, three of which are known sibling species. However, we studied the fourth species, *C. setivalva*, just recently, and its phylogenetic relationships with the other species of this complex were detected for the first time in this work. Of the utmost importance is localization of a unique species *C. novosibiricus* near the cam cluster. We described this species recently and found that its uniqueness was in that its chromosome arm combination assigned it to the cytocomplex cam, whereas the banding morphology and pattern rather ascribes it to *thummi.* The phylogenetic tree demonstrates exactly this borderline position of the species in question between the two cytocomplexes cam and thu (Figure 3).

As for the clusters of banding sequences of sibling species in this tree, note two important aspects. First, only 10 species of the 12 forming the cluster plu (cytocomplex thu) are sibling species; *C. behningi* and *C. nuditarsis* do not belong to this group according to distinctions in their morphological characteristics, although displaying similar banding sequences in certain chromosome arms. Consequently, clustering of these two species with plu indicates their cytogenetic relatedness; however, in no way suggests their attribution to the gr. *plumosus*. Generally, this group of sibling species is very young and, consequently, retained many relationships with other members of the cytocomplex *thummi* (*C. annularius*, *C. anthracinus*, *C. cingulatus*, etc.).

Second, note also the cluster comprising sequences of two recently studied groups of sibling species—gr. *decorus* (Nearctic species) and gr. *obtusidens* (Palearctic species). So far, their putative close phylogenetic relationships were suggested from similarity in certain banding sequences and morphological traits of larvae (fluviatilis type of ventral tubules). The ArmsACDEF tree provides a solid confirmation for this hypothesis. In addition, this phylogenetic tree demonstrates that the gr. *riihimakiensis* actually exists.

2. The phylogenetic trees built using sequences of one of the five chromosome arms in question (A, C, D, E, and F trees) demonstrate that evolutions of banding sequences of individual arms were relatively independent (trees not shown). Although the main clusters evident in the overall phylogenetic tree—plu, abe, rii, dec, obt, pil, mat, cam, and pst—remain in the individual trees, their relative arrangement and size change. Note that single-arm trees detect a rather large group of species carrying homologous sequences in the same chromosome arms, whereas homology of the five arms, as detected by the ArmACDEF tree, is a singular situation. Presence of homologous (basic) sequences in chromosome arms of different species indicates their common origin and make their further analysis most intriguing.

3. Phylogenetic trees for individual chromosome arms A, E, and F of *Chironomus* species and for these three arms together were constructed earlier using either overlapping inversions (Keyl, 1962; Martin, 1979) or computer analysis (Wülker *et al.*, 1989; Shobanov and Zotov, 2001). The authors found certain association between the sequence evolution and evolution of species and cytocomplexes. Our data demonstrate a good correlation between evolution of sequences and divergence of many sibling groups as well as individual cytocomplexes.

On the other hand, we demonstrated that evolutions of banding sequences in individual arms were relatively independent, resulting in a specific patterns of clustering in each chromosome arm.

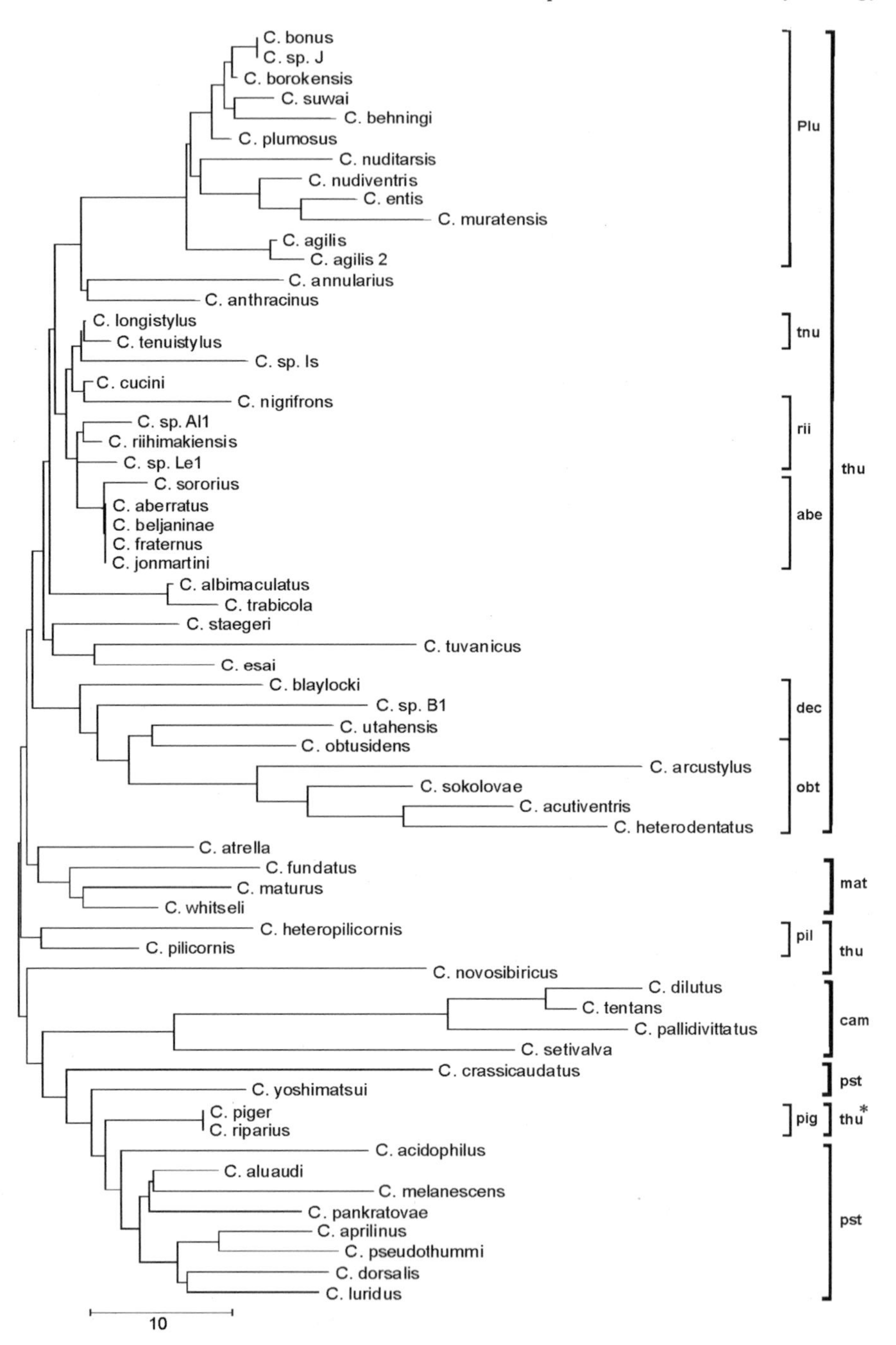

Figure 3. The ArmsACDEF phylogenetic tree of 63 Chironomus species based on banding sequences of five chromosome arms. Brackets show clusters of cytocomplexes (cam, mat, pst, and thu; rightmost) and groups of sibling species (abe, dec, obt, pig, pil, plu, sta, and tnu; to the right).

4. CONCLUSION

The described measure for detecting differences in banding sequences is a particular case of the measure we used for comparing DNA sequences (Gusev *et al.*, 2001), when four, not two, repeat types in S_1 and S_2 (direct, symmetric, direct complementary, and symmetric complementary) were considered. Note also that the measure used lacks any particular suggestions on the evolutionary mechanisms; therefore, it is applicable to other types of chromosome rearrangements and their combinations, not only to inversions.

The breakpoint analysis used in the work appeared very convenient and efficient in studying chromosome evolution basing on changes in banding pattern of polytene chromosomes. It is evident that this method will be equally efficient for analysis of inversions at the level of DNA.

ACKNOWLEDGEMENTS

Authors thank Prof. N.A. Kolchanov for initiation and incessant support of this work. The work was partially supported by the Russian Foundation for Basic Research (grant No. 03-07-96837); grant PCB RAS (No. 10.4); the Siberian Branch of the Russian Academy of Sciences (integration projects Nos. 145 and 119); Russian Federal Research Development Program Investigations and Development in Priority Directions of Science and Technology (contract No. 56/2001); INTAS (project No. 21-2382); Basic Research Programm of RAS (grant No. 12.4).

PART 4. SYSTEM COMPUTATIONAL BIOLOGY

INTEGRATIVE ANALYSIS OF GENE NETWORKS USING DYNAMIC PROCESS PATTERN MODELLING

A. FREIER[1]*, M. LANGE[2], R. HOFESTAEDT[1]
[1]Bioinformatics Group, Faculty of Technology, Bielefeld University, Germany,
e-mail: afreier.or.hofestae@techfak.uni-bielefeld.de; [2]Plant Genome and Resource Center,
IPK Gatersleben, Germany, e-mail: mlange@ipk-gatersleben.de
* *Corresponding author*

Abstract In this article, a novel object-oriented modeling approach in the field of biochemical network modeling is presented. Molecular objects are modeled conceptually using object classes, internally based on the standard object models Java and ODMG. Objects and object networks are composed automatically using data integration. In combination with that, a specific view concept based on access paths has been implemented to model biochemical processes from integrated databases directly. Together with the application of graphical methods, networks are computed by the system. Each step of the workflow can be executed using a server and a graphical interface implemented with Java.

Key words: network modeling, data integration, database views, object databases

1. INTRODUCTION

Our goal is the computational construction and analysis of gene controlled metabolic networks (Russel, 1996). Biochemical processes involve a huge number of interconnected nodes, whereat efficient methods and tools are needed to support the access to different data sources, to integrate data retrieved from these sources and to apply common analysis methods, e.g. graph theory, mathematical simulation, and visualization. This will give us an overview of topology and dynamics of cellular process networks and the occurrence of biochemical objects, e.g. the state of metabolic systems under the conditions of metabolic diseases.

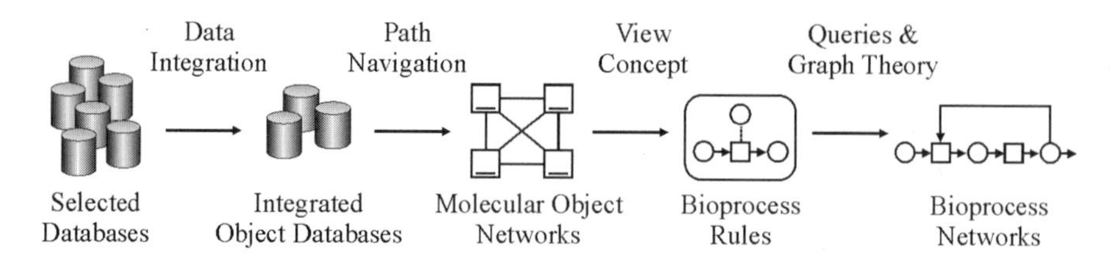

Figure 1. Process pattern modeling: mining biochemical networks from integrated data.

Today, more and more internet databases are published providing selected molecular data concerning metabolism, gene networks and their application (Baxevanis, 2001). Actually, several systems for molecular database integration (Etzold *et al.*, 1996; Stevens *et al.*, 2000; Haas *et al.*, 2001; Freier *et al.*, 2002) are already used to get efficient access to distributed databases. At the same time, information systems for modeling and visualization of regulative molecular networks are presented (Kolchanov *et al.*, 1998; Waugh, 2000; Glass and Gierl, 2002; Goesmann *et al.*, 2002). Still, even systems specialized at the same topic (e.g. gene networks) show differences in data modeling and in information content. Actually, existing systems implement object services providing a previously defined object structure, where methods of application's exclusively are specialized to. Thus, a processing of user-specific objects is not possible. The main idea of the iUDB (Individually Integrated Molecular Databases) system is to provide a data independent toolbox for object database implementation, data integration, network modeling and analysis in application to genome data. Figure 1 shows the workflow accomplished by the system.

2. METHODS AND ALGORITHMS

Three different models have been combined in our approach. In association with KDD (Knowledge Discovery in Databases) (Han, 1999) the initial step in our workflow starts with preprocessing the input data. To integrate data into our object database, we prefer the relational database model. iUDB has been enabled to access any data source, for which a JDBC (Java Database Connectivity) driver is available for. Like that, the data integration mechanism becomes independent from its content.

For computational modeling of our biological data types, we are using the object-oriented modeling paradigm. As we know, common methods (e.g. UML) and tools from software development are already available here. The user himself can select typical concepts including object classes, complex data types, inheritance, and object references modeling the object type. Note that a network is described, if at least each object class refers to or is referred by other object classes. In the domain of gene networks, we are modeling, e.g. "Enzyme", "Pathway", and "Gene" classes. The object specification will be internally mapped to an ODMG database (The object database..., 1997).

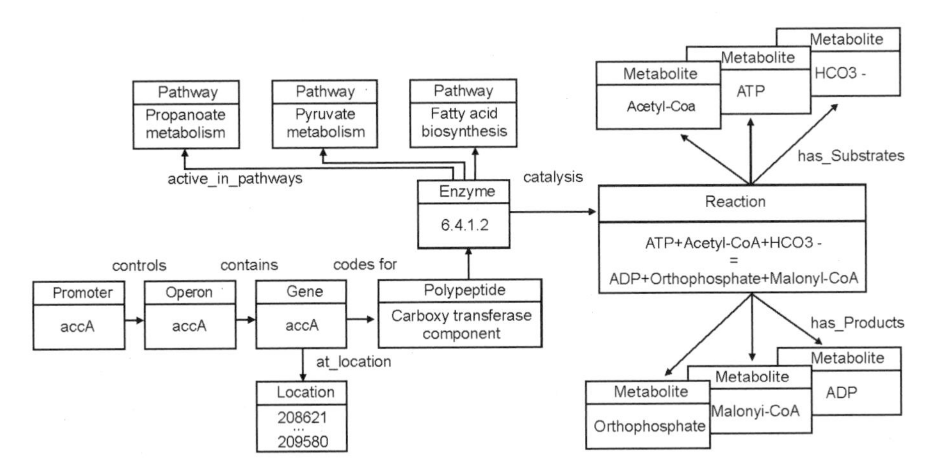

Figure 2. Example: molecular object network.

However, by adding classes to the scheme, we include all structural information needed for the analysis of cell processes, which are bioprocesses (e.g. metabolic reactions), knowledge (relationships and logical conclusions between objects) as well as the physiology of biological structures (e.g. sequences, cells and compartments). An example of object networks is shown in Figure 2. According to this example, the classes "Pathway", "Enzyme", "Reaction", "Metabolite", "Promoter", "Operon", "Gene", "Polypeptide", and "Location" have been modeled by the user. The interestingness of the approach is the fact, that all objects and their references, e.g. all things shown in the figure should be entered by the system automatically using, e.g. data integration. What we can do now is to navigate through access paths $p = (c, \{a_1, a_2, \ldots, a_n\})$. Starting at "Promoter", we can use the path (Promoter, {controls, contains, is_coding_for, is_active_in}) to retrieve all Pathways influenced by a given promoter.

In order to model bioprocesses, a third model is used. In the center of the approach, there is a rule based production system consisting of rules $r = (S, P, I, L)$ used, where r itself represents the topology of a bioprocess, containing four elements. The sets S and P contain input and output objects, I is a set of influences and L specifies the location of the process described. To obtain these rules from a database, a specific view concept has been developed. Starting at a defined root class, each rule pattern (or rule type) t access paths to each class, where S, P, I, and L should be mined from.

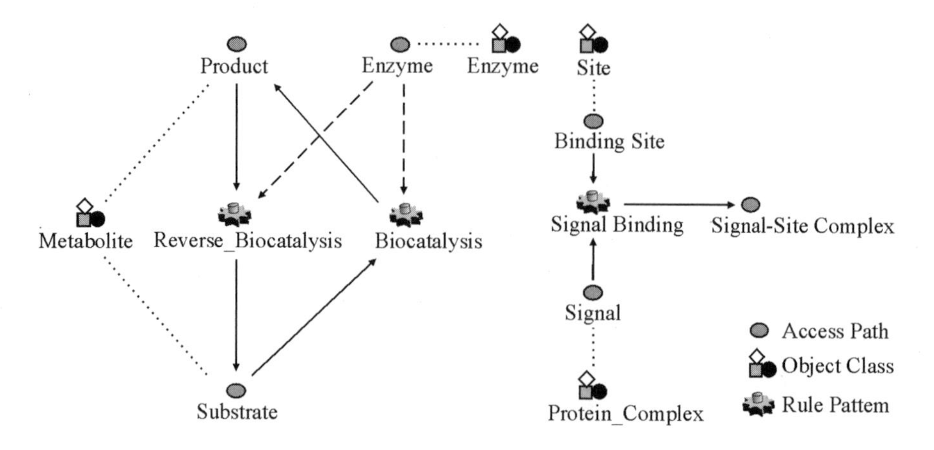

Figure 3. Bioprocess pattern examples.

A model $M = (T, R, N)$ holds the recent information, where T is a set of patterns, R is a set of rules mined by application of T against the database and N are the networks found by rule clustering (e.g. by application of graph-based searches). In Figure 3, two examples are displayed.

The left part of the Figure shows two rule types modeling metabolic reactions. "Bio-Catalysis" rules transform "Substrate" into "Product", influenced by "Enzyme", whereby "Substrate" and "Product" are different access paths to the "Metabolite" class and "Enzyme" is an access path to the "Enzyme" class. Note that the path specification is not displayed in the Figure. The rule type "Reverse Bio-Catalysis" describes reversible reactions, meaning products are transformed into substrates.

The next difference is selective. The user specifies class properties, in which dependency rules should be created or not. For the current example this property could be, e.g. a hypothetical class attribute "reversibility", set to values similar to "reversible" or "not reversible". In the second example in Figure 3, a signal binding rules are modeled. Access paths lead to all object classes involved in the process (Protein Complex and Binding Site). Be stating that an aggregation of both classes is not part of the database scheme, the output of the process cannot be specified directly. In short, the idea is to aggregate classes with each other.

This means for our example, that the access path "Signal-Site Complex" is composed by the two access paths "Signal" and "Binding Site". The application of both paths will be the elements of the aggregation. It is obvious that in our approach we finally combine and not merge different standard models for different applications.

3. RESULTS AND DISCUSSION

The main result of our work is the implementation of the models and methods discussed in section 2 in the iUDB server program. In Figure 4 an overview of the system is given.

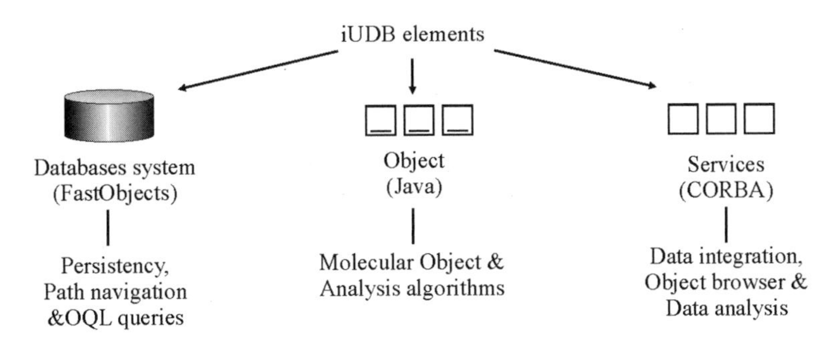

Figure 4. iUDB system elements and their implementation.

All object classes and algorithms have been implemented in Java. Platform independent object access, analysis methods (e.g. advanced queries), and data integration capabilities are provided as CORBA (OMG, 1996) services. Internally, we are using an ODMG conform object database (FastObjects), providing Persistency, Path navigation and OQL (Object Query Language) database queries. The graphical user interface of iUDB has been implemented in Java Swing. According to the tasks in Figure 1, it contains modules for data source management, object integration, bioprocess views and network analysis.

3.1 Data preprocessing

Data preprocessing includes the classification (Baxevanis, 2001) and management of all available data sources. iUDB contains a module for browsing data sources and data source management. In Figure 5, we can see the class editor, where networks of object classes can be modeled interactively. Also, there is an integration dialog, which enables the user to construct complex objects from a prepared list of data sources. For each object attribute mappings to different data sources, e.g. BRENDA (Schomburg *et al.*, 1999), can be defined. Furthermore, objects can be reengineered from data source tables. However, we see the advantage of the object-oriented approach here: to obtain complex objects, complex join operation must be carried out by standard database systems. At the same time, the number of elements in the class diagram is much less than, e.g. in the same scheme modeled with relational databases. Summarized, the result of data preprocessing are integrated object databases with a user-defined data structure.

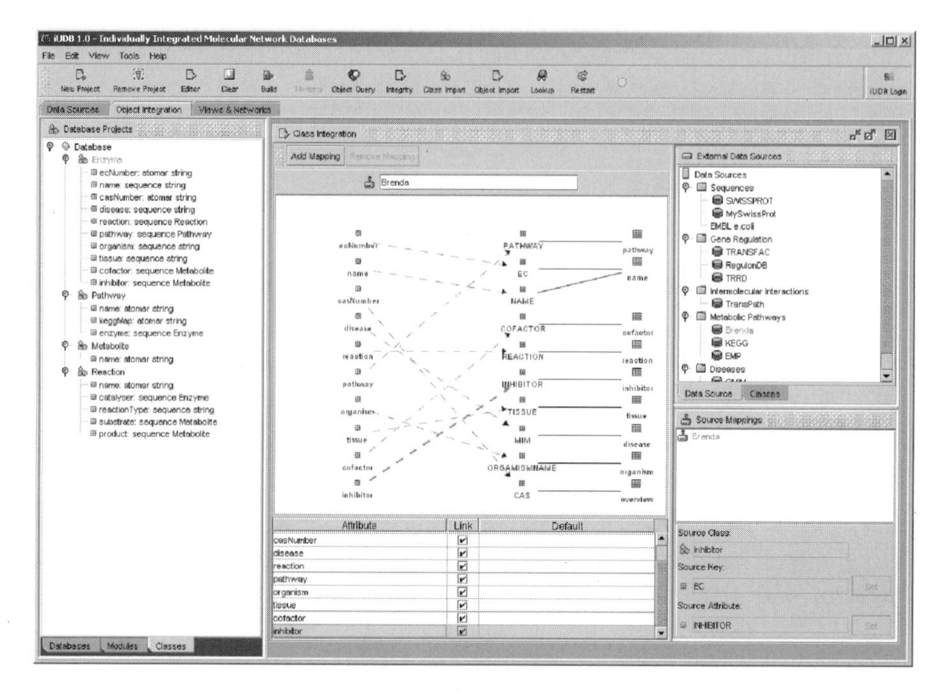

Figure 5. Data integration: retrieving complex objects from relational database sources.

3.2 Bioprocess views

As we have seen in section 2, the combination of object approach and rule system is useful to "animate" the object networks integrated in our database. The conceptual design of biochemical networks using our view concept dramatically reduces modeling time, compared to create processes or transformation scripts, e.g. relational database views, by hand. The concept of access paths exhaustingly uses database references and could be implemented alternatively an enumeration of joins in SQL databases. Because of the lower complexity of operations and first measurements, we expect a performance lead over standard SQL databases.

For efficiency, we snapshot rule views defined by the user by adding bioprocess rules to the database directly and interconnect them with the related objects. So far, the resulting data is a bipartite graph consisting of objects of the object network, interconnected additionally by bioprocess rules. To improve query performance, we interlink objects and rules in a bi-directional way. In Figure 6, we can see the view editor module enabling the user to design bioprocess views interactively.

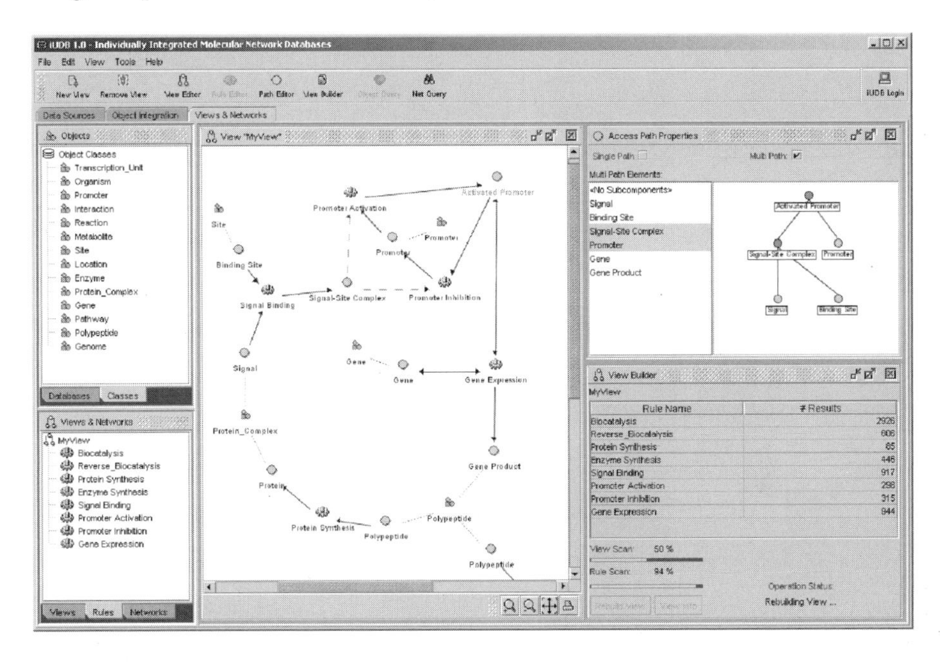

Figure 6. Modeling networks of bioprocess patterns using a specific view concept.

3.3 Network analysis

The analysis of integrated data using bioprocess patterns generates rules R for all patterns found in the database. They can be interconnected transitively as, e.g. pathways, by matching the identity (OID) of database objects they refer to. All rules of a network or pathway found are a subset of the rule set R. Networks can be stored in the system and loaded as a pre-processed network later. Actually, we implemented the following analysis methods: transitive closure (object environment); search for pathways between given objects; computation of object interactivity profiles; traversal of reaction paths; typical set operations, e.g. union, difference and intersection which are necessary for network comparison. Here the capabilities of the underlying ODMG compliant object database system is used. Besides, the interactive pathway construction is possible and enhanced graphical methods will be implemented in the future development. Figure 7 shows the application of the system to the Glycolysis pathway. For each object in the database, the user can select consuming and producing rules to add them to the network.

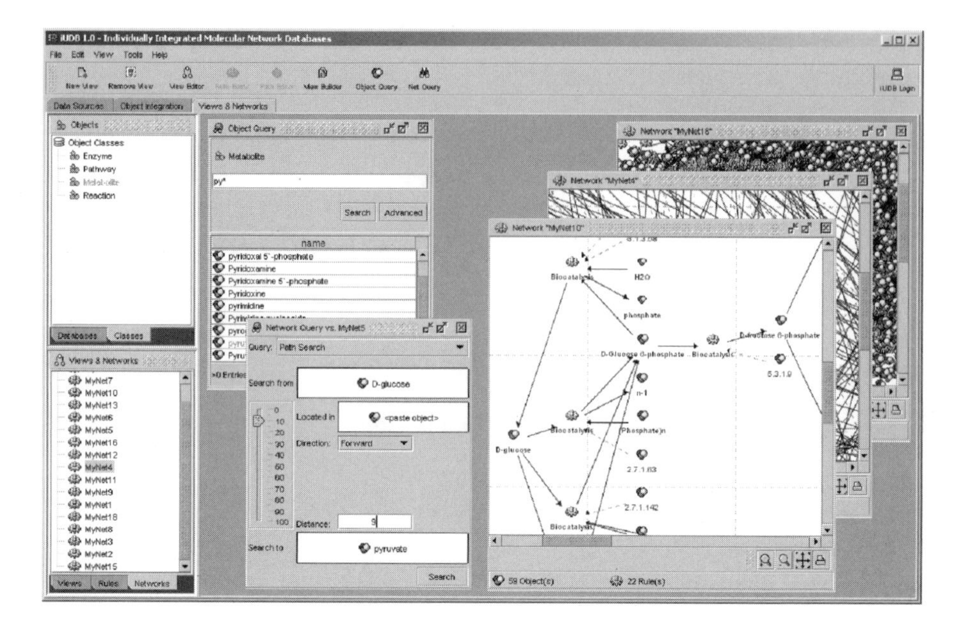

Figure 7. Applied graph theory: interactive analysis of biochemical objects and networks.

4. DISCUSSION

The idea of database integration applied to network analysis is not a new one. Moreover, it is a necessary task. Still, the growing number and the evolution of molecular databases demands adaptive systems able to integrate new data sources, build new integrated databases, and apply suitable visualisation tools dynamically. With iUDB, we have developed a novel system supporting the object-oriented modelling and analysis of gene network and metabolic data. For data modelling, we used standard object models, which have been established in software development and already applied in bioinformatics. Finally, the result is the interactive and automatic computation of bioprocess networks and pathways. In the near future, we will to continue the workflow with specialised applications, e.g. graphical analysis, visualisation and simulation. The direct transformation of networks computed by iUDB into Petri-nets is an example. Actually, a first version of the system is available under the URL: http://tunicata.techfak.uni-bielefeld.de.

ACKNOWLEDGMENTS

This work is supported by the German Research Council (DFG).

DEVELOPMENT AND ANALYSIS OF MODELS OF GENETIC AND METABOLIC NETWORKS AND SIGNAL TRANSDUCTION PATHWAYS IN THE GENENET SYSTEM

E.A. ANANKO[1,3]*, K.A. LOKTEV[1], N.L. PODKOLODNY[1,2,3,4]

[1]*Institute of Cytology & Genetics, Siberian Branch of the Russian Academy of Sciences, prosp. Lavrentieva 10, Novosibirsk, 630090 Russia, e-mail: eananko@bionet.nsc.ru;* [2]*Institute of Computational Mathematics & Mathematical Geophysics, Siberian Branch of the Russian Academy of Sciences, prosp. Lavrentieva 6, Novosibirsk, 630090 Russia, e-mail: pnl@bionet.nsc.ru;* [3]*Novosibirsk State University, ul. Pirogova 2, Novosibirsk, 630090 Russia;* [4]*Ugra Research Institute of Information Technologies, ul. Mira 151, Khanty-Mansyisk, 628011 Russia*

* *Corresponding author*

Abstract The GeneNet system was designed to construct and analyze information models of complex molecular genetic systems. The GeneNet system allows manifold data on the structure–function organization of gene and metabolic networks and signal transduction pathways to be accumulated in the context of a unified approach and to be visualized as interactive graphs. The GeneNet data are used to model the network dynamics and search for optimal means for control and correction of gene network behavior. The GeneNet system now contains descriptions of 37 gene networks regulating various functions of organisms. The information stored in the system was collected by annotating experimental publications. The academic version of GeneNet is available at http:// wwwmgs.bionet.nsc.ru/mgs/gnw/genenetworks.shtml.

Key words: gene networks, signal transduction pathways, metabolic pathways, regulation of gene expression, visualization

1. INTRODUCTION

In recent years, particular attention has been paid to the functions of complex molecular genetic systems and their computer reconstruction and analysis. This interest brought about numerous databases that accumulate data not only on genes and proteins, but also on various relationships between complex biological molecules. For example, the databases of metabolic networks are KEGG (Kanehisa and Goto, 2002), WIT (Overbeek *et al.*, 2000), and EcoCyc (Karp *et al.*, 2002); the databases of gene networks and regulatory pathways are RegulonDB (Salgado *et al.*, 2001), BIND (Bader *et al.*, 2001), InterActiveFly (Brody, 1999) SPAD (http:// www.grt.kyushu-u.ac.jp/spad/), CSNDB (Takai-Igarashi and Kaminuma, 1999), HOX Pro (Spirov *et al.*, 2002), TRANSPATH (Krull *et al.*, 2003), BioCarta (http://www.biocarta.com/), etc. Most of these databases deal with individual biological objects or particular aspects of body activities. Therefore, it is quite difficult to reconstruct a comprehensive model of cell operation, to say nothing about the operation of the entire organism, using these scattered data.

We developed a unique technology that allows for searching for and accumulating manifold data on gene and metabolic networks, signal transduction pathways, and other complex molecular genetic systems. The information is then used for computer reconstruction of the maximally possible integral patterns of operation these systems under various conditions. This technology has been implemented in the GeneNet system (Ananko *et al.*, 2002; Kolchanov *et al.*, 2002a).

Specific of the GeneNet system is that it deals with complex gene networks, each of them regulating a particular biological function (Kolchanov *et al.*, 2000; 2002c). Using a unified approach implemented in the GeneNet system, manifold data on any components of such systems and relationships between these components of any complexity level can be organized as formalized descriptions (Kolpakov *et al.*, 1998). GeneNet allows for viewing the structure–function organization of gene networks in a graphic representation. The data collected by annotating scientific papers are further used to construct dynamic models and search for optimal means of control and correction of gene network behavior. In addition, GeneNet provides for logical analysis of the information stored.

Now, GeneNet contains descriptions of the structure–function organization of 37 gene networks regulating various biological processes, including cell cycle, some types of homeostasis in the body, tissue and organ morphogenesis, response of the organism to external effects, etc.

The academic version of GeneNet is available at http:// wwwmgs.bionet.nsc.ru/mgs/gnw/genenetworks.shtml.

2. THE BASICS OF GENENET

It is well known that operation of any gene network regulating a particular activity of the entire body involves many components, namely, genes that constitute the core of a network, proteins that perform structural, transportation, enzymatic, and regulatory functions, and low-molecular-weight components, such as metabolites, signal molecules, and energy-connected cell components (Kolchanov *et al.*, 2000). The interactions between these components are quite complex. Using an object-oriented approach (Booch, 1991), we divided the gene network components into the elementary structures and elementary processes. The elementary structures include several classes of objects: (1) genes, (2) proteins and protein complexes, (3) RNAs, and (4) small molecules. The elementary processes include two types of interactions: (1) reactions and (2) regulatory events (Kolpakov *et al.*, 1998). Being a highly flexible system, GeneNet allows users to add easily new classes of objects. Each class of objects is described in a separate GeneNet database table and characterized with a set of parameters specific of this class (Ananko *et al.*, 2002). This set of parameters can also be extended if necessary. Some properties of biological objects and reactions are determined only for a specific gene network they belong to.

3. STRUCTURE OF THE GENENET SYSTEM

GeneNet contains the following units:

1. An application server that provides connections between all the system components.
2. A database (Ananko *et al.*, 2002) that stores information on biological objects and reactions and the system data related to the operation of the server and gene network editor. A relational version of the GeneNet database is implemented in the Oracle9i environment. To access the GeneNet database using a standard web browser, an SRS (Sequence Retrieval System) portal was developed. The current release of the GeneNet SRS version of June 1, 2003 comprises 16 interlinked tables and over 17 000 entries, including descriptions of more than 1000 genes, 1600 proteins, and 3500 interactions between gene network components. The information stored in the database was collected by annotating more than 2000 scientific publications.
3. The GeneEd graphic editor for creating, visualizing, and editing graphic representations of gene networks. GeneEd is also used to create new objects in the database, edit the properties of the existing objects, support various types of decomposition of complex gene networks, and provide

authorized access to the database (Loktev *et al.*, 2002). In GeneEd, gene networks can be represented in two ways: as a hierarchical tree of relations between objects and as a two-dimensional graph. The hierarchical tree of elements (Figure 1*a*) allows a user to manipulate multicomponent objects, i.e. create, copy, delete, remove, and select objects. This hierarchical tree provides a clear picture of hierarchical interrelations between objects in complex systems. The friendly graphic representation of a gene network (Figure 1*b*) allows users to work interactively with the program. In these graphs, all graphic elements strictly correspond to specific objects and depend on the object attributes as defined in the database. The editor allows users to create schemes of any complexity. In addition, GeneEd allows for creating nested hierarchical schemata, i.e. when a detailed graph of a system component is unavailable at a particular level but can be viewed as separate subschemata (Figure 1*c*).

4. GN_Viewer is a client application that allows users to view graphic representations of gene networks in the Internet using a standard web browser. GN_Viewer also supports several types of decomposition for gene networks. For example, a user can select the data obtained for a particular organism, cell type, or under specific external effects.
5. A unit for logical analysis that allows users to construct gene networks on query to database and displays the results in graphics. For example, one can construct a gene network that includes elementary objects and interactions located in a specified neighborhood of a selected object. Figure 2*a* shows an example of the human protein HSF-1 gene network constructed on query to database. It is also possible to construct a gene network showing a path from one specified element to another (Figure 2*b*). In addition, GeneNet has the option to search for closed cycles in a graph and analyze the graph connectedness.

4. CERTAIN SPECIFIC FEATURES OF THE STRUCTURE–FUNCTION ORGANIZATION OF GENE NETWORKS

Logical analysis of the GeneNet data revealed certain specific features of the structure–function organization of gene networks.

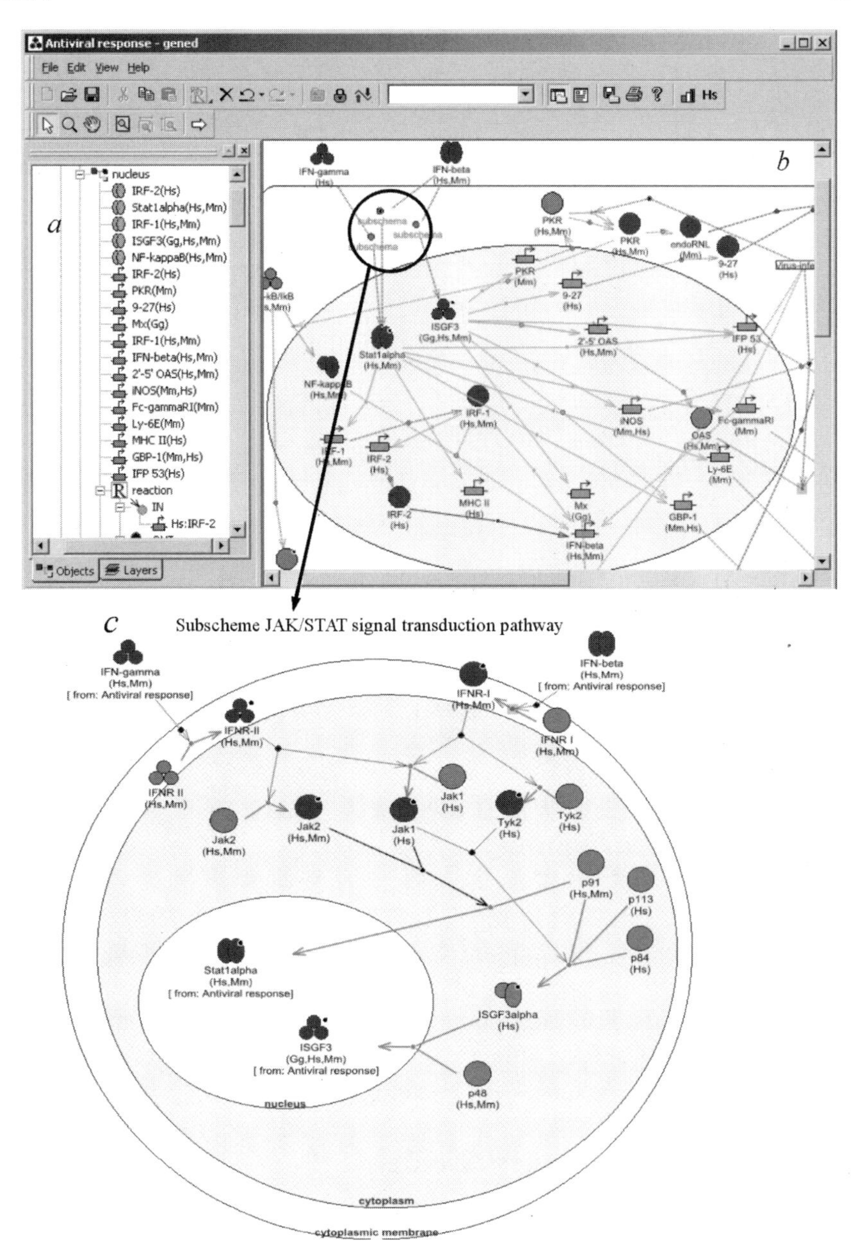

Figure 1. Two ways of a gene network representation in the GeneNet system. The example is a fragment of a gene network of the antiviral response: (*a*) a fragment of the hierarchical tree of interrelations between objects with the expanded list of objects localized to the cell nucleus; (*b*) a fragment of a two-dimensional graph of the structure–function organization; and (*c*) activation of the subscheme JAK/STAT signal transduction pathway by clicking the special-type indirect reaction subscheme. The central circle is the cell nucleus; peripheral circle, the cytoplasm. In the schemes, small dark circles denote proteins; rectangles, genes; and arrows, interactions between the gene network components.

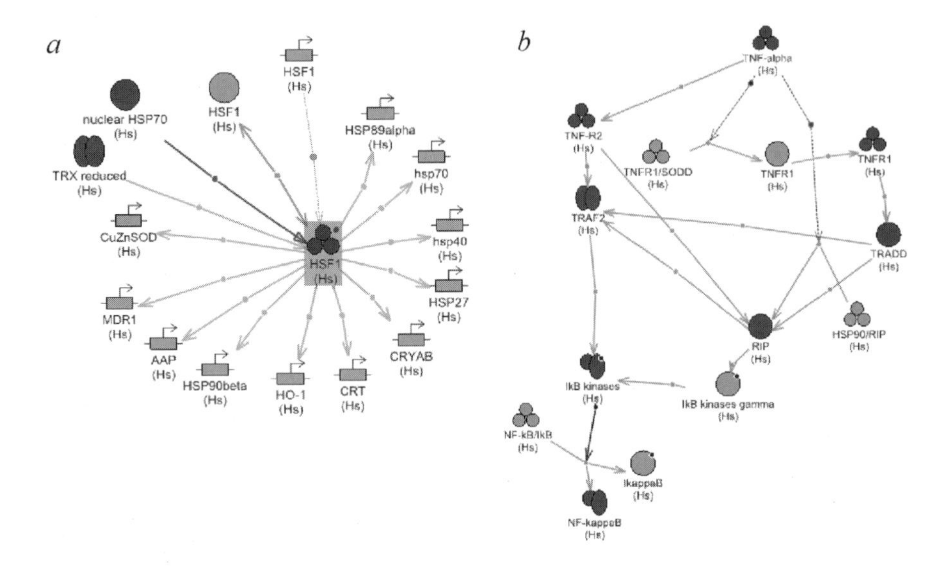

Figure 2. Examples of logical analysis within the GeneNet system: (*a*) construction of gene network involving all the components related to human HSF-1 protein (with a specified neighborhood amounting to 1); and (*b*) the gene network reflecting the pathway connecting human TNF-α and NF-κB proteins. Circles indicate proteins; square boxes, genes; and arrows, interactions between components of the gene network.

4.1 Key regulators and a cassette pattern of gene activation

Each gene network has key regulators (Kolchanov *et al.*, 2000). As a rule, the key regulators are transcription factors that activate a cassette of genes responsible for optimal operation of this network. For example, in the gene network of erythrocyte differentiation and maturation, this key regulator is the transcription factor GATA-1 (Figure 3*a*); in the gene network of the antiviral response, transcription factors STAT1 and ISGF3 (Figure 3*b*).

4.2 Closed regulatory circuits

Closed regulatory circuits are necessary components of any gene network. The gene networks regulating homeostasis of physiological processes in the body typically have regulatory circuits with a negative feedback. For example, in a gene network regulating the intracellular cholesterol concentration, the transcription factor SREBP is activated at low cholesterol concentrations. SREBP induces expression not only of the genes of enzymes involved in cholesterol synthesis, but also of the genes responsible for cholesterol uptake through a cell membrane.

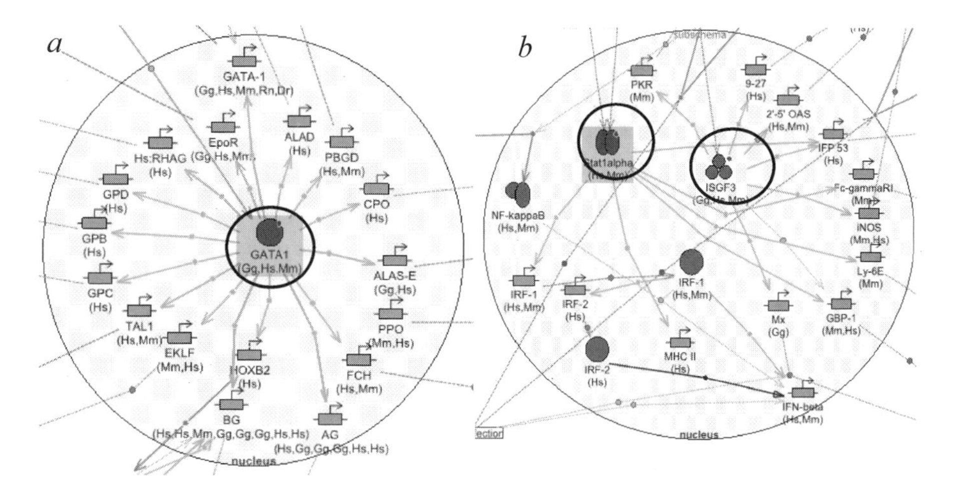

Figure 3. Examples of key regulators of gene networks: (*a*) GATA-1 is the key regulator of the gene network of erythrocyte differentiation and maturation; and (*b*) STAT1 and ISGF3 are the key regulators of the gene network of antiviral response. In the scheme, the circles denote proteins; rectangles, genes; and arrows, interactions between the gene network components.

At high cholesterol concentrations, the negative feedback inhibits activation of SREBP and, hence, both the synthesis of cholesterol and its uptake from extracellular space decrease drastically (Figure 4*a*).

Positive feedback circuits are typical of fast and irreversible processes, for example, cell differentiation or morphogenesis. Positive feedbacks also provide the cascade pattern of gene activation. For example, in the gene network of erythrocyte differentiation and maturation, the positive feedback circuit provides self-activation of a gene encoding GATA-1 protein, the key transcription factor of this network (Figure 4*b*).

5. PROSPECTS FOR GENENET DEVELOPMENT

In future, particular attention will be paid to the logical analysis unit in order to study the gene networks more efficiently at various levels: at a molecular genetic level, in the cell, and the entire body.

Now, a new standard for descriptions of elementary interactions is being developed. Several new features will allow users to input qualitative and quantitative data on dynamics of the processes and to generate automatically the specifications of gene network models in an XML-like format to load subsequently the gene network dynamics into the modeling system. It is planned to integrate more tightly the TRRD (Kolchanov *et al.*, 2002b) and GeneNet databases (Ananko *et al.*, 2002).

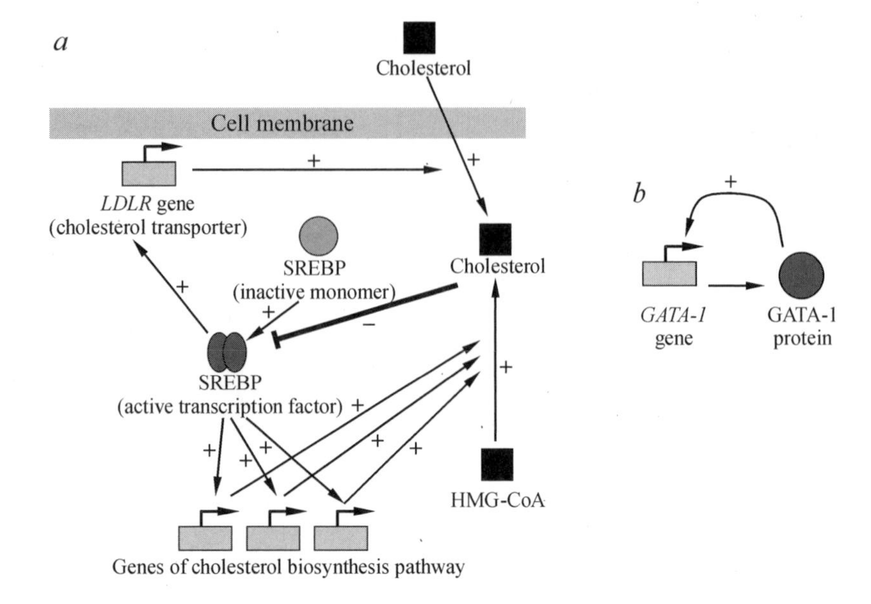

Figure 4. Closed regulatory circuits: (*a*) negative feedback regulatory circuits responsible for maintenance of intracellular cholesterol homeostasis; and (*b*) a positive feedback regulatory circuit that provides self-activation of *GATA-1* gene, coding for the key transcription factor of the gene network of erythrocyte differentiation and maturation.

ACKNOWLEDGMENTS

The authors thank N.A. Kolchanov for fruitful discussions; E.V Ignatieva, O.A. Podkolodnaya, and I.S. Stepanenko for checking the information; Yu.P. Guryeva, A.N. Kudryavtseva, N.S. Logvinenko, E.A. Nedosekina, A.L. Proscura, I.I. Turnaev, and V.V. Suslov for annotating the publications; I.V. Lokhova for bibliographical support; D.A. Grigorovich for system administration; I.V. Filippova and G.B. Chirikova for translation of the paper into English. The work was supported in part by the Russian Foundation for Basic Research (grants Nos. 01-07-90376-в, 02-07-90359, 03-07-96833-p2003, 03-07-96833, 03-04-48506-a, 03-07-06077-мас, 03-01-00328); Russian Ministry of Industry, Science, and Technologies (grant No. 43.073.1.1.1501); the Siberian Branch of the Russian Academy of Sciences (integration projects Nos. 119 and 10.4); NATO (grant No. LST.CLG.979816).

EXTENSION OF CELL CYCLE GENE NETWORK DESCRIPTION BASED ON PREDICTION OF POTENTIAL BINDING SITES FOR E2F TRANSCRIPTION FACTOR

I.I. TURNAEV[1]*, D.Yu. OSHCHEPKOV[1,2], O.A. PODKOLODNAYA[1,2]
[1]Institute of Cytology & Genetics, Siberian Branch of the Russian Academy of Sciences, prosp. Lavrentieva 10, Novosibirsk, 630090 Russia, e-mail: turn@bionet.nsc.ru;[2]Novosibirsk State University, ul. Pirogova 2, Novosibirsk, 630090 Russia
* *Corresponding author*

Abstract Annotation of scientific papers using the GeneNet technology allowed for reconstruction of the gene network regulating the cell cycle (the G1/S transition). To extend the network, we searched for potential sites for E2F transcription factor binding in promoter regions of genes whose expression depends on the cell cycle phase, however, any experimental evidence for the presence of E2F binding sites is lacking. New potential target genes for E2F factor were found. Products of these genes are involved in DNA replication and cell cycle regulation as well as in chromatin compacting, repair, and apoptosis. Combination of GeneNet-compiled data with the results of prediction of new potential E2F sites in regulatory regions enriched the reconstructed cell cycle gene network with new regulatory linkages.

Key words: Cell cycle regulation, site recognition, E2F sites, gene network

1. INTRODUCTION

Cell cycle is one of the basic processes supporting vital functions of organisms. Entering into the cell cycle and its progression depend on expression and activation of the cell cycle genes, which are arranged in a composite and highly integrated network (Kohn, 1999). Activity of E2F transcription factor (TF) plays a key role in this gene network. E2F TF binding sites were detected in promoters of genes whose products are essential for nucleotide synthesis, DNA replication, and progression of the cell cycle (Wells *et al.*, 2000). Transcription of the genes regulated by E2F factor increases in the G1 phase of cell cycle, reaches its maximum in the late G1 or S phases, and decreases in the G2 and M phases (DeGregori *et al.*, 1995; Muller *et al.*, 1997). This dynamics is mediated by the activity of E2F TF, which binds to its sites in promoters of the corresponding genes and regulates their transcription with respect to the cell cycle phases. The most complete detection of the target genes for E2F TF may contribute to a more complete description of the processes regulating the cell cycle progress.

For this purpose, we reconstructed in this work the gene network of cell cycle regulation (G1/S transition) by the computer technology GeneNet (Kolchanov *et al.*, 2002). To extend this network, we searched for potential binding sites of E2F TF in promoter regions of the genes whose expression depends on the cell cycle phase, however, any experimental evidence for the presence of E2F binding sites is lacking. New potential target genes for E2F factor were found. Products of these genes are involved in DNA replication and cell cycle regulation as well as in chromatin compacting, repair, and apoptosis. Combination of GeneNet-compiled data with the results of prediction of new potential E2F sites in regulatory regions enriched the reconstructed cell cycle regulation gene network with new regulatory linkages.

2. MATERIAL AND METHODS

The gene network of cell cycle regulation (G0/G1-S) was reconstructed by the GeneNet technology (Kolchanov *et al.*, 2002). This approach allows for description of the structure–function arrangement of gene networks with a unified format data representation. For the description of gene networks, an object-oriented approach is used, in which all the network components may be divided into two groups: (i) elementary structures (such as genes, proteins, mRNAs, and metabolites) and (ii) elementary events or interactions between elementary structures (such as reactions and regulatory events).

The section Cell Cycle G0/G1-S of GeneNet database was developed by annotation of more than 100 scientific papers and contains description of the gene network regulating cell transition from the G0/G1 to S phase of the cell cycle. This section contains information on 162 objects (41 genes, 18 mRNAs, 101 proteins, 23 transcription factors, and 2 external signals), 200 linkages and regulatory events, and 6 processes. The information with references to the literature sources is accessible at the address http://wwwmgs.bionet.nsc.ru/mgs/gnw/genenetworks.shtml/. The E2F sites were recognized by the SITECON method (Oshchepkov *et al.*, this issue). The method is based on determination of conservative context-dependent conformational and physicochemical properties of short DNA sequences and uses the published data on 38 conformational and physicochemical properties of dinucleotides (http://wwwmgs.bionet.nsc.ru/mgs/gnw/bdna/).

3. RESULTS

3.1 Cell cycle regulation gene network (G1/S transition): reconstruction in the GeneNet database

Analysis of elements and linkages of this gene network revealed some *characteristic features of the cell cycle regulation gene network* that are common to the gene networks (Kolchanov *et al.*, 2002):

(i) Presence of inputs for reception of external signals from both activators and repressors, and their signal transduction pathways. In the case of cell cycle regulation gene network, these are the signals from mitogens (for example, PDGF, EGF, or IGF) acting via the corresponding signal transduction pathways, in particular, through the MAP-kinase cascade, one of the main pathways for activation of cell proliferation (Figure 1; Lukas *et al.*, 1996; Cook *et al.*, 1999), and the signals from antimitogens (for example, TGF-beta; Florenes *et al.*, 1996; Hu *et al.*, 1999).

(ii) Presence of the central transcription factor E2F/Dp and a cassette of genes controlled by this factor. In a quiescent cell, transcription of the genes from this cassette is directly repressed by E2F/Dp/pRB TF (Figure 2*a*, *c*).

GFs act on a cell by the Gfs/MAPK cascade/AP1/cyclin D pathway. As a result activities of cyclin D/cdk4,6 and later, cyclin E/cdk2 kinases are increased (Figure 1; Lukas *et al.*, 1996). These kinases phosphorylate pRB in two steps; the hyperphosphorylated pRB loses its ability to bind to E2F1/DP1; and thereby, the derepressed E2F1/DP1 complex starts activating transcription of the genes it regulates (Figure 2*b*, *c*; Muller *et al.*, 1997; Dyson, 1998; Lundberg, Weinberg, 1998). Therefore, the repression of the gene cassette under consideration by E2F/Dp/RB TF changes for the activation of this cassette by E2F/Dp TF; accordingly, the cell transits from G0 to the cell cycle.

(iii) Regulatory circuits with feedbacks, both positive and negative, providing sustainable operation of the gene network. In the case of cell cycle regulation gene network, these are, first, conjugated regulatory circuits with a positive feedback (Muller *et al.*, 1997; Dyson, 1998; Lundberg and Weinberg, 1998), each of them contributing to the enhancement of E2F1/DP1 TF activity and, thus, to the enhancement of transcription of the gene cassette it controls. Second, these are the circuits with a negative feedback inactivating transcription of the gene cassette controlled by E2F/DP TF (Figures 3 and 5; Xu *et al.*, 1994). Mutually coordinated operation of these circuits provides a fast transcription activation (G1/S) and repression (G2, M) of the cell cycle genes.

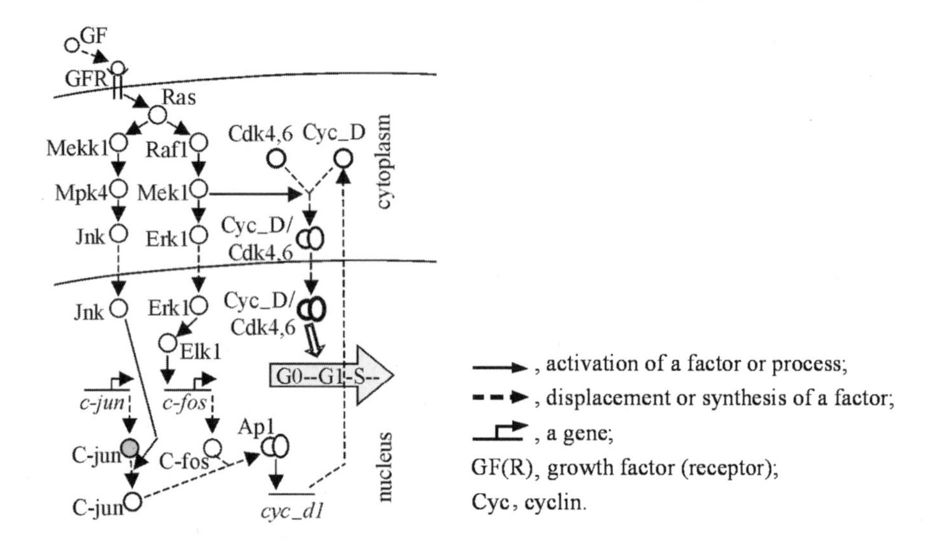

Figure 1. Activation of cyclin D1 by growth factors via the MAPK cascade.

Operation of the cell cycle regulation gene network is sustained by the following processes:

(i) Regulation of gene transcription. The first step in cell transition from G0 to G1 is the activation cyclin D gene transcription by AP1 TF via the GFs/MAPK cascade/AP1/cyclin D pathway (Figure 1). Further activation of cyclin D/cdk4,6 kinase (in the G1 phase) increases the activity of E2F1/Dp1 TF and, thus, enhances transcription of the genes of cyclins D and E and the E2F itself within the gene cassette activated by this TF (Figures 2*b* and 3). Further activation of cyclin E/cdk2 results in attaining the maximum activity of E2F1/DP1 TF (G1/S transition) and, respectively, in transcription of the gene cassette activated by this factor (Muller *et al.*, 1997; Dyson, 1998; Lundberg and Weinberg, 1998). Therefore, the genes whose products activate E2F1/DP1 TF and that are also present in the gene cassette activated by this factor form a set of conjugated positive feedbacks (G1/S) including the genes, such as E2F1, cyclin D, cyclin E, CDC25A, and c-Myc (the most essential of the genes are shown in Figure 3). During the late S phase, E2F1/DP1 TF, binding to its sites in gene promoters, is phosphorylated by cyclin A/cdk2 kinase. As a result, the phosphorylated E2F/DP1 factor is detached from the corresponding DNA sites (Figure 3), thereby ceasing transcription activation of the gene cassette controlled by this factor (Xu *et al.*, 1994).

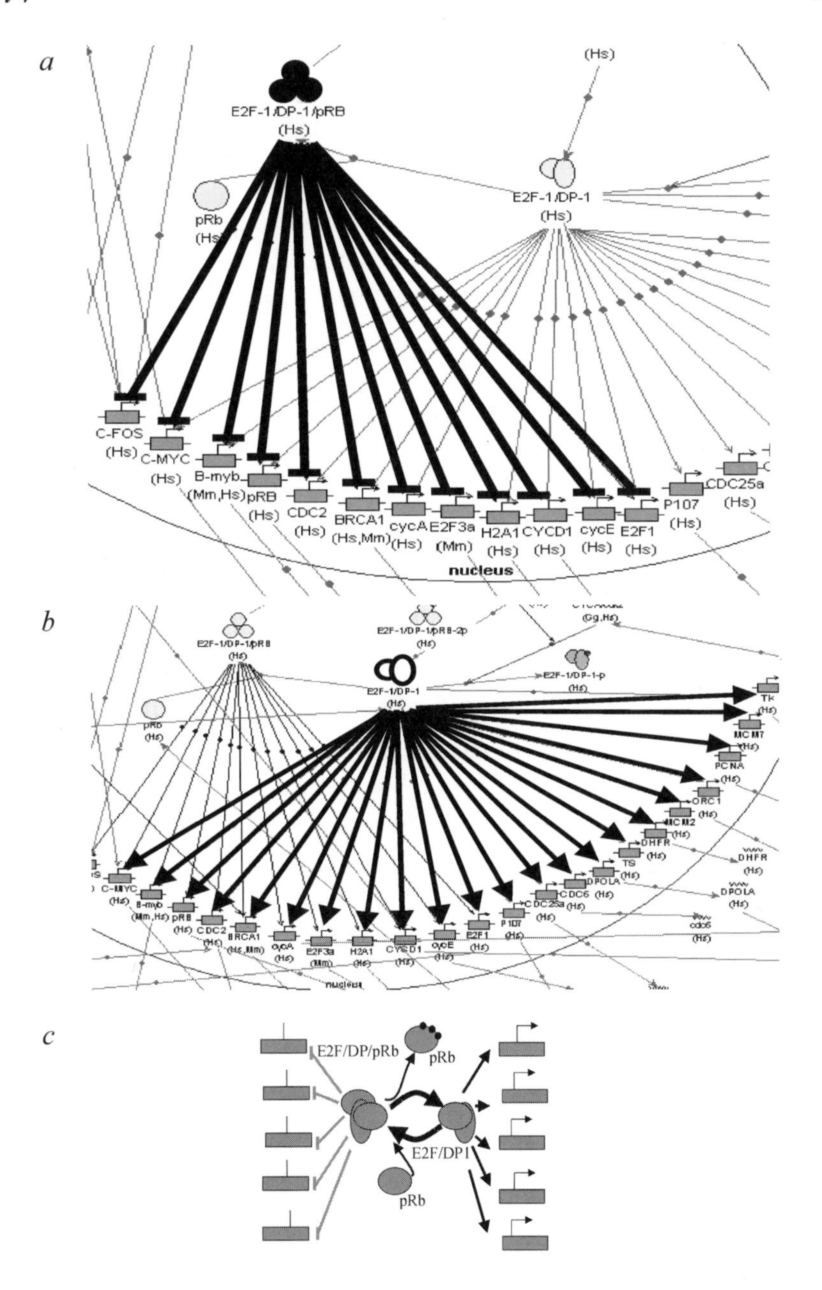

Figure 2. Regulation of the gene cassette by E2F TF: (*a*) repression of the gene cassette by E2F/DP/RB TF; (*b*) activation of the gene cassette by E2F/DP TF; and (*c*) E2F-mediated activation and repression of genes.

(ii) Post-transcription events (except for degradation of proteins). Through the MAPK cascade and Mek1 kinase therein, the growth factors initiate assembly of the cytoplasmic complex cyclin D/cdk4,6 followed by its transfer to the nucleus, thereby facilitating activity of cyclin D/cdk4,6 (Figure 1; Cheng *et al.*, 1998). In the G0 phase in the nucleus, the cyclin E/cdk2 complex is inactivated through its binding to P27/Kip1 factor. A high concentration of P27/Kip1 is characteristic of G0, while a low concentration is characteristic of a cycling cell (Rivard *et al.*, 1996). In the cell cycle G1 phase, the cyclin D/cdk4,6 complex is accumulated in the nucleus, where it binds P27 and, thereby, increases the amount of free active cyclin E/cdk2 complexes.

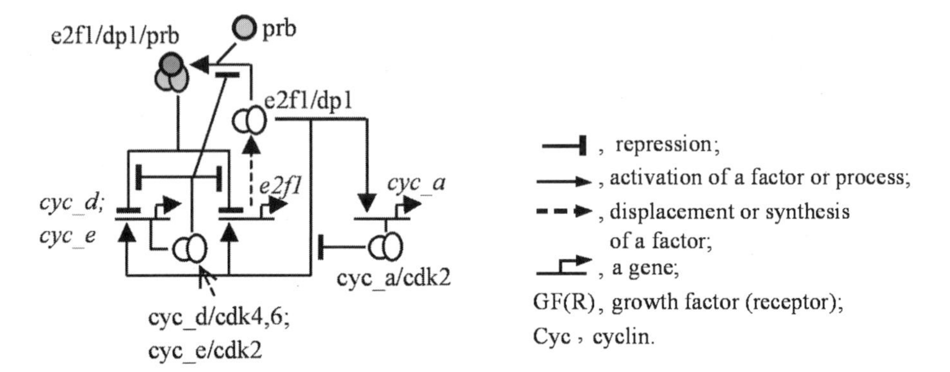

Figure 3. Layout illustrating the main feedback circuits providing the G0,G1/S transition.

(iii) Cell cycle-dependent degradation of proteins. Some proteins regulating the cell cycle may degrade, both constitutively and by the alternative pathways depending on the cell cycle phase. For example, the activities of cyclin D–dependent CDK4,6 kinases beyond the cell cycle are repressed by phosphorylation of cyclin D at Thr-286 by GSK-3beta kinase followed by release of phosphorylated cyclin D from the nucleus to cytoplasm and its fast ubiquitin-dependent degradation (Figure 4). Activation of the PI3-K/Akt signal transduction pathway by growth factors (such as IGF) inactivates GSK-3beta kinase (the early G1). This results in accumulation of cyclin D/CDK4,6 complex in the nucleus during the G1 phase. In the S phase, the activity of GSK-3beta kinase is restored; correspondingly, the cyclin D-dependent kinase activity of CDK4,6 decreases (Diehl *et al.*, 1998). In a similar way, phosphorylation of P27 protein at Thr-187 by cyclin E/cdk2 kinase results in an ubiquitin-dependent degradation of P27, increasing the concentration of free cyclin E/cdk2 complex (Sheaff *et al.*, 1997).

An increase in activity of the cyclin E/cdk2 complex is self-limited by a negative feedback, and autophosphorylation of cyclin E by cyclin E/cdk2 kinase serves as a signal for a fast ubiquitin-dependent degradation of this complex (S/G2; Singer *et al.*, 1999). E2F1 is protected against degradation when it is bound to RB (G0, the early G1; Hofmann *et al.*, 1996). Upon unbinding, E2F1/Dp1 (in G1/S) degrades. Apparently, concentration of free E2F1/Dp1 increases (G1/S) when transcription and synthesis rates of E2F1 exceed its degradation rate and decreases when the degradation proceeds more rapidly than synthesis of new E2F1 protein (S/G2).

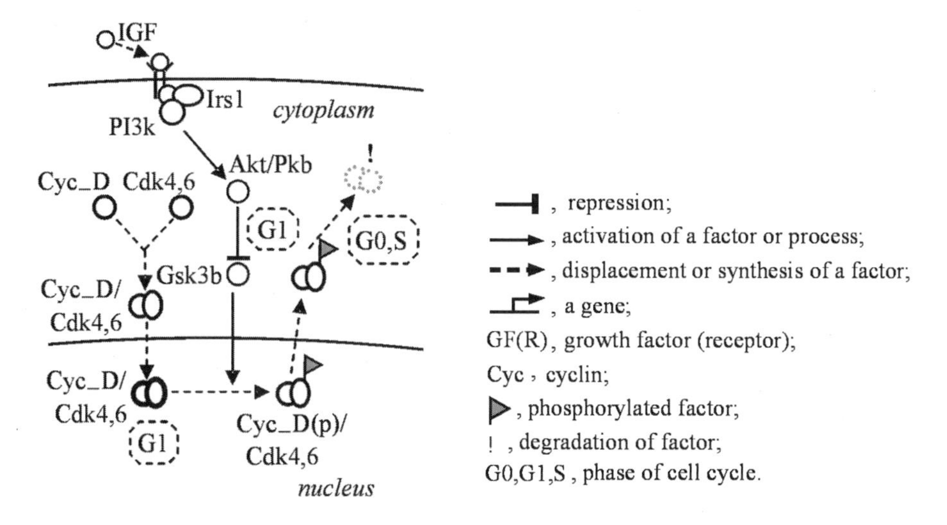

Figure 4. Cell cycle-dependent degradation of cyclin D.

Therefore, the reconstructed cell cycle gene network is rather a comprehensive description of how this process is regulated. However, analysis of regulatory elements of this gene network demonstrated that many linkages in the network remain open and, hence, cannot provide its sustainable operation. As E2F TF is among the key regulators of the cell cycle gene network, detection of a most complete set of E2F sites in regulatory regions of the genes whose products control the cell cycle or are regulated during the cell cycle may contribute greatly to the knowledge on dynamic regulation of this process. In this connection, we searched for the potential binding sites of E2F TF in promoter regions of such genes.

3.2 Recognition of TF E2F binding sites

The E2F sites were searched for in promoter regions (from –300 to +100 with respect to the transcription start) of 79 genes regulating the cell cycle and of the genes involved in cell cycle working processes wherein E2F binding sites were not yet found experimentally. E2F binding sites were detected in 40 of the studied promoter regions (Table 1).

Table 1. The genes with promoters carrying potential E2F binding sites

Function/Gene and species	Number of recognized sites	Function/Gene and species	Number of recognized sites
G0/G1 switch regulatory protein		*POLE2*, human	3
FosB, mouse *	2	*DPOLB*, cow	1
c-Fos, mouse *	1	*DPOL*, mouse	2
Elk-1, human	1	*DPOLA*, human*	3
Fos, human *	1	*Thymidine kinase*, human	1
G1/S transition regulation		*Thymidine kinase*, mouse*	2
E2F1, hamster	5	*Flap endonuclease-1*, mouse	1
E2F2, human *	2	Nucleotide metabolism	–
E2F6, mouse	1	*Cad*, hamster*	1
CDK2, human	1	DNA compacting	–
CDK4, mouse *	1	*Histone H2A.X*, human	2
CDC7, mouse	2	Tumor suppressor	–
CDC25A, human *	1	*ARF*, mouse	1
c-Myc, rat *	1	*p27/Kip1*, mouse*	1
N-myc, mouse *	3	*p53*, human *	1
B-myb, human *	1	*p18INK4c*, human	2
G1/S and G2/M transition regulation		*p19INK4d*, human	1
cyclin A1, human	1	*BRCA2*, human	1
Wee1, human *	4	Differentiation	
G2/M regulation		*c-Rel*, chicken	3
PLK, human	1	*CDK9*, human	2
14-3-3 sigma, human	1	*c-Myb*, human	1
DNA replication		*IRF-1*, mouse	1
POLD1, human	1	Cytokinesis regulation	
POLD2, human	1	*Survivin*, mouse	1

* The genes with promoters with previously confirmed presence of potential E2F binding sites (Kel *et al.*, 2001).

4. DISCUSSION

Information on the detected potential binding sites for E2F/Dp TF allowed us to add new regulatory linkages to the reconstructed gene network. For example, the detected potential E2F sites in promoters of early response genes (*c-Fos*, *FosB*, *Elk-1*, *c-Myb*, *B-Myb*, *c-Myc*, and *N-myc*) suggest the presence of positive feedback regulatory circuits activating E2F gene and the early response genes (the early G1 phase) (Figure 5).

The presence of potential E2F sites in promoters of the genes encoding cyclin-dependent kinases (*CDK2*, *CDK4*, and *CDC7*) suggests the occurrence of additional positive feedbacks operating in the G1/S phase (Figure 5). Occurrence of such sites in promoters of inhibitors of cyclin-dependent kinases (*p18INK4c*, *p19INK4d*, and *p27Kip1*) implies new negative feedback circuits and suggest the mechanism underlying involvement of these genes in checkpoints (certain points where cell cycle stops and does not enter the next phase until all the processes of the preceding phase are completed; for example, a cell does not start mitosis until DNA replication is completed; Hartwell and Weinert, 1989).

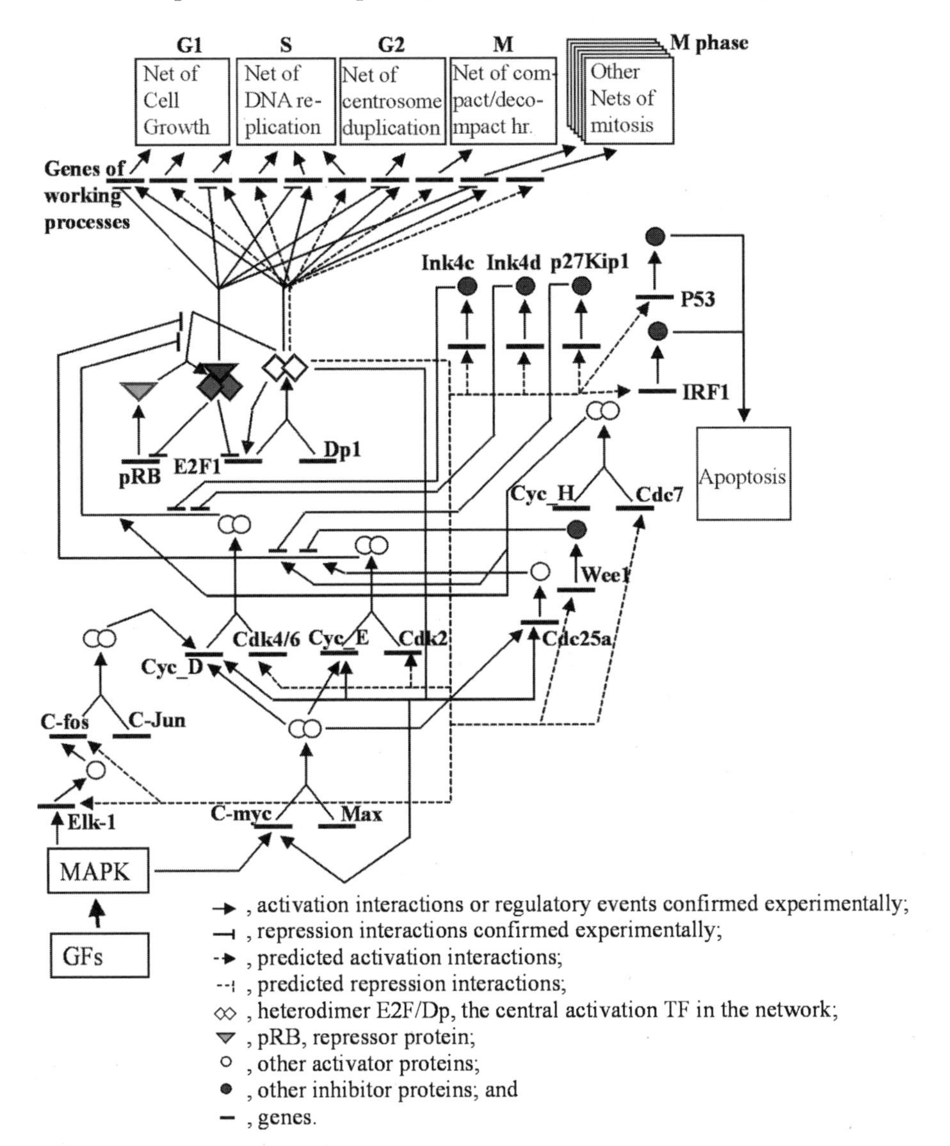

Figure 5. Layout of the gene network regulating eukaryotic cell cycle according to experimental data supplemented with the hypothetical regulatory interactions predicted by recognition of E2F sites.

In this way, the mechanism of E2F activity self-limitation when the activity exceeds a certain threshold value may be realized. In addition, the detection of potential E2F sites in *p53*, *BRCA2*, and *IRF-1* genes, whose products are involved in both the cell cycle cessation and activation of cell death program (apoptosis), suggests occurrence of additional negative feedback circuits limiting of E2F TF activity and blocking the cell cycle (Figure 5).

Thus, the potential binding sites for E2F transcription factor detected in the work allows us to enrich considerably the reconstructed cell cycle regulation gene network with additional regulatory linkages, which enhance sustainability of its operation.

5. CONCLUSION

A comprehensive approach to reconstruction of gene networks based on combination of experimental data with the results of recognition of E2F TF binding sites allowed us to extend the description of cell cycle regulation gene network and describe in more detail the mechanisms providing a sustainable dynamics of the cell cycle. The next step of our work will involve chemical kinetic approach to mathematical simulation of the gene network regulating the cell cycle progression.

ACKNOWLEDGEMENTS

The authors are grateful to N.A. Kolchanov, E.A. Ananko, E.V. Ignat'eva, and L.V. Omel'yanchuk for fruitful discussions of the study and to G.B. Chirikova for translating the manuscript into English. The study was supported in part by the Russian Foundation for Basic Research (grants Nos. 01-07-90376-в, 02-07-90359, 03-07-96833-p2003, 03-07-90181-v, 03-07-96833, 03-04-48506-a, 03-04-48469-a, 03-07-06077-мас, 03-01-00328); Russian Ministry of Industry, Science, and Technologies (grant No. 43.073.1.1.1501); grant PCB RAS (No. 10.4); the Siberian Branch of the Russian Academy of Sciences (integration project No. 119); NATO (grants Nos. LST.CLG.979816 and PDD(CP)-(LST.CLG 979815)); and US Department of Energy (grant No. 535228 CFDA 81.049).

MATHEMATICAL SIMULATION OF DYNAMICS OF MACROPHAGE GENE NETWORK ACTIVATED BY LIPOPOLYSACCHARIDES AND/OR INTERFERON-g

E.A. NEDOSEKINA[1,2]*, E.A. ANANKO[1,2], L. MILANESI[3], V.A. LIKHOSHVAI[1,2,4], N.A. KOLCHANOV[1,2]

[1]*Institute of Cytology & Genetics, Siberian Branch of the Russian Academy of Sciences, prosp. Lavrentieva 10, Novosibirsk, 630090 Russia, e-mail: nzhenia@bionet.nsc.ru;* [2]*Novosibirsk State University, ul. Pirogova 2, Novosibirsk, 630090 Russia;* [3]*CNR Institute of Biomedical Technology (CNR-ITB), via Fratelli Cervi, 93, 20090 Segrate (Milan), Italy;* [4]*Ugra Research Institute of Information Technologies, ul. Mira 151, Khanty-Mansyisk, 628011 Russia*

* *Corresponding author*

Abstract Macrophage activation is an important component of the immune response. The gene network controlling activation of macrophages comprises a large number of genes, proteins, and metabolites interacting with one another and regulating a variety of reactions. Using the GeneNet technology (Kolchanov *et al.*, 2000), we developed a formalized description of the gene network regulating macrophage activation stimulated by lipopolysaccharides and (LPS) and interferon-γ (IFN-γ). In the concept of generalized chemical kinetic simulation approach (Likhoshvai *et al.*, 2001), the mathematical model describing dynamics of this gene network was constructed basing on the information accumulated in GeneNet. This model allows the data on the structure–function organization of gene network, mechanisms of individual reactions, and the known values of static parameters and dynamic characteristics to be taken into account. The current version of this model comprises description of 306 elementary processes involving 36 genes, over 100 proteins and protein complexes, and other substances. The model allows specific operation patterns of the gene network in norm and during several pathologies to be investigates and concentrations of various compounds a quiescent and an activated cells to be calculated.

Key words: gene networks, database, mathematical simulation, computer model, macrophage activation

1. INTRODUCTION

Macrophages perform various regulatory and effector functions in the body, such as, phagocytosis, regulation of the immune and anti-inflammatory responses, regeneration of injured tissues, etc. Various pathologies, such as rheumatoid arthritis (Kinne *et al.*, 2000), atherosclerosis (Gough *et al.*, 1999), multiple sclerosis (Martino *et al.*, 2000), and Alzheimer's disease (Smits *et al.*, 2000), are accompanied by impaired macrophage function.

Upon activation, macrophages change their morphology, intensify phagocytosis and production of various compounds—active radicals (NO, (H_2O_2), and O_2^-), enzymes (cyclooxygenase-2, inducible nitric oxide synthase iNOS, and lysozyme), membrane proteins (antigens of histocompatibility complex, various receptors, and cell adhesion molecules), etc. Synthesis and secretion of a set of cytokines (IL-1, IL-6, IL-10, IL-12, IFN-α, IFN-β, and TNF-α) increases too.

Activation of macrophages is intensely studied. It is assumed that the gene network regulating this process comprises hundreds of genes, RNAs, proteins, and nonprotein compounds. However, the mechanisms underlying the function of these cells and well as the causes of the related pathologies yet remain vague enough. Quantitative characteristics of molecular processes (reaction constants and intracellular concentrations of various substances) require further studies; in addition, the available data have been obtained under various conditions, in different cell lines, and a diversity of preparations, which could not but influence the results. Mathematical simulation, requiring prior collection of experimental information and its analysis, may assist in solving some of these problems. Calculations using the model can aid gaining new information on operation of gene networks.

Our goal was to construct a mathematical model describing macrophage activation (Nedosekina *et al.*, 2002) in the concept of generalized chemical kinetic simulation approach (Likhoshvai *et al.*, 2001). For this purpose, we have collected and compiled the relevant data and, using the GeneNet technology (Kolchanov *et al.*, 2000), created a formalized description of the gene network of macrophage activation by lipopolysaccharides (LPS) of bacterial cell wall and interferon-γ (IFN-γ), synthesized by T cells of the immune system.

2. RESULTS AND DISCUSSION

2.1 Gene network of macrophage activation

The GeneNet technology (Kolchanov *et al.*, 2000) was used for describing the gene network of macrophage activation in a formalized manner. The structure–function organization of macrophage gene network activated by LPS and IFN-γ was reconstructed through analyzing published experimental data and is available in the Internet at http://wwwmgs.bionet.nsc.ru/systems/MGL/GeneNet/ (Nedosekina *et al.*, 2002). The latest version comprises data on over 400 various components of this gene network, including about 130 proteins, 36 genes, and over 200 reactions and regulatory stimuli.

NFκB, IRF-1, and Stat1α are the key transcription factors of the gene network considered (Figure 1). These factors enhance transcription of the largest set of the genes under study compared with other transcription factors, including genes encoding cytokines (IL-1β, IL-6, IL-12p35, IL-12p40, TNF-α, IP-10, IFN-α, IFN-β, etc.), enzymes (COX-2 and iNOS), membrane proteins (ICAM1), and other transcription factors (ICSBP and IRF-2). However, other transcription factors—Pu.1, USF-1, c-Jun, CREB, NF-IL6, AP-1, and ISGF3—are involved in activation of this gene network (Figure 1).

Transcription factors require coactivators for their work, as, for example, in regulation of transcription of class II histocompatibility complex genes. Induced expression of the genes, such as MHCII IA^b and MHCII IE^b is impossible without the factor CIITA (Class II TransActivator), which does not bind to DNA directly, but is necessary for the work of transcription factors (Herrero *et al.*, 2001). Expression of the gene encoding CIITA depends, in turn, on the transcription factors Stat-1α, USF-1, and IRF-1.

The cytotoxic molecule NO (Figure 2) plays an important role in a valid function of the macrophage. The enzyme iNOS is involved in NO synthesis; L-arginine, molecular oxygen, and NADPH are the substrates; and tetrahydrobiopterin, FAD, and FMN, the cofactors. Upon macrophage activation, NO production increases drastically. However, a mere elevation in iNOS expression is insufficient—increases in the amounts of substrates and cofactors are required as well. For example, IFN-γ activates the enzyme GTP cyclohydrolase I (GCHI) (Figure 2), necessary for synthesis of tetrahydrobiopterin (H4B). The model developed describes the pathway of synthesis of the cofactor H4B too.

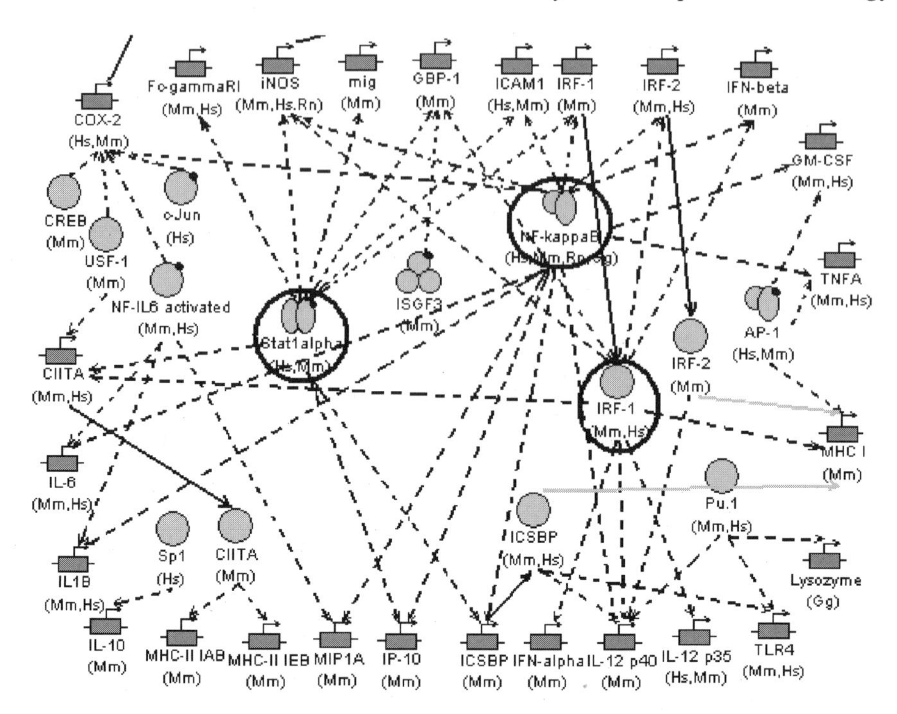

Figure 1. A fragment of the gene network of macrophage activation: the cell nucleus (filled ovals represent proteins, here, transcription factors and the transactivator CIITA; filled rectangles, genes; arrows, reactions and regulatory stimuli; and the key transcription factors are encircled).

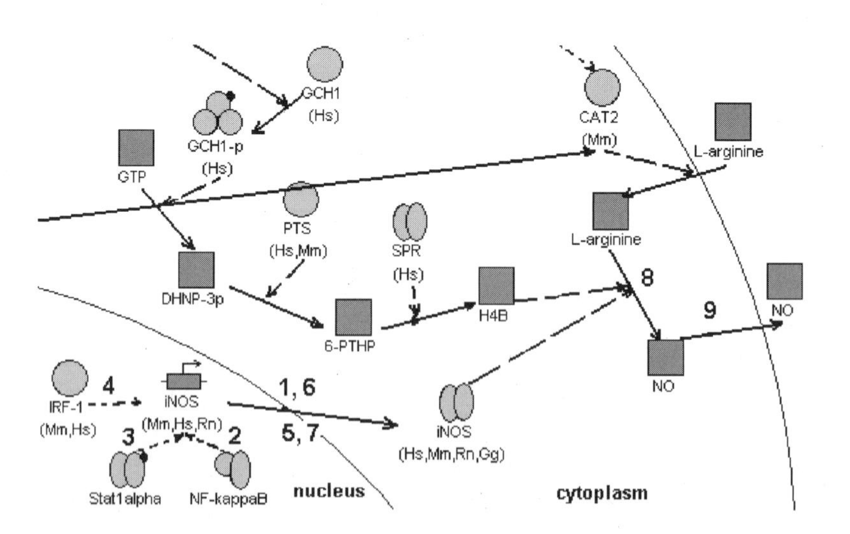

Figure 2. A fragment of the gene network of macrophage activation: production of NO and its cofactor tetrahydrobiopterin (rectangles indicate low-molecular-weight compounds).

Cationic amino acid transporter 2 (CAT2) is responsible for concentration of the substrate L-arginine (Figure 2). Expression of CAT2 is also stimulated when the macrophage cell is activated by LPS or IFN-γ. However, a simultaneous effect of these activators decreases the concentration of CAT2 on the background of an elevated level of the corresponding mRNA. Nevertheless, the mechanism underlying these processes is yet unknown (Kakuda *et al.*, 1999).

2.2 Mathematical simulation of the gene network of macrophage activation

A generalized chemical kinetic simulation approach (Likhoshvai *et al.*, 2001) was used to construct a mathematical model of the macrophage activation gene network. The basic principle used in development of the GeneNet database is description of elementary structures and objects. The generalized chemical kinetic simulation approach uses the same principle; therefore, a scheme of the gene network fits well into this concept. Each elementary process is described as an independent structural unit of the model using nonlinear differential equations. For example, when simulating NO production, it is necessary to take into account not only the enzymatic reaction of NO production, but also the rate of iNOS production, that is, the rate of constitutive synthesis of iNOS mRNA (with basic transcription factors); enhancement of the mRNA synthesis by transcription factors NFκB, Stat1α, and IRF-1 (Figure 1); degradation of the mRNA; and rates of synthesis of the protein iNOS and its degradation. For this purpose, the following differential equations are required (equations are numbered as the links in Figure 2):

1. Constitutive synthesis of mRNA:
   $$\frac{d[\text{iNOS\_mRNA}]}{dt} = k_s$$, where k_s is the constant of iNOS mRNA constitutive synthesis;
2. Enhancement of the synthesis by transcription factor NFκB:
   $$\frac{d[\text{iNOS\_mRNA}]}{dt} = \frac{K_1[\text{NF}\kappa\text{B}]K_g}{K_2 + [\text{NF}\kappa\text{B}] + K_g}$$, where K_1, K_2, and K_g, constants of biochemical reactions in this (2) and the next two (3–4) reactions;
3. Enhancement of the synthesis by transcription factor Stat1α:
   $$\frac{d[\text{iNOS\_mRNA}]}{dt} = \frac{K_1[\text{Stat1}\alpha]K_g}{K_2 + [\text{Stat1}\alpha] + K_g};$$

4. Enhancement of the synthesis by transcription factor IRF-1:
   $$\frac{d[\text{iNOS\_mRNA}]}{dt} = \frac{K_1[\text{IRF1}]K_g}{K_2 + [\text{IRF1}] + K_g};$$
5. Degradation of mRNA:
   $\frac{d[\text{iNOS\_mRNA}]}{dt} = -k_d[\text{iNOS\_mRNA}]$, where k_d, degradation constant of iNOS mRNA;
6. Synthesis of the enzyme iNOS:
   $\frac{d[\text{iNOS\_protein}]}{dt} = k_s$, where k_s, constant of iNOS synthesis;
7. Degradation of the enzyme iNOS:
   $\frac{d[\text{iNOS\_protein}]}{dt} = -k_d[\text{iNOS\_protein}]$, where k_d, constant of iNOS degradation;
8. Enzymatic reaction of NO formation (a simplified equation):
   $$\frac{d[\text{NO}]}{dt} = \frac{K_1[\text{Arginine}][\text{iNOS}][\text{T4B}]}{K_2 + [\text{Arginine}][\text{iNOS}] + [\text{Arginine}][\text{T4B}] + [\text{iNOS}][\text{T4B}]};$$
9. Transfer of NO into the intercellular space:
   $\frac{d[\text{NO}_{\text{outside}}]}{dt} = k_{tr}[\text{NO}_{\text{in\_cell}}]$, where k_{tr}, reaction constant of the transfer.

A formalized description of gene network operation—the mathematical model—is constructed of such equations. In addition to these equations, the gene network regulating NO production includes also the reactions producing the coenzyme tetrahydrobiopterin, activation/synthesis pathways of transcription factors, etc. The current version of this model comprises description of 306 elementary processes involving 36 genes, over 100 proteins, mRNAs, low-molecular-weight compounds, protein complexes, and the intermediate compounds. The mathematical tools contain 153 dynamic variables and about 450 various constants.

The key stage in constructing a model is verification of the values of its parameters (constants of biochemical reactions). For this purpose, we specified the initial approximations of the model's parameters basing on analysis of the relevant literature data. At the next stage, we searched for the optimal values of these parameters that would provide the best compliance with the experimental data (Ratushny *et al.*, this issue). This problem was solved numerically by an optimization approach oriented to use of the database of dynamic data and base of the model operation scripts under various external stimuli (Likhoshvai *et al.*, 2002).

Taking into account the experimental data on dynamics of NO production (Chen *et al.*, 1995), we optimized the constants of the reactions described above.

Comparison of the results obtained using the model and the experimental data, which we used while optimizing the parameters, are exemplified by dynamics of NO concentration upon activation of macrophages by lipopolysaccharides (Figure 3).

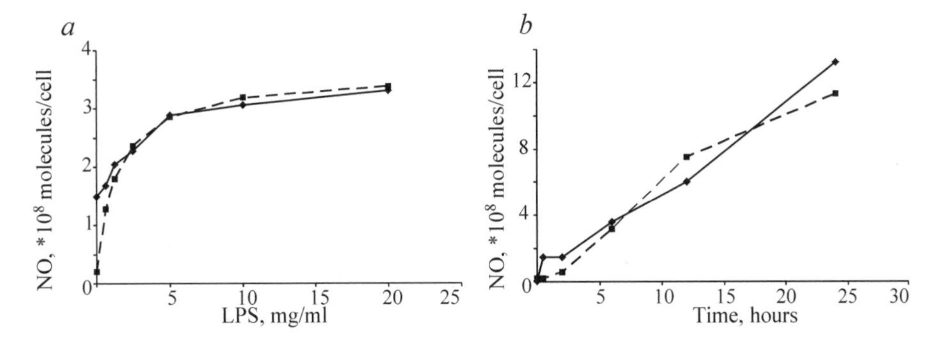

Figure 3. Concentration dynamics of produced NO depending on (*a*) LPS concentration and (*b*) the time since LPS activation: firm line, experimental data (Chen *et al.*, 1995); dash line, calculations using the model.

The curve in Figure 3*a* demonstrates that the NO production reaches a plateau with increase in LPS concentration. Presumably, this results from the limited number of LPS receptors on the cell surface, preventing further boosting of the cell response at high LPS concentrations.

The shape of the curve in Figure 3*b* is not evident, as the experimental data end on hour 24 after the LPS activation. However, calculations using the model show that the NO production peaks 48 h after the activation and then decreases. These results were obtained when macrophage activation by lipopolysaccharides at a concentration of 10 μg/ml was simulated. We also simulated the activation using the concentrations of 10 ng/ml (Figure 4, broken line) and 0.1 ng/ml (Figure 4, dash line). At an LPS concentration of 10 μg/ml, the NO production returns to a level exceeding the initial concentration by 20% approximately on day 14 after the stimulus (Figure 4, firm line); at 10 ng/ml, by 1–2 days earlier; and at 0.1 ng/ml, already on day 7.

The equations describing NO concentration dynamics demonstrate that it is also important to take into account the concentration dynamics of transcription factors. Comparison of the results obtained using the model and the experimental data, which we used when optimizing parameters of the model, for mRNA of the transcription factor ICSBP is shown in Figure 5.

We also studied syntheses of major histocompatibility complex proteins. The results obtained for relative levels of CIITA and MHCII IA^b mRNAs upon induction with IFN-γ are shown in Figure 6*a*, *b*.

Calculations of the synthesis dynamics of the protein MHCII IA^b under the same conditions are shown in Figure 6*c*. These graphs demonstrate that the maximal synthesis rate of both the protein and its mRNA are reached by 24–48 h; however, the rate of CIITA mRNA production by that time is already slowed.

Unlike the previous kinetic patterns of cell component concentrations considered, a pronounced elevation in the content of the protein MHCII IA^b is observed only after 48 h of the IFN-γ action, and the increase continues to 72 h.

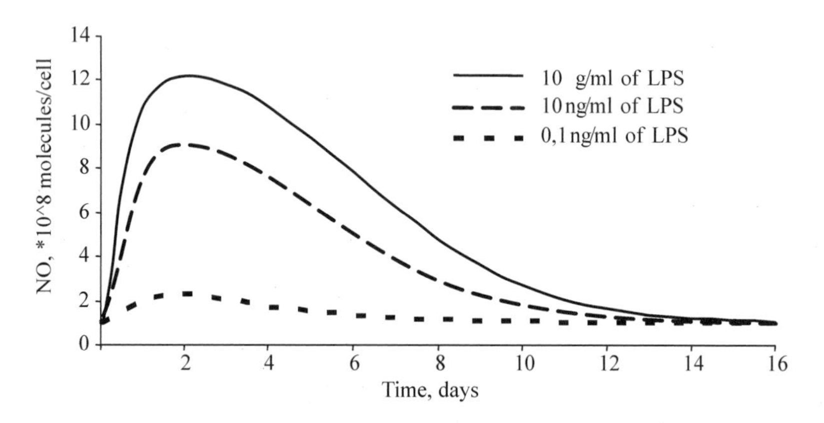

Figure 4. Concentration dynamics of produced NO in response to activation by lipopolysaccharides (results of mathematical simulation).

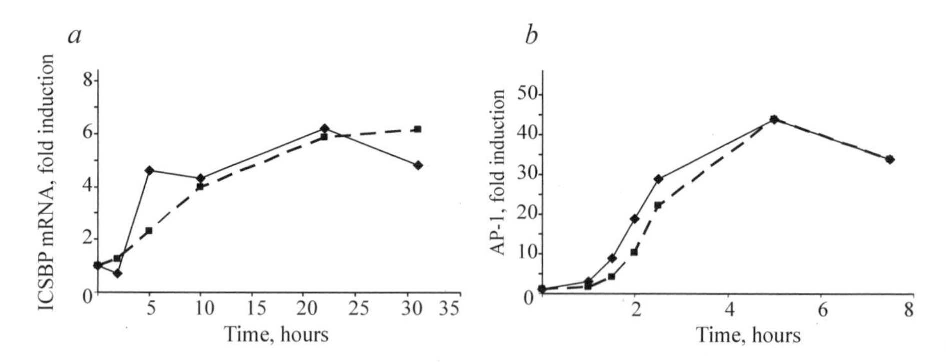

Figure 5. Concentration dynamics of (*a*) mRNA for transcription factor ICSBP and (*b*) AP-1 protein depending on the time since LPS-caused activation: firm line, experimental data (Hambleton *et al.*, 1996; Kantakamalakul *et al.*, 1999); dash line, calculations using the model.

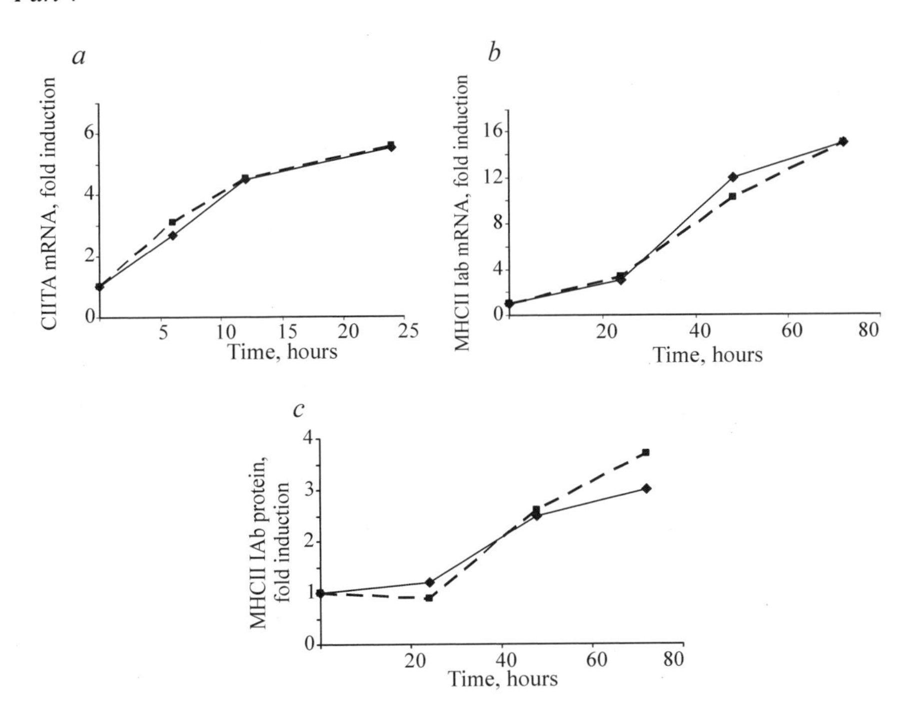

Figure 6. Induction of expression of MHCII genes—dependence of the concentrations of gene network components on the time since IFN-γ-caused activation: (*a*) concentration dynamics of CIITA mRNA; (*b*) concentration dynamics of MHCII IA^b mRNA; (*c*) concentration dynamics of MHCII IA^b protein; firm line, experimental data (Herrero *et al.*, 2001); dash line, calculations using the model.

3. CONCLUSION

The information on activation of the macrophage gene network under the effects of LPS and IFN-γ accumulated in the system GeneNet may be useful when planning experiments, for obtaining a more integrate view of this gene network, clarifying the roles of its individual components, and constructing mathematical models. The current version of the model allows dynamic characteristics of the components of this gene network to be calculated under certain specified effects of external stimuli.

This model can be used for solving a variety of topical problems, such as (a) searching for optimal characteristics of compounds that would influence the gene network in a prespecified manner; (b) testing alternative hypotheses on missing components and/or regulatory links; (c) studying the effects of mutations and specific patterns of gene network behavior in the case of certain pathologies; (d) solving the problems of searching for optimal stimuli to compensate for impairments of gene network operation; and (e) calculating concentrations of various components in a quiescent and an activated cells.

ACKNOWLEDGMENTS

This work was supported by the Russian Foundation for Basic Research (grants Nos. 01-07-90376-в, 02-04-48802, 02-07-90359, 03-07-96833-р2003, 03-07-96833, 03-04-48506-а, 03-07-06077-мас, 03-01-00328); Russian Ministry of Industry, Science, and Technologies (grant No. 43.073.1.1.1501); grant PCB RAS (No. 10.4); the Siberian Branch of the Russian Academy of Sciences (integration project No. 119); NATO (grant No. LST.CLG.979816).

COMPUTER DYNAMIC MODELING OF THE GENE NETWORK CONTROLLING INTRACELLULAR CHOLESTEROL HOMEOSTASIS

A.V. RATUSHNY[1,2]*, E.V. IGNATIEVA[1,2], V.A. LIKHOSHVAI[1,2,3]
[1]*Institute of Cytology & Genetics, Siberian Branch of the Russian Academy of Sciences, prosp. Lavrentieva 10, Novosibirsk, 630090 Russia, e-mail: ratushny@bionet.nsc.ru;* [2]*Novosibirsk State University, ul. Pirogova 2, Novosibirsk, 630090 Russia;* [3]*Ugra Research Institute of Information Technologies, ul. Mira 151, Khanty-Mansyisk, 628011 Russia*
* *Corresponding author*

Abstract An adequate computer dynamic model of the gene network regulating cholesterol homeostasis that would allow for investigation of its components (RNAs, proteins, their complexes, metabolites, etc.) is demanded for a deeper insight and a more integral understanding of this intricate gene network, study of its various operation modes and the underlying mechanisms, assessment of effects of mutations on its operation, and choosing of the optimal strategies for influencing this system while solving particular problems, for example, therapeutic. In this work, a computer dynamic model of operation of the gene network controlling intracellular cholesterol homeostasis, regulation of its synthesis in the cell, and intake of the blood plasma cholesterol is constructed. The model is described in terms of elementary processes—biochemical reactions—and is fitted to the relevant published experimental data. The optimal set of model's parameters is determined. The results obtained are compared with the corresponding experimental data.

Key words: computer dynamic modeling, gene network, cholesterol biosynthesis, homeostasis

1. INTRODUCTION

Constituents of biological membranes contain various sets of lipids, which are individual for different organisms, tissues, cells, and cell organelles.

Cholesterol, an amphipathic lipid that is a precursor of various other steroids (corticosteroids, sex hormones, bile acids, and vitamin D), is an essential structural component of mammalian cell membranes. Cholesterol is synthesized in many tissues from acetyl-CoA. It is also supplied to the tissues within low density lipoproteins (LDL), housing the major part of blood plasma cholesterol, taken up by the cells via LDL receptors. Free cholesterol is removed from the tissues with the help of high density lipoproteins. In various pathologies, the level of cholesterol in blood plasma is the factor causing atherosclerosis of vital arteries of the brain, heart muscle, and other organs (Murray *et al.*, 1988).

The overall lipid composition in cell membranes and, in particular, cholesterol content, is regulated according to a homeostatic mechanism, which allows for maintaining the level of lipids in stringent limits and at an optimal concentration. Cholesterol and fatty acids self-regulate their own syntheses in mammalian cells according to a negative feedback mechanism.

For a wider study and better understanding of the tangled nonlinear gene network regulating intracellular cholesterol homeostasis, study of its various operation modes and the underlying mechanisms, assessment of effects of mutations on its operation, etc., a computer dynamic model describing operation of this gene network was built in this work. The model is described in terms of elementary processes—biochemical reactions. A set of parameter values for the model allowing a credible compliance of the calculation results with published experimental data to be reached was determined by numerical experiment using the genetic algorithm.

1.1 Cholesterol biosynthesis and regulation of its intracellular content by a negative feedback mechanism

All the cells retaining the nucleus are capable of synthesizing cholesterol. Cholesterol biosynthesis is localized to microsomes and cytosol. Acetyl-CoA is the sole source of all the carbon atoms forming the cholesterol molecule (Murray *et al.*, 1988). The main stages of cholesterol biosynthesis are qualitatively represented in Figure 1 form the GeneNet database (http://wwwmgs.bionet.nsc.ru/systems/mgl/genenet/). Promoters of all the sterol-regulated genes contain a regulatory region—the sterol regulatory element (SRE)—and are subjected to a stringent transcriptional control. In mammalian cells, the family of membrane-bound transcription factors—sterol regulatory element-binding proteins (SREBPs)—governs the cholesterol content in cell membranes (Brown and Goldstein, 1999). SREBPs controls expression of over 20 genes, whose products are involved in syntheses of cholesterol and unsaturated fatty acids and their intake into the cell (Goldstein *et al.*, 2002).

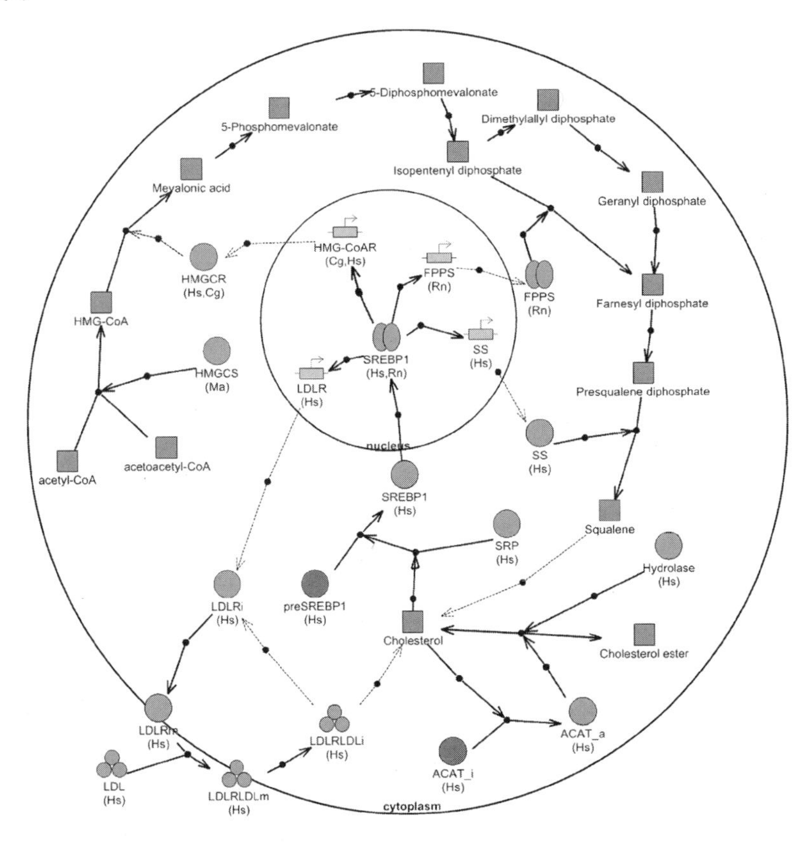

Figure 1. Gene network controlling intracellular cholesterol homeostasis (picture from the GeneNet database; http://wwwmgs.bionet.nsc.ru/systems/mgl/genenet).

Immediately after being synthesized on endoplasmic reticulum (ER) membranes, SREBP forms a stably bound complex with SREBP cleavage-activating protein (SCAP; Sakai *et al.*, 1997). This is inactive SREBP form, as this transcription factor has no access to the nucleus. SCAP performs two functions, namely (1) it is sensitive to cholesterol content in ER membranes and (2) it is involved in transportation of SREBPs from ER to the Golgi apparatus, housing the proteases that release (cleave) transcriptionally active *N*-terminal SREBP domain in a two-stage process, wherefrom the domain reaches the nucleus to activate transcription of sterol-regulated genes (Nohturfft *et al.*, 2000; Goldstein *et al.*, 2002).

Thus, transportation of the complex formed of sterol regulatory element binding protein (SREBP) cleavage activating protein (SCAP) and SREBP (SCAP–SREBP) from ER to the Golgi apparatus is the determining event (limiting stage) in regulation of cholesterol homeostasis in mammalian cells, following the negative feedback pattern. Recent publications (Brown *et al.*, 2002; Yang *et al.*, 2002) provide new insights into how an integral ER protein,

SCAP, mediates this process. Conformation of the SCAP protein is sensitive to the content of cholesterol in ER membranes. When the cholesterol content in ER membranes is high, the SCAP conformation permits the SCAP–SREBP complex to bind tightly to insulin-induced gene (INSIG) retention protein yet remaining in ER. When the cholesterol content in the cell is low, SCAP changes its conformation in such a way that the complex SCAP–SREBP dissociates from INSIG retention protein to insert into common coat protein (COP) II vesicles (Gimpl *et al.*, 2002). Within COP II vesicles, the complex SCAP–SREBP is transported into the Golgi apparatus via a classical secretory pathway (Espenshade *et al.*, 2002). An essential moment in regulation of cholesterol synthesis is a rather quick response of the cells to intracellular cholesterol content, which is unexplainable by SREBP-based regulation of transcription of sterol-regulated genes. Such a short-term regulation found its explanation in existence of bicyclic cascade system involved in regulation of HMG-CoA reductase, the key enzyme of cholesterol synthesis (Beg *et al.*, 1984; 1986). HMG-CoA reductase and the enzyme inhibiting its activity, reductase kinase, may exist in two (active or inactive) reversible states. Transition from one state into the other proceeds as a result of phosphorylation–dephosphorylation reaction. The active HMG-CoA reductase is the nonphosphorylated enzyme; inactive, phosphorylated; and reductase kinase, *vice versa.*

2. METHODS AND ALGORITHMS

A generalized chemical kinetic approach (Likhoshvai *et al.*, 2001) is used in this work to simulate gene network dynamics. Formalization exploits a block-diagram principle, i.e. each process is represented by an individual block and described separately of other processes. Block is the modeling unit element whose formal structure is completely characterized by the three following vector components: (1) V, the list of dynamic variables; (2) P, list of constants; and (3) F, form of the right part of the system $dV/dt = F(V, P)$, specifying the law of variation of the values of dynamic variables in the time domain. The following equation is applicable to formal description of the processes occurring in a gene network, such as mono-, bi-, and trimolecular reactions; mono and multisubstrate enzymatic reactions; regulation of transcription and translation; etc.:

$$\frac{dV_q}{dt} = A_q\, k \frac{1 + \sum_{i=1}^{n} \prod_{j=1}^{k_i} (k_{1i} V_{\sigma(i,j)})^{h_{ij}}}{\alpha + \sum_{i=1}^{m} \prod_{j=1}^{l_i} (k_{2i} V_{\delta(i,j)})^{h_{ij}}},$$

where V_q is the concentration of qth component involved in the process; q, number of the components involved (reagents, regulators, and products); $\sigma(i,j)$ and $\delta(i,j)$, specified functions determining the index of a component involved from the known values of i and j. Here, k has the dimensionality of (mol/l)/sec; k_{1i} and k_{2i}, $(\text{mol/l})^{-1}$; and the rest constants are dimensionless.

The model is constructed from blocks described using the above equation, and is actually a system of ordinary differential equations, where kinetics of each component equals sum of the instantaneous rates of their changes in all the processes where they are involved. To simulate external factors, discrete expressions changing values of dynamic variables according to a certain law at particular time points (Likhoshvai *et al.*, 2001) are used.

Evolutionary approach to fitting computer models. An evolutionary algorithm was used to fit the computer dynamic model of the gene network in question to experimental data. This algorithm is realized as follows. At the initial stage, a population of individuals, each carrying the simulated gene network with individual set of its parameters, is generated in a random manner. Then, reproduction within this population is modeled. Each individual is capable of generating a specified number of descendants that differ from the initial individual by one or several mutations, i.e. randomly changed parameters. A certain number of most fitted individuals are selected at each stage of evolution. Fitness is considered as a degree of compliance of the calculation results obtained using the model to the relevant published experimental data. The specified minimized functional f may be formally represented in different ways, for example, as

$$f = \sqrt{\sum_i \left(v_i^t - v_i^e\right)^2} \quad \text{or} \quad f = \sum_i \left(\frac{v_i^t}{v_i^e} + \frac{v_i^e}{v_i^t} - 2\right),$$

where v_i^t is the ith value of a certain characteristics obtained by numerical calculations using the computer dynamic model (for example, concentration of particular gene network component under certain conditions); v_i^e, ith experimental value. Then, the described processes of reproduction, mutation generation, and selection are repeated for the new population, and so on. While realizing this algorithm for simulation of evolution, it is possible to impose various limitations, for example, introduce various patterns of mutation generation (different probabilities for particular parameters to be changed, different number of mutations per one generation, etc.), specify the numbers of individuals selected and descendants generated by one individual, etc.

3. COMPUTER DYNAMIC MODEL OF THE GENE NETWORK

The current version of computer dynamic model of the gene network regulating homeostasis of intracellular cholesterol comprises 80 gene network components (of them, 57 are dynamic variables) and 139 processes.

Figure 2 shows the bipartite graph representing the model of gene network regulating intracellular cholesterol homeostasis. Circles mark substances; square boxes, processes; and arrows, link components involved in processes with the corresponding products. Firm lines indicate that substance is changed in the corresponding process; dash arrows, that substance influences the process rate but is not consumed.

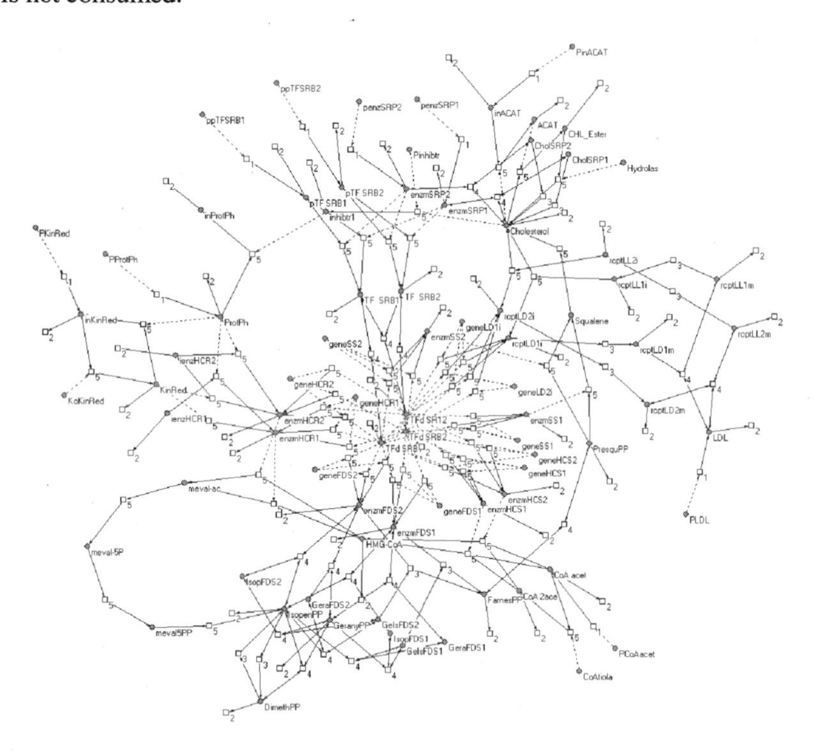

Figure 2. Bipartite graph of the computer model of the gene network controlling intracellular cholesterol homeostasis: circles, dynamic variables; square boxes, processes; (1) constitutive synthesis; (2) utilization or degradation; (3) monomolecular reaction; (4) bimolecular reaction; and (5) generalized Michaelis–Menten reaction (Likhoshvai *et al.*, 2001). For graph visualization, the program package Pajek was used (Batagelj and Mrvar, 1998).

The model was fitted to the relevant published experimental data. Enzymatic reactions of this gene network have been studied rather comprehensively, and the parameter values of enzymatic kinetics are available in experimental works;

many of them are compiled in electronic databases, in particular, EMP (http://www.empproject.com/) and BRENDA (http://brenda.bc.uni-koeln.de/). The values for parameters of macroprocesses were chosen taking into account the related biological data on characteristic rates of actual processes. For example, we considered the translation rate constant equaling 0.1 sec^{-1}. This is slightly lower than the maximally possible limit, which can be assessed taking into account that the A sites of adjacent ribosomes in a polysome cannot occur closer than 20–60 nt due to steric limitations and the average elongation rate amounts to 3–10 codons/sec (Spirin, 1986). Values of the parameters of the model lacking in the literature, were fitted by numerical experiment. In this process, the values of parameters were chosen so that the behavior of the system complied maximally with the available published data on dynamic characteristics of its behavior.

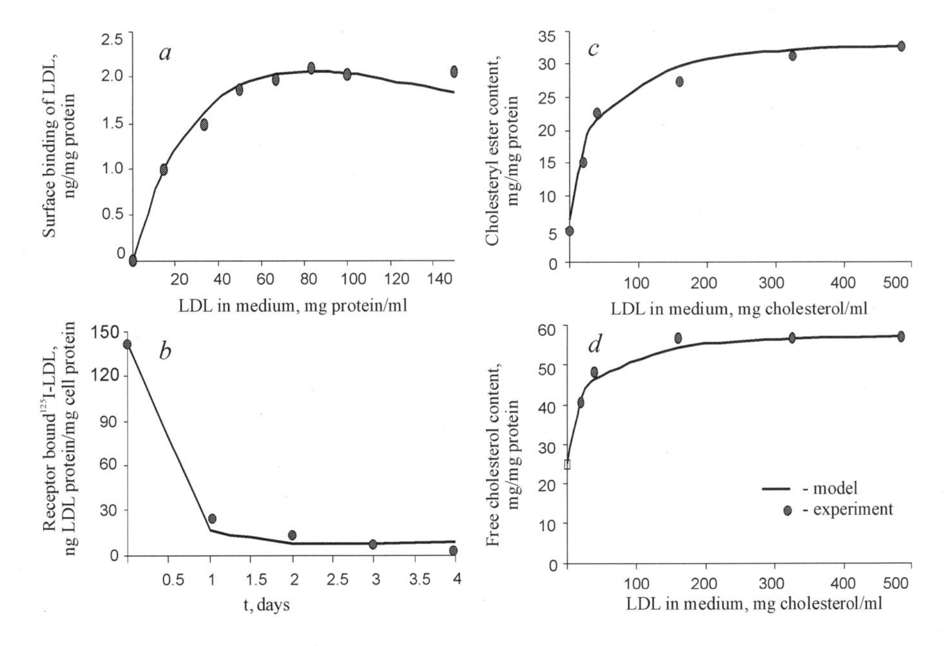

Figure 3. Comparison of the calculations made using the model with experimental data: circles, experimental data from (*a*) Brown and Goldstein (1979) on binding of LDL on the cell surface depending on LDL content in the blood plasma; (*b*) Goldstein *et al.* (1977) on LDL receptor synthesis when cells were transferred from the medium lacking LDL to the medium containing LDL at a certain concentration;

and (*c* and *d*) Goldstein *et al.* (1975) on the effect of increase in LDL concentration on the contents of cholesterol esters and free cholesterol, respectively, in normal fibroblasts; firm lines, simulations.

For example, the following data from the monograph by Klimov and Nikul'cheva (1999) were used for assessing several parameters of the model: (1) LDL concentration in blood plasma of healthy fasting adults amounts to $200 \div 300$ mg/dl; (2) average numbers of nonesterified and esterified cholesterol molecules per one LDL particle is 475 and 1310, respectively; (3) LDL halflife in blood of healthy individuals is about 2.5 days; correspondingly, the degradation rate constant is approximately 3.2×10^{-6} sec^{-1}; (4) the total number of LDL receptors per one cell at 37°C amounts on the average to $15\,000 \div 70\,000$; (5) the lifespan of LDL is approximately $1 \div 2$ days; correspondingly, the constant of LDL utilization equals on the average 7.7×10^{-6} sec^{-1}; (6) the period of LDL receptor recycling is about 20 min; etc. When using quantitative data obtained under more complex experimental conditions, we applied a script construction technique. Its essence lies in a maximally possible reconstruction of experimental protocol using the model followed by comparison of the quantitative characteristics obtained *in silico* with the corresponding experimental data. For example, Figure 3 shows a series of experiments by Brown and Goldstein (1979) and Goldstein *et al.* (1975; 1977) used to fit a number of unknown parameters of the model by applying the evolutionary algorithm. As a result, a plausible compliance of the modeling results with the experimental data was achieved.

4. CONCLUSION

Thus, the constructed computer dynamic model has been fitted to the corresponding experimental data obtained in experiments with human fibroblasts. The model developed is a convenient tool for research into various quantitative dynamic aspects of the intricate molecular system controlling cholesterol synthesis in the cell and its intake from the blood plasma. The work by Ratushny *et al.* (this issue) describes application of this model to quantitative analysis of the effects of various mutations on operation of the gene network regulating intracellular cholesterol homeostasis.

ACKNOWLEDGMENTS

The work was supported in part by the Russian Foundation for Basic Research (grants Nos. 01-07-90376-в, 02-04-48802, 02-07-90359, 03-07-96833-p2003, 03-07-96833, 03-04-48506-a, 03-07-06077-мас, 03-01-00328); Russian Ministry of Industry, Science, and Technologies (grant No. 43.073.1.1.1501); grant PCB RAS (No. 10.4); the Siberian Branch of the Russian Academy of Sciences (integration projects Nos. 119 and 145); NATO (grants Nos. PDD(CP)–(LST.CLG 979815) and LST.CLG.979816).

AN INVESTIGATION OF THE STRUCTURAL STABILITY OF *DROSOPHILA* CONTROL GENE SUBNETWORK IN COMPUTER EXPERIMENTS

A.V. GALIMZYANOV*, R.N. TCHURAEV
Institute of Biology, Ufa Research Centre, Russian Academy of Sciences, prosp. Oktyabrya 69, Ufa, 450054 Russia, e-mail: galim@anrb.ru
** Corresponding author*

Abstract We have studied the stability of *Drosophila melanogaster* gene subnetwork consisting of maternal coordinate and gap genes through computer experiments with a mathematical model developed by the method of generalized threshold models. According to numerical assessments, the *Drosophila* embryo is heterogeneous from the standpoint of gene network parametric stability in nuclei along the anterior–posterior axis. Our experimental results show that fluctuations in the action thresholds of regulatory substances lower the network parametric stability to a considerably greater extent than fluctuations in the unit intensities of RNA/protein syntheses and degradation. The less is the difference between concentration maximum and minimum values of a certain morphogen (gradient local amplitude) at compartment boundaries, the higher is the model sensitivity to fluctuations in the threshold values of morphogen concentrations. The lowest parametric stability is found at the boundaries between compartments. This can be attributed to gene alternative functioning in the network. The structure of the *Drosophila* gene network can neutralize mutation phenotypic manifestations. The test procedure of gene network models for parametric stability can be used to identify key genes that make the greatest "contribution" to parametric stability disturbances of the system and to determine the embryo's compartment boundaries where the gene network displays of low parametric stability.

Key words: control gene networks, *Drosophila melanogaster*, gap genes, stability, computer modeling

1. INTRODUCTION

During mathematical investigation of real gene networks controlling ontogenetic processes, it is necessary to evaluate the structural, and particularly parametric, stability of the models constructed. Such evaluation, on the one hand, is used to justify the supposition that modes of ontogenesis are not much sensitive to random fluctuations in parameters; on the other hand, it is intended to ascertain the adequacy of the model to normal processes of development of a particular organism under study. Stability of a gene network at the level of model is tested through evaluating the "sensitivity" of its operation modes to random fluctuations in kinetic parameters over relatively wide intervals of values (Tchuraev and Ratner, 1975). Here, we imply such parameters as unit intensities of transcription and translation, degradation coefficients of transcripts and polypeptides, threshold concentrations of regulatory proteins, and so on. Action thresholds of regulatory proteins and their complexes depend mainly on the degree of affinity for the corresponding sites. This degree of affinity, as well as other kinetic coefficients, is determined in the primary nucleotide sequences of the corresponding sites. Thus, fluctuations in the values of kinetic parameters are related to fluctuations in the primary structure of DNA genome molecules. Let us think that the evaluation of sensitivity of the model's operation modes to random fluctuations in the parameters over relatively wide intervals of values reflects both the parametric and, to some extent, structural stabilities of the model.

Many years ago, a problem for molecular genetic systems of gene expression control was posed that involved an investigation of their parametric stability (Tchuraev and Ratner, 1975; Ratner and Tchuraev, 1978). The problem was solved for the system controlling λ phage development taken as an example. This approach was later on applied to test the parametric stability of the subsystem controlling the flower morphogenesis of *Arabidopsis thaliana* (Tchuraev and Galimzyanov, 2001a). On the basis of a chemical kinetic model, similar problems were also examined (von Dassow *et al.*, 2000). The authors studied the parametric stability of a *Drosophila* gene network consisting of N segment polarity genes. The experimental results gave the possibility to derive a concept of stable ontogenetic modules, i.e. such regulatory subsystems that are capable of preserving their characteristic behavior under temporal changes in input signals. This paper gives formulation and solution of the problem on parametric stability evaluation of *Drosophila* gene subnetwork consisting of maternal coordinate and gap genes, which forms a basic level in the early ontogenesis control system of the organism in question. Computer experiments were performed with the use of the previous model (Tchuraev and Galimzyanov, 2001b) constructed, in our opinion, by a more adequate method of generalized threshold models (Tchuraev, 1991) in its realization as a computer program (Galimzyanov, 2000).

2. OBJECTS AND METHODS

2.1 Control gene network

The gene network controlling *Drosophila* early ontogenesis (Dr-CGN) consists of maternal coordinate, gap, pair-rule, segment polarity and homeotic genes and forms a hierarchic system, with some elements of hierarchy related to feedback cycles (Tchuraev and Galimzyanov, 2001b). Low-level genes are affected by genes from the upper level as well as by some genes of their own level. This provides a cascade-like operation of genes in time, so that homeotic genes, for example, are activated later than the segment polarity genes. Together with the morphogens Bicoid (Bcd), Caudal (Cad), and Nanos (Nos), the genes *giant* (*gt*), *hunchback* (*hb*), *knirps* (*kni*), *Krüppel* (*Kr*), and *tailless* (*tll*) of the gap class form an upper (basic) level of the network. Just after formation of a zygote, successive gene expression begins with translation of maternal mRNA of the pre-early genes *bcd*, *nos*, cad^{mat}, and hb^{mat}, located at the egg's ends. Proteins synthesized on them are irregularly distributed along the embryo's anterior–posterior axis and specify the spatial patterns of gap-gene expression. Gradients of Bcd and Nos protein concentrations along the anterior–posterior axis have an exponential form (Driever and Nüsslein-Volhard, 1988); the Cad^{mat} protein concentration, a near-linear form (MacDonald and Struhl, 1986); and Hb^{mat} protein concentration, a rectilinear form in the embryo's first third and an exponential form in the rest part of the egg (Tautz, 1988). Gap genes are transcribed in the embryo's N overlapped compartments and control activity of each other as well as of genes at the lower cascade level, i.e. pair-rule genes. Pair-rule genes are expressed within seven alternate stripes seen in the embryo. They code for transcriptional factors that determine segment polarity gene expression. Homeotic genes, the latest group of the genes, specify the type of each segment.

2.2 Mathematical and software tools

The method of generalized threshold models (GTM) takes into account specific features of control processes at the molecular level makes it possible to obtain both qualitative and quantitative patterns of the dynamics of gene networks. The principal idea of the method consists in dividing the molecular system of encoding polymers (DNA, RNA, and proteins) and metabolites (m-systems) into the control and controlled subsystems. In this case, the former subsystem is described in terms of discrete mathematics and the latter, in terms of the theory of differential equations with particular right parts. On the basis of the GTM formalism, the system controlling gene expression is represented as a network of genetic blocks, where input variables are concentrations of regulatory substances for the corresponding genes and outputs are concentrations of their products. Information microstructure of a genetic block (Figure 1) includes discriminators, finite automata, a combination scheme, and time delay element—discrete elements forming in combination the control signal $u_j(t)$ depending on the concentration of regulatory substances. The control variable $u_j(t)$ exerts control over the operation of execution units, i.e. the mechanisms of mRNA and protein synthesis, and is specified by a set of piecewise linear differential equations (Tchuraev, 1991).

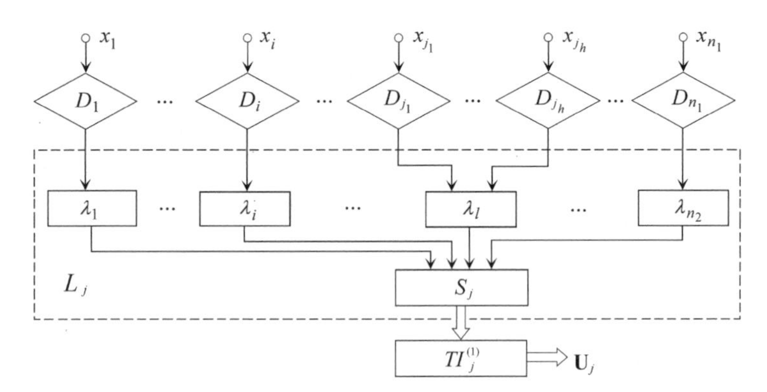

Figure 1. Information microstructure of a genetic block: $x_i(t)$ is concentration of the regulatory substance; D_i, discriminators. The discriminator transforms continuous variables $x_i(t)$ into discrete binary variables $e_i(t)$ in such a way that $e_i(t) = H\,[x_i(t) - Pij\,]$, where $H\,(\upsilon)$ is the Heaviside function, Pij is the threshold value of concentration, $\lambda_1, \ldots, \lambda_i, \ldots, \lambda_l, \ldots, \lambda_{n_2}$ are the designations of "microelements", elementary subautomata of the whole automaton L_j, S_j is the designation of a combinative element (combinator of binary signals), $TI_j^{(1)}$ is the delay element, and $\boldsymbol{U}_j$ is the control vector.

To investigate *in silico* the operation dynamics of complex real genetic regulatory systems, we have developed a software package Analyzer of the Gene Network Dynamics (AGENDY; Galimzyanov, 2000), realizing the GTM formalism as a computer program within the methodology of object-oriented programming. The package makes it possible to find the stationary states of control gene networks at different values of parameters, obtain kinetic curves for molecular components (RNA, proteins) and patterns of gene activities in time, and discriminate between the hypotheses on gene interactions.

2.3 Schemes of computer experiments

For the parametric stability, we tested a fragment of the Dr-CGN model consisting of the genetic blocks that correspond to maternal coordinate (*bicoid*, *caudal*, and *nanos*) and gap (*giant*, *hunchback*, *knirps*, *Krüppel*, and *tailless*) genes (Figure 2). Previously, kinetic curves were obtained in the model for molecular components (RNA, proteins) at different initial parameters corresponding to different concentrations of BCD, NOS, Cad, and HB morphogens in nuclei along the egg's anterior–posterior axis (Tchuraev and Galimzyanov, 2001b). Calculations were made for each nucleus of a hundred located on the control line. This control line consists of two parts, takes a position of the lateral equator (Reinitz and Sharp, 1995), and determines coordinates of the nuclei relative the dorsal and ventral cavities.

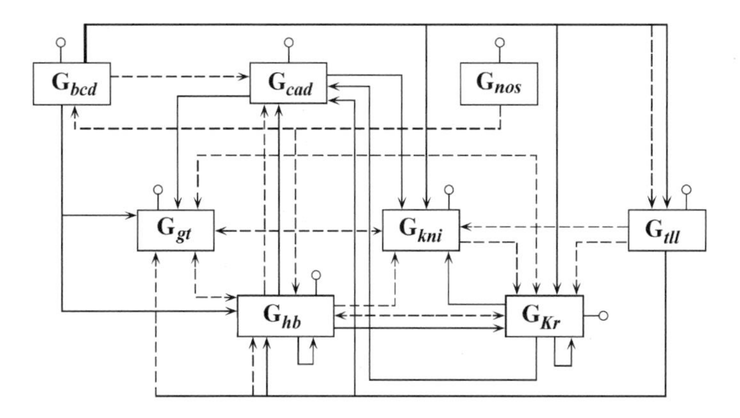

Figure 2. Scheme of the model gene network controlling *Drosophila* early ontogenesis. The model includes synthesis blocks G1, G2, ..., G8, i.e. blocks synthesizing protein products of the genes *bicoid*, *caudal*, *nanos*, *giant*, *hunchback*, *knirps*, *Krüppel*, and *tailless*. Each synthesis block is described as a genetic block in terms of the GTM formalism. The Dr-CGN model is comprehensively described in (Galimzyanov and Tchuraev, 2002). Genetic blocks are symbolized by branch-bearing rectangles; information relations among genetic blocks are marked with either solid (positive regulation) or dotted (negative regulation) lines. Thick solid lines are used in the special case when one and the same protein acts as both a repressor and an activator.

Four types of computer experiments were performed. In experiments 1 and 2, the values of kinetic parameters were selected within the limits of *l*-percent ($l = 10$, 20, 30, and 40%) deviation from the values of parameters in the basic set, when the Dr-CGN model was operating in a normal mode. In experiment 1, we varied unit intensities of gene transcription and mRNA translation as well as coefficients of RNA and protein degradations. In experiment 2, we varied only threshold values of regulatory protein concentrations. Experiment 3 combined the conditions of experiments 1 and 2. In Experiment 4, the values of kinetic parameters were selected at random over wide intervals. With the aid of computer, 500 random choices were made for each type of experiment.

As controls, we took the model patterns of gap gene expression. Four types of computer experiments were performed. Dynamics of the gene network in each of the egg's one hundred nuclei at a randomly selected set of parameters was compared with the dynamics of the gene network in the same nucleus at the basic set of parameters, i.e. in a normal mode. Comparison involved the totality of genetic blocks in the active state. Stability of the model to fluctuations in the values of kinetic parameters was evaluated according to the average of normal modes of the gene network operation in all the egg's nuclei of the total number of random choices. The parametric stability coefficient C_{PS} was determined by the equation $C_{PS}=\frac{1}{n}\sum_{i=1}^{n} c_i$, where n is the length of the control line (n = 100 nuclei) and c_i, percentage of normal modes of gene network operation in the ith nucleus of the total number of random choices (parametric stability local coefficient).

3. RESULTS AND DISCUSSION

The values of coefficient C_{PS} found in experiments 1, 2, and 3 are listed in Table 1. In experiment 4, the parametric stability coefficient amounted to 73.2%. Figure 3 shows the curves for evaluating the model's parametric stability at different initial data.

Table 1. Parametric stability coefficients of the model

Experiment, No.	Coefficient C_{PS} (%) at different l			
	l = 10%	l = 20%	l = 30%	l = 40%
1	97.8	93.0	88.6	83.5
2	76.2	59.8	51.0	45.3
3	74.2	57.0	47.5	41.1

Thus, the following results are obtained:

1. In experiments 1, 2, and 3, the model's parametric stability decreases with an increase in the deviation percentage of the values of kinetic parameters as compared to those of the basic set. This happens because of an "unbalance" in the system with regard to such characteristics as the relation between the rates of RNA/protein syntheses and time of

activation of the controlled genes, impairing formation of normal gene expression patterns along the egg's anterior–posterior axis.

2. In experiment 4, coefficient of the model's parametric stability (a random choice of values from a wide interval) is similar to that at a 10% deviation from normal modes in experiments 2 and 3. During experiment 1, the model operates in a normal mode in nuclei in the control line in the majority of cases. Experimental conditions 2 and 3, with deviations of the thresholds from the norm exceeding 30%, simulate a critical effect on the behavior of the system, when the majority of nuclei show less than 50% normal modes from the total number of choices. Thus, fluctuations in thresholds of regulatory protein activities tend to lower the model parametric stability much greater than fluctuations in coefficients of RNA/protein synthesis and degradations of gap genes.
3. In all the experiments with *Drosophila* gene subnetworks, the nuclei that form the control line have compartments different from each other in their quantitative (the number of variants with normal operation modes) and qualitative (the form of a curve fragment) parametric stability characteristics. In general, the boundaries between compartments correspond to the boundaries between expression regions of genes *gt* and *tll*, *tll* and *gt*, *gt* and *hb*, *hb* and *Kr*, *Kr* and *kni*, *kni* and *gt*, and *gt* and *tll*, respectively, found in the model. This can be attributed to gene alternative operation in the subnetwork. *This suggests that Drosophila embryo is heterogeneous from the standpoint of gene network parametric stability in the nuclei along the egg's anterior–posterior axis.*
4. The curves for evaluating the parametric stability obtained in experiments 2 and 3 are qualitatively similar—with points close to compartment boundaries displaying maximum and minimum values of the parametric stability local index. The regions of 10–15 *n.n.*, 38–45 *n.n.*, 60–70 *n.n.*, and 80–85 *n.n.* contain negative peaks (Figure 3*c*). These experimental results suggest that under fluctuations in the values of regulatory substance thresholds, shape of the parametric stability curve is qualitatively determined by concentration gradients of individual morphogens. Consequently, the less is the difference between concentration maximum and minimum values of a certain morphogen (gradient local amplitude) at compartment boundaries, the higher is the model sensitivity to fluctuations in the threshold values of morphogen concentrations.

Note here that distribution of concentrations of maternal-gene protein products with a necessary gradient along the egg's anterior–posterior axis is important for forming the normal patterns of Dr-CGN gene expression. Since we assumed that the gene networks controlling ontogenetic processes displays a high degree of parametric stability, the first and second results of our computer experiments serve as additional verification of the adequacy of the model to *Drosophila* early ontogenesis. The third and fourth results are of prognostic significance and count in favor of *inferred* properties of the *Drosophila* gene network during its early ontogenesis.

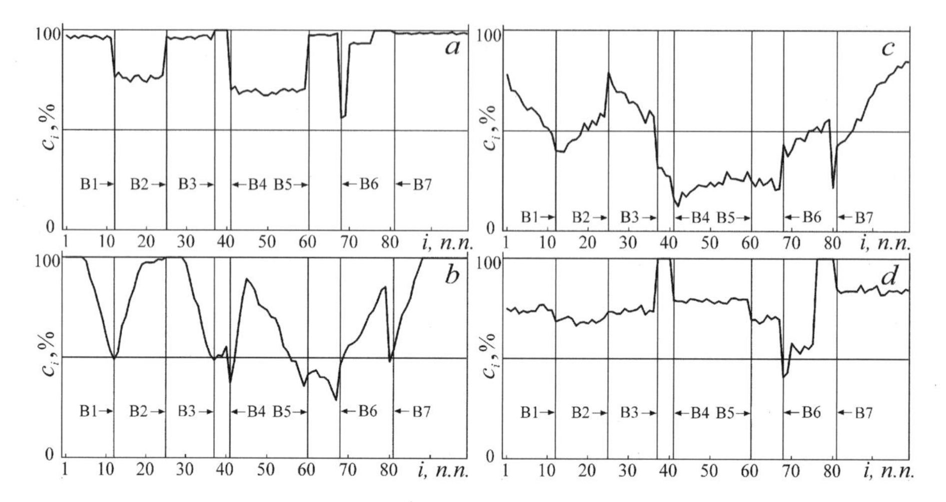

Figure 3. Curves for evaluating the model's parametric stability in the experiments: (*a*) E(1, 30%); (*b*) E(3, 10%); (*c*) E(3, 30%); and (*d*) E(4, –). E(*k*, *l*) is the type of experiment, where *k* is the number of experiment *l*, deviation percentage; *n.n.*, number of nucleus (in each figure, *n.n.* 1–25 correspond to embryo's anterior part; 80–100, posterior part; c_i, local index of the model parametric stability; B1, B2, B3, B4, B5, B6, and B7, boundaries between expression regions of the genes *gt* and *tll*, *tll* and *gt*, *gt* and *hb*, *hb* and *Kr*, *Kr* and *kni*, *kni* and *gt*, and *gt* and *tll*, respectively.

Application of the computational procedures described above not only permits the limits of the ranges of kinetic parameter values where the model is parametrically stable to be found, but also ensures a more thorough analysis aimed at distinguishing the genes in the gene network that make the greatest "contribution" to parametric stability disturbances (i.e. are unreliable elements, so to say). In addition, it is possible to determine the boundaries of embryo's compartments where the gene network shows a lower parametric stability. Of interest is a correlation of unreliable elements with the key genes of the network, because, on the one hand, the operation of key genes should be more protected against fluctuations, and, on the other, they enter into relations with many other genes in the network, which seemingly lowers the degree of stability of the whole network. Correlations of this kind for other gene networks might be by no means as trivial as those studied in our paper.

4. CONCLUSION

Comparison of the data on parametric stability with other control gene networks in eukaryotes—*Drosophila* (von Dassow *et al.*, 2000) and Arabidopsis (Tchuraev and Galimzyanov, 2001a)—and the prokaryote, λ phage development control system (Tchuraev and Ratner, 1975; Kananyan *et al.*, 1981), suggests the following conclusions:

1. Gene networks that control the development of organisms possess a high degree of parametric stability;
2. DNA sites providing specific binding of regulatory proteins allow synonymous substitutions with respect to ontogenetic processes; and
3. Neutralization of mutation phenotypic manifestations is possible particularly owing to high organization of the structure in the control gene network.

Further steps in studying gene networks that control ontogenetic processes, particularly, their stability, may imply both the extension of the previously constructed models (involvement of genes from some other classes) and accounting for cell–cell interactions and compartment formation characteristic of eukaryotic organisms. Noteworthy that the method of generalized threshold models has a specific field of application, and its further development requires the spatial heterogeneity of real systems would be taken into account. Our research seems to facilitate the construction of more sophisticated models aimed at understanding of ontogenetic processes and mechanisms.

ACKNOWLEDGEMENTS

We are grateful to N.A. Kolchanov, V.A. Likhoshvai, I.E. Toto, and J. Reinitz for critical discussion of the data given in the paper reviews. We also thank I.P. Mazola for her valuable assistance in preparing the manuscript.

MODELING PLANT DEVELOPMENT WITH GENE REGULATION NETWORKS INCLUDING SIGNALING AND CELL DIVISION

H. JÖNSSON[1,2], B.E. SHAPIRO[3], E.M. MEYEROWITZ[1], E. MJOLSNESS[4,1]*

[1]Division of Biology, California Institute of Technology, Pasadena, CA 91125; e-mail: henrik or meyerow@caltech.edu; [2]Department of Theoretical Physics, Lund University, Lund, Sweden; [3]Jet Propulsion Laboratory, California Institute of Technology, Pasadena, CA 91109; e-mail: bshapiro@jpl.nasa.gov; [4]Department of Information and Computer Science, University of California, Irvine, CA92697; e-mail: emj@uci.edu

* *Corresponding author*

Abstract The shoot apical meristem of *Arabidopsis thaliana* is an example of a developmental system which can be modeled at genetic and mechanical levels provided that suitable mathematical and computational tools are available to represent intercellular signaling, cell cycling, mechanical stresses, and a changing topology of neighborhood relationships between compartments. In this paper, we present a simplified dynamical 2-dimensional model of a growing plant. Cells in the shoot grow and proliferate, while the number of stem cells at the apex stays constant due to differentiation into tissue cells. Cell types are defined by protein concentrations within the cells, and the dynamics of the differentiation follows from a gene regulation network, which includes intercellular signals.

Key words: arabidopsis, shoot apical meristem (SAM), Cellerator, computer modeling

1. INTRODUCTION

Developmental systems in biology are complex multicellular systems thatrequire multiple tools to be fully understood. In this paper, we show how a mathematical model of biological components can be used to simulate qualitative behavior of a growing plant. The model consists of cells, which grow and proliferate, mechanical interactions between cells, and an underlying genetic network describing the dynamics of the cell states. Important for the dynamics is the availability of signaling between neighboring cells.

The model is applied to the shoot apical meristem (SAM), from which the complete aboveground adult plant is derived. Cells in the SAM retain the ability to divide throughout the life of the shoot, while differentiation of these cells into mature cell types balance the size of the SAM, which stays close to constant throughout shoot life. Cell fates are dependent on the cell positions, and signaling between neighboring cells is believed to play a major role in the differentiation process.

Simulations result in a growing plant, where the SAM is pushed upwards while the stem of the plant is expanding. The stability of the model and its underlying assumptions are discussed along with recent data suggesting a more complicated genetic network with feedback control for regulating the stem cell population within the SAM.

2. METHODS AND ALGORITHMS

2.1 The Shoot Apical Meristem

Consider the shoot apical meristem (SAM) of *Arabidopsis thaliana*, which is a model organism among plants (The Arabidopsis Genome Initiative, 2000). The SAM is the source of the complete aboveground part of the organism. It forms during embryogenesis and retains a nearly constant size and shape from germination, throughout the life of the shoot that it is producing. Among its products are secondary and higher order SAMs that produce branches. The SAM contains a dynamically stable spatial pattern of meristematic regions, despite cell division that causes individual cells or their daughters to move into different regions (Figure 1*a*; Meyerowitz, 1997). The central zone (CZ) is at the very apex, the peripheral zone (PZ) is on the sides, and the rib meristem (RIB) is in the central part of the meristem.

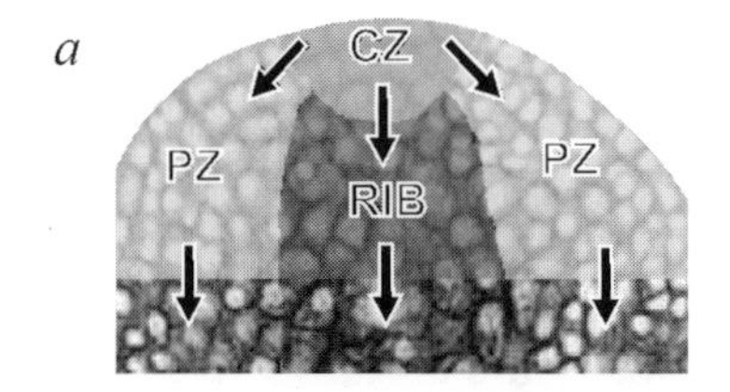

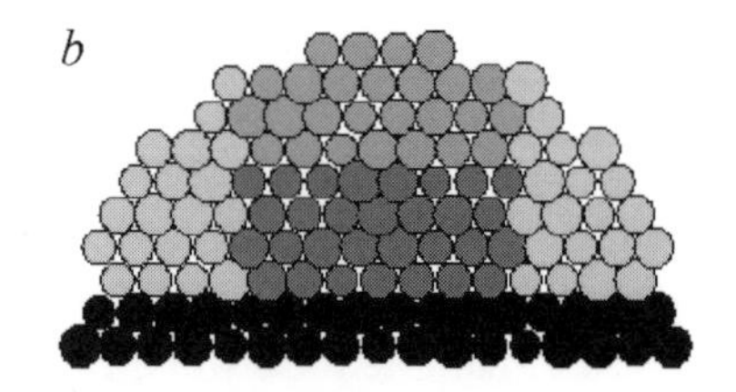

Figure 1. Meristem cell regions: (*a*) section of a meristem where meristematic cell regions are indicated by gray scales. CZ is central zone; RIB, rib meristem, and PZ, peripheral zone. The arrows show the displacement and differentiation path for cells; (*b*) initial configuration for a simulation run where cell gray scales correspond to different protein concentrations.

The slowly dividing central zone cells are thought to be the ultimate stem cell population and provide cells for the maintenance of the meristem. The peripheral zone is where new leaf or flower primordia are initiated, while the rib meristem provides cells for the formation of the stem. Each of these zones has zone-specific gene expression, where for example the central zone is the domain of expression of the *CLAVATA3* gene (Fletcher *et al.*, 1999).

It is known that cell positions are more important than cell lineages for the cell fates in the SAM (Scheres, 2001). Experiments where a single cell is marked show that its descendents can end up as parts of different tissues (Jegla and Sussex, 1989). In the root meristem, laser ablation studies have shown that if a cell is separated from the more mature cells, it remains arrested in a not fully differentiated state (van den Berg *et al.*, 1995). These features are used as basic assumptions in the model presented in this chapter.

In recent experiments, mutant phenotypes have been used to find the roles of different genes in the development of the SAM, to define the mechanisms of communication between different groups of meristematic cells, and to find the patterns of gene regulatory interactions that define the expression domains of the genes. For example, the *CLAVATA3* expression domain has been shown to be partly regulated by the activities of the *CLAVATA1* and *WUSCHEL* genes, as well as by its own activity (Fletcher *et al.*, 1999; Brand *et al.*, 2000; Bowman and Eshed, 2000).

2.2 The Underlying Model

In previous work we have introduced a mathematical framework for gene regulation networks combined with cell signaling (Marnellos and Mjolsness, 1998), and the "Cellerator" package for automatic model generation from reaction relationships (Shapiro *et al.*, 2001) and regulatory relationships along with cell division (Shapiro and Mjolsness, 2001). These tools may be combined to produce models capable of simultaneously representing transcriptional regulation, intercellular signaling, cell division, and mechanical deformation as appropriate to a developmental model. For this study, the model framework is implemented in a C++ program, which is used for the SAM simulations.

Generalizing from Marnellos and Mjolsness (1998) and Mjolsness *et al.* (1991), we use the combined gene regulation and cell–cell signaling dynamics:

$$\frac{d}{dt}v_a(t) = \frac{1}{\tau_a}\left[g(u_a + h_a) - \lambda_a v_a\right], \tag{1a}$$

where

$$u_a(t) = \sum_b T_{ab} v_b(t) + \sum_{I \in Nbrs} \Lambda^I \sum_b \hat{T}_{ab} v_b^I(t) + \\ + \sum_{I \in Nbrs} \Lambda^I \sum_b \sum_c \tilde{T}_{ac}^{(1)} \tilde{T}_{cb}^{(2)} v_c(t) v_b^I(t). \tag{1b}$$

Here v denotes the protein concentrations within a cell. The matrix T represents an intracellular gene regulation network, $\hat{T}$ is an intercellular network, and $\tilde{T}^{(1)}$ and $\tilde{T}^{(2)}$represent a more detailed intercellular signaling network which separates the connection of receptors and ligands ($\tilde{T}^{(2)}$) from the connection of receptors and nuclear pathway target genes ($\tilde{T}^{(1)}$). The parameter h is used to tune the basal expression level, while λ determines the degradation rate and τ sets a time scale for the reaction. The function $g(x)$ is a sigmoid function, which is able to vary the final output, from an almost linear, to an on-off behavior of the gene expression.

A dynamical neighborhood relation is used to describe the intercellular signaling (Λ in equation (1)). In this case, a simple connection matrix, $\Lambda \in \{0,1\}$, is used to describe if cells are neighbors ($\Lambda = 1$), or not, ($\Lambda = 0$). A pair of cells is defined as neighbors if the distance between them is less than a threshold value, proportional to the radii sum, such that only nearest neighbor cells are connected. Since the cells are moving and dividing, the neighborhood connection matrix is updated at each time step of the simulation. Cell shapes are approximated as spheres, and a simple model for cell growth and cell division is added, which can be chosen from a variety of published models (Goldbeter, 1991; Gardner *et al.*, 1998; Shapiro and Mjolsness, 2001). Mechanical interaction between cells is modeled by a softly truncated spring force between cell centers, with a relaxing distance typically set to the sum of the radii of the interacting cells. The cell movement, rather than the acceleration, is proportional to the force, to simulate a highly viscous media (Shapiro and Mjolsness, 2001). While the repelling force is modeled as a standard spring force, the adhesion force is truncated to a given width and strength, reflecting that there is no adhesion between cells that are far apart. The connection matrix Λ is also used for optimizing the calculation of the mechanical interaction, only applying the truncated spring force for neighboring cells.

2.3 The Simulated SAM Network

A model where the cell–cell signaling is the main driver of cell differentiation is defined. Cells that initially correspond to stem cells in the central zone have ability to change state into peripheral or rib meristem cells, when they are neighbors to these cells. Also peripheral zone and rib meristem cells can differentiate when they are neighbors to cells of the stem.

Four genes are introduced as markers of different cell types in the SAM. An intracellular winner-take-all network is introduced (Figure 2*a*) such that only one of the genes is highly expressed in each cell. This unique expression is achieved by a network in which each gene promotes its own expression, while it represses the expression of the other genes. A cell and its descendents will usually end up in a state where the gene with the highest initial concentration is expressed, while the other genes are not. The cells are initiated with different expressions in the different meristemic regions as shown in Figure 1*b*.

An intercellular network is also introduced as shown by the dashed lines in Figure 2*b*. The intercellular network introduces a repression of a selected gene in neighboring cells, together with promotion of its own expression. The result is that a cell can change state if it is neighbor to a region of more "mature" cells. The intercellular signaling is driving the dynamical differentiation of cells, from central zone cells into peripheral zone and rib meristem cells, and from pz/rib cells into cells of the plant stem.

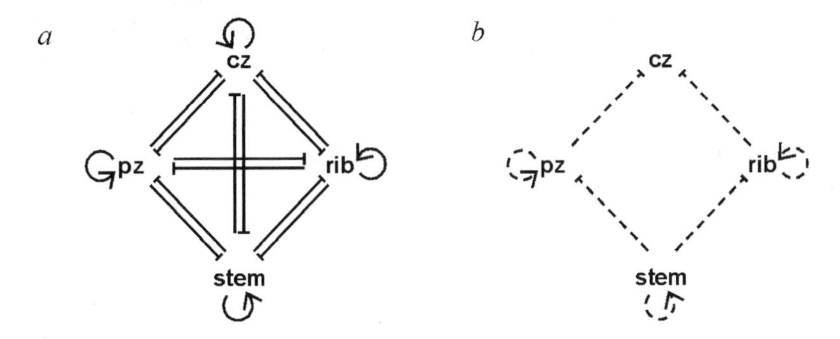

Figure 2. Gene interaction network. Arrows (↑) represent upregulation, which is implemented as a positive entry in the corresponding *T*-matrix in equation 1, while barred lines (┬) represent downregulation (negative *T*-parameter): (*a*) solid lines represent the intracellular network; (*b*) dashed lines show the intercellular interactions between neighboring cells.

Cell growth and division is stopped as the cells become part of the stem. This is implemented by using the gene expression within a cell to control the growth parameters. No difference in growth or proliferation rate is implemented for different meristematic regions, although it is again straightforward to control these parameters using the protein concentrations within cells (Jönsson *et al.*, 2002).

3. RESULTS AND DISCUSSION

The dynamical behavior of the simplified SAM model is shown in Figure 3. The cells are marked with different gray scales representing the protein concentrations. The peripheral zone protein corresponds to the lightest gray, while the central zone and rib meristem proteins are marked in two darker shades of gray. The cells where the stem marking protein is high in concentration are colored as black cells at the bottom. Figure 3*a* shows the simulation close to the initial configuration of Figure 1*b*. Some of the cells are already in the process of changing state, which is detected by the darker gray cells at the boundaries between different regions.

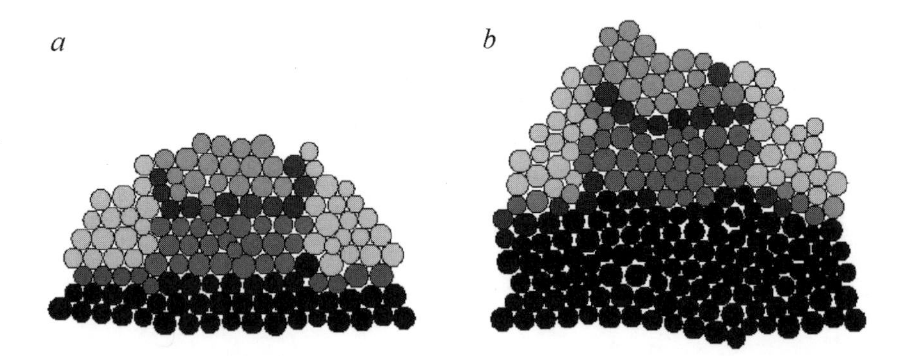

Figure 3. Result of cell division, intercellular signaling, and intracellular gene regulation network dynamics. Different gray shades correspond to protein concentrations related to the marker genes: (*a*) start of simulation (t = 300 in arbitrary units); (*b*) result of long-term dynamics (t = 1700).

In Figure 3*b*, a later time point is shown, where a number of cell divisions have occurred. Also, a number of cells have differentiated, resulting in almost constant gene expression regions in the SAM although the individual cells have changed state. In the simulation, the plant grows as the number of cells in the stem increases despite that these cells do not divide.

3.1 Stability

As discussed in section 2.1, the gene expression regions of the SAM are quite stable. The gene network is also well designed resulting in a self-organization of the domains. If for example the SAM is bisected, it can form two functioning SAMs with characteristic cell domains (Steeves and Sussex, 1989). The simple model simulated in this paper, where expression domains in the SAM stay constant during growth by an increase of cells due to cell divisions balanced with a decrease due to differentiation, does not reflect this stability. This is illuminated in longer simulations, and Figure 4 shows statistics of the sizes of the gene expression domains for a number of runs at late time points. In each run, initial cell sizes and individual cell cycle periods are varied slightly (cf Figure 1*b*). It can clearly be seen that the variation of the region sizes increases the longer the simulations run.

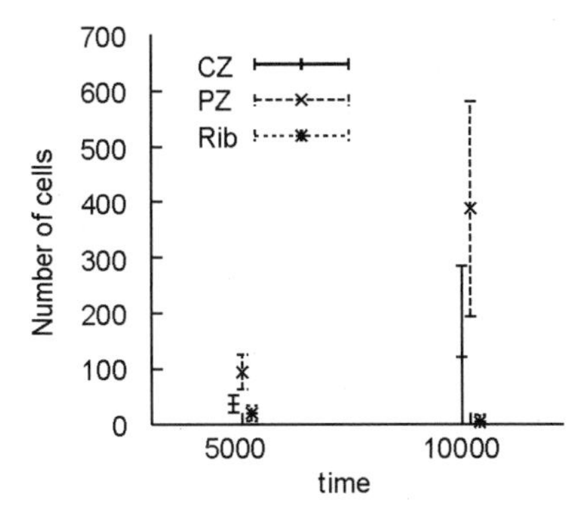

Figure 4. Statistics of the domain sizes for 10 simulations with slightly different initial configurations. The mean and standard deviation (bars) of the distribution is shown at two different time points, showing the increased variation the longer a simulation runs. The time unit is arbitrary (cf. Figure 3*b* where t = 1700). CZ is central zone; RIB, rib meristem, and PZ, peripheral zone.

The instability might be explained by a simple analysis of the model, where the dynamics of the volume, V, of an expression domain is described as in equation (2).

$$\frac{dV}{dt} = \alpha V - \beta V^{2/3}. \tag{2}$$

The first term represents the volume increase due to cell division, proportional to the volume, while the second term is the decrease due to cells changing state at the border towards another domain, proportional to $V^{2/3}$. The equation has two fixed points ($dV/dt = 0$) at $V_1^* = 0$, and $V_2^* = (\beta/\alpha)^3$. The second fixed point is the interesting one, but it is instable, which means that a small deviation from it will result in that the volume either decrease or increase away from the fixed point value. It is possible to tune parameters to stay close to the fixed point for quite a while, but in the long run, it is inevitable that the cell domain either disappears, or grows to infinite size. The more troubling behavior of the model is that it will never self-organize into meristematic regions, but can only maintain regions that are initiated at the start of a simulation.

In section 2.1, we described recent experiments that suggest a feedback network between genes expressed in different domains of the SAM. This data suggests a regulation of the stem cell region that might better explain the stability of the stem cell region. In a static simulation, we have also shown that it is possible to create a stable, self-organizing stem cell domain in the SAM using a regulatory network based on this data (Jönsson *et al.*, 2003). Although the new data provides clues for the stability discussion it does not answer all questions, and the stability of the SAM regions remains to a large extent an open problem. The simulations and analysis described in this paper can be seen as an argument for a more complicated set of feedback controls on gene regulation.

3.2 Outlook

We have shown how computer simulations based on a multicellular mathematical model of a developmental system can be used to help qualitative reasoning, in the case of a developmental system. We have addressed the question of cell differentiation of plant cells, and in our simulations, qualitative features of a growing plant are achieved. However, the lack of long-time stability and self-organization of the meristematic regions in the model indicates a more advanced system of interacting genes to create the stable expression regions in the SAM.

ACKNOWLEDGMENTS

The research described in this paper was carried out, in part, by the Jet Propulsion Laboratory, California Institute of Technology, under contract with the U.S. National Aeronautics and Space Administration. Further support came from the Whittier Foundation, the ERATO Kitano Symbiotic Systems project, and the California Institute of Technology President's Fund. HJ was in part supported by the Knut and Alice Wallenberg Foundation through the Swegene consortium.

THE GLOBAL OPERATION MODES OF GENE NETWORKS DETERMINED BY THE STRUCTURE OF NEGATIVE FEEDBACKS

V.A. LIKHOSHVAI[1,3,4]*, S.I. FADEEV[2,3], Yu.G. MATUSHKIN[1,3]
[1]Institute of Cytology & Genetics, Siberian Branch of the Russian Academy of Sciences, prosp. Lavrentieva 10, Novosibirsk, 630090 Russia, e-mail: likho@bionet.nsc.ru; [2]Sobolev Institute of Mathematics, Siberian Branch of the Russian Academy of Sciences, prosp. Koptyuga 4, Novosibirsk, 630090 Russia;[3]Novosibirsk State University, ul. Pirogova 2, Novosibirsk, 630090 Russia; [4]Ugra Research Institute of Information Technologies, ul. Mira 151, Khanty-Mansyisk, 628011 Russia
* *Corresponding author*

Abstract Any gene networks contain positive and negative feedbacks regulating their operation. This work investigates the interrelation between operation modes of gene networks and their structure–function organization. Four types of negative feedback – based regulatory mechanisms were considered. Basing on numerical experiments, the patterns allowing for prediction of a phase-plane portrait of gene network operation from the structural graph of the gene network were determined. The results reported may find application in the fields of biotechnology and pharmacogenetics, in particular, for constructing gene networks with prespecified properties, searching for optimal strategies for control of gene networks, constructing "intellectual" genetic programs providing correction of parameters of organism's functions, etc.

Key words: hypothetical gene network, mathematical model, computer model, regulation, negative feedback, positive feedback, stationary solutions, stability, theory of gene networks

1. INTRODUCTION

The ability to self-reproduce and survive efficiently under changing conditions of the environment is a unique property of the living systems.

Gene networks, underlying performance of the majority of vital functions, are responsible for realization of this singular ability.

Gene networks are complex functional objects comprising a certain set of physical objects differing in their nature and complexity (functional sites in genomic DNA, genes, RNAs, proteins, various protein complexes, low-molecular-weight compounds, etc.). All these objects are united into a space–dynamic integer via processes of synthesis/degradation as well as active/passive transfer of substances and energy. Proteins are essential elements of the gene networks, as they provide normal progress of the majority of processes. Indeed, it is because a certain part of the proteins synthesized is involved in repression or activation of expression of particular genes, the gene networks acquire the ability to self-regulate and response adequately to changes in external conditions (Kolchanov, 1997).

Analysis of mathematical and computer models demonstrates the essential role of positive and negative feedback mechanisms in generation of individual portraits of stationary and/or cyclic modes of gene network operation (Thomas, 1995; Edwards and Glass, 2000; Elowitz and Leibler, 2000; Gardner *et al.*, 2000). In this work, we are investigating dynamic properties of the gene networks with exclusively negative feedback regulatory mechanisms using numeric simulation of hypothetical gene networks (HGNs).

We are considering four types of negative feedback regulatory mechanisms. For these types, criteria are formulated that allow for prediction of the possible diversity of HGN operation modes from the HGN structural graph without numerical calculations. The results reported may find application in the fields of biotechnology and pharmacogenetics, in particular, for constructing gene networks with prespecified properties, searching for optimal strategies for control of gene networks, constructing "intelligent" genetic programs providing correction of parameters of organism's functions, etc. Intelligent systems are capable of providing a maximally balanced correction of the parameters of the body functions (impaired, for example, due to a genetic defect) to bring them back to an individual norm.

2. MODELS OF HYPOTHETICAL GENE NETWORKS

2.1 Systems of equations describing operation dynamics of hypothetical gene networks

Changes in concentrations of certain set of substances characterize operation of the gene network in the time domain. Biochemical processes together with processes of active and passive transfer of substances and energy form the background of gene network operation. Granting this, it is possible to describe gene networks in terms of systems of ordinary differential equations with the rational right parts (Savageau, 1985) and concentrations of various substances as variables. In HGNs, only proteins—products of synthesis of the corresponding genetic elements (GE)—are considered as the objects changing in the time domain. Taking into account that genes are expressed (proteins are synthesized) in multistep processes—transcription, splicing and translation—composed of a considerable number of quickly progressing elongation stages, let us use systems of equations with retarded arguments of the following form as models of protein synthesis in HGN:

$$\frac{dx_i(t)}{dt} = \sum_{j \in G_i} A_j\left(x_r(t-\tau_{j,i}) \middle| r \in R_j\right) - \beta_i x_i(t),\ i = 1, 2, \ldots, n. \tag{1}$$

Here, n is the number of various proteins; x_i, protein concentrations; β_i, rate constant of the ith protein degradation; G_i, set of the ascribed numbers of genetic elements wherefrom the ith protein is synthesized; $\tau_{j,i}$, retardation time necessary for synthesis of the ith protein from the moment its synthesis from the jth genetic element commenced; $A_j\left(x_r(t-\tau_{j,i}) \middle| r \in R_j\right)$, rational expression specifying operation mechanism of the jth genetic element; and R_j, set of the ascribed numbers of proteins whose concentrations appear in $A_j\left(x_r(t-\tau_{j,i}) \middle| r \in R_j\right)$ as arguments (Likhoshvai *et al.*, 2003).

2.2 Representation of HGN with oriented graphs

Let us consider various proteins within HGN as graph nodes. Let us also consider that the graph has the arc (u, v) if and only if the protein corresponding to u is contained in the regulatory complex controlling operation activity of the genetic element that encodes the protein corresponding to the node v. Let us name the resulting directed graph the HGN graph. Figure 1 exemplifies structural graphs of hypothetical gene networks.

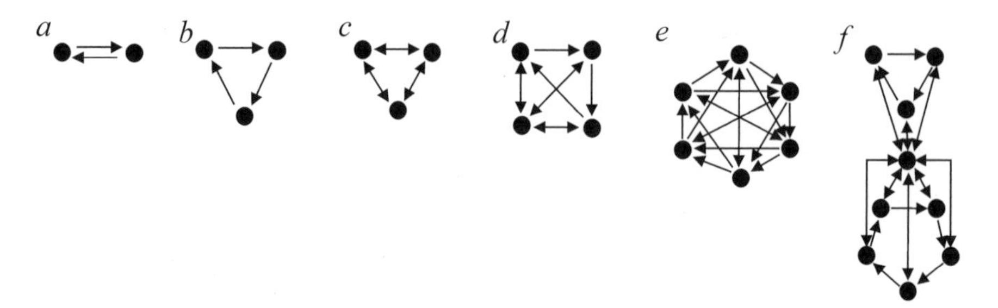

Figure 1. Examples of structural graphs of hypothetical gene networks (*a, c*) with stationaries; (*b, e*) limit cycles; (*d*) stationarie and limit cycle; and (*f*) stationarie and quasi-cycle. For simplicity, the oppositely directed arcs in *c*, *d*, *e*, and *f* are indicated by bidirectional arrows.

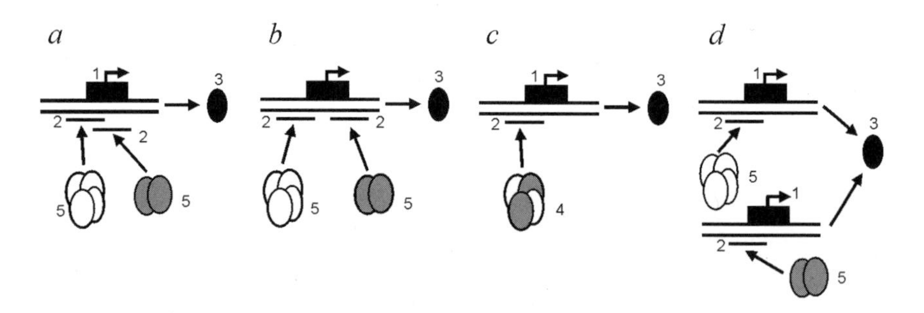

Figure 2. Layouts illustrating the four types (*a–d*) of negative feedback activity regulation of genetic element operation in gene networks. Designations: (1) expression initiation region; (2) regulatory site; (3) protein synthesized; (4) heteromultimer; (5) homomultimer.

2.3 Hypothetical gene networks of classes 1–4

Hypothetical gene networks of arbitrary construction represent a very wide class of theoretical objects, as virtually all the nonnegative rational functions may stand as functions A_j (Likhoshvai *et al.*, 2003). In this work, we are limiting ourselves to consideration of only negative feedback mechanisms. We also omit from consideration the HGNs where proteins directly inhibit the activity of their own genetic elements. Let us further examine four types of regulatory mechanisms of all the diversity of possible regulatory mechanisms. To the first type, sketched in Figure 2*a*, we will assign the regulatory mechanisms that satisfy the following conditions: (i) the regulators are homomultimeric proteins; (ii) inhibition is competitive; and (iii) each protein is encoded by single GE. The HGNs where all the regulatory mechanisms are of the first type are ascribed to class 1. Systems of the following form are the mathematical models of such HGNs:

$$\frac{dx_i(t)}{dt} = \frac{\alpha_i}{1 + \sum_{k \in D_i} \delta_{i,k} x_k^{\gamma_{i,k}}(t - \tau_{i,i})} - \beta_i x_i(t),\ i = 1, \ldots, n. \tag{2}$$

Here and below in equations (2)–(5), D_i denotes the set of ascribed numbers of proteins forming activity regulators of synthesis of the *i*th protein. To the second type (layout is shown in Figure 2*b*), we will assign the negative feedback mechanisms meeting the following supplementary conditions: (i) the regulators are homomultimeric proteins; (ii) inhibition is not competitive; and (iii) each protein is encoded by single GE. The HGNs of class 2 will be constructed using only second type regulatory mechanisms. Systems of the following form represent mathematical models of these HGNs:

$$\frac{dx_i(t)}{dt} = \frac{\alpha_i}{\prod_{k \in D_i} (1 + \delta_{i,k} x_k^{\gamma_{i,k}}(t - \tau_{i,i}))} - \beta_i x_i(t),\ i = 1, \ldots, n. \tag{3}$$

The HGNs of class 3 will be constructed using only third type regulatory mechanisms. These are the mechanisms meeting the following supplementary conditions: (i) the regulators are heteromultimeric proteins; (ii) not more than one regulator corresponds to each GE; and (iii) each protein is encoded by single GE (Figure 2*c*). The following systems represent mathematical models of class 3 HGNs:

$$\frac{dx_i(t)}{dt} = \frac{\alpha_i}{1 + \delta_i \prod_{k \in D_i} x_k^{\gamma_{i,k}}(t - \tau_{i,i})} - \beta_i x_i(t),\ i = 1, \ldots, n. \tag{4}$$

Class 4 comprises the HGNs whose regulatory mechanisms belong to the fourth type, i.e. (i) the regulators are homomultimeric proteins; (ii) not more than one regulator corresponds to each GE; and (iii) each protein is encoded by more than one GE (Figure 2*d*). Mathematical models of class 4 HGNs are the following systems:

$$\frac{dx_i(t)}{dt} = \sum_{k \in D_i} \frac{\alpha_{i,k}}{1 + \delta_{i,k} x_k^{\gamma_{i,k}}(t - \tau_{k,i})} - \beta_i x_i(t),\ i = 1, \ldots, n. \tag{5}$$

The types introduced are constituent units of regulatory processes actually existing in nature. For example, interaction of repressor(s) with specific site(s) resulting in disguise of the site for RNA polymerase is a typical mechanism of transcription repression in prokaryotes. In this process, sites for binding repressors may be solitary and multiple (Figure 1*a*–*d*); sites may sterically overlap (Figure 2*a*) or be disjoint (Figure 2*c*); repressors may be homomultimeric (Figure 2*a*, *b*, *d*) or heteromultimeric (Figure 2*c*); and transcription initiation sites may be multiple (Figure 2*d*). Numerous examples of the situations listed are available in the EcoCyc database (http://biocyc.org). Although the types introduced can be rarely met in nature just as they are, it is of interest to study properties of HGNs in which regulatory mechanisms are of one of the types listed above. Knowledge of these properties will allow for further investigation of more general constructions.

Specific feature of the types introduced is unambiguous determination of the model (2)–(5) by the graph. Therefore, we will speak below that a HGN of classes 1–4 is constructed according to graph G, if G represents the structural graph of the HGN.

To clarify maximally the role of the HGN structural part in forming the dynamic properties, let us minimize the parametric diversity in models (2)–(5). For this purpose, let us impose the following limitations on the parameters:

$$\tau_{k,i} = \tau, \qquad \alpha_i = \alpha_{i,k} = \alpha, \qquad \delta_i = \delta_{i,k} = 1, \qquad \gamma_{i,k} = \gamma,\ \beta_i = 1. \tag{6}$$

The HGN with limitations (6) will be considered as a canonical hypothetical gene network (cHGN).

3. DYNAMIC PROPERTIES OF CANONICAL HYPOTHETICAL GENE NETWORKS

Numerical analysis demonstrated that two groups of values of the parameters α and γ are evident in the parametric diversity of each cHGN. Characteristic of the first group of values, L, is only one stable stationary point of the gene network, i.e. if the pair $(\alpha, \gamma) \in L$, there is no need in numerical calculations to find out the qualitative operation portrait of the HGN. Specific feature of the region L is rather low value of one of the parameters (or both). Therefore, it is natural to name the region L the region of low values of the parameters α and γ. For example, for the cHGN of class 1, constructed according to graphs shown in Figure 1*a*, *b*, only one stationary point $x_i = 2\alpha/(\sqrt{4\alpha+1}+1)$ exists for $\gamma = 1$ and any $\alpha > 0$. It is easy to demonstrate that the stationary uniqueness retains even when $\gamma < 1$. The name of region L is justified by the fact that $\gamma \leq 1$ are the lowest γ values within the range allowed, as this parameter cannot possess negative values according to the sense of problem. Occurrence of the region L was detected numerically for a considerable number of variants of structural graphs, including those shown in Figure 1.

If (α, γ) are beyond the region L, the form of structural graph along with the accepted values of parameters becomes an essential factor forming cHGN properties. To put it otherwise, we cannot state that cHGN has only one stationary point. The answer to the question on its limit properties requires numerical analysis of the corresponding model for specified (α, γ). Here, it cannot be excluded that the gene network will behave in a different way at other (α, γ) values. Nevertheless, numerical calculations distinguished another group of parameters H where the form of structural graph is the sole factor forming the limit modes of cHGN operation.

For example, at $\alpha = 3$, $\gamma = 2$, the structural graph in Figure 1*a* forms exactly two stable stationaries ($x_1 = 2.618$, $x_2 = 0.382$ and $x_1 = 0.382$, $x_2 = 2.618$) of the corresponding cHGN. The cHGN of class 1 constructed according to the graph in Figure 1*c* has three different stationaries ($x_1 = 4.506$, $x_2 = x_3 = 0.2340$; $x_2 = 4.506$, $x_1 = x_3 = 0.2340$; and $x_1 = 4.506$, $x_1 = x_2 = 0.2340$) at $\alpha = 5$, $\gamma = 2$; however, only one ($x_1 = x_2 = x_3 = 1$) at $\alpha = 3$, $\gamma = 2$. Similar situation is observed with the cHGNs of classes 2–4 constructed according to the graph in Figure 1*c*. At $\alpha = 5$, $\gamma = 2$, the cHGN constructed according to the graph in Figure 1*b* has only one attractor, which is a limit cycle. The cHGNs of classes 1, 2, and 4 constructed according to the graph in Figure 1*e* have two alternative modes, each representing a limit cycle, at $\alpha = 5$, $\gamma = 5$; the cHGN of class 3, only one limit cycle.

When the structural graphs shown in Figure 1*d*, *f* are used, the corresponding cHGNs are capable of existing in several qualitatively distinct states: reach stationary under one set of initial conditions or operate in a mode of continuous oscillations of protein concentrations under another. In the first case, the oscillatory mode formed for each class is most likely of cyclic character; in the second, quasi-cyclic (Likhoshvai *et al.*, 2001; Likhoshvai and Matushkin, 2002b). Characteristic of the region H is a rather high value of one of the parameters (or both). Therefore, we named the region H the region of high values of the parameters α and γ.

In a sense, the regions L and H are extreme subsets of the overall diversity of parameter values. This means that if (α, γ)$\in L$, than the limit cHGN behavior is rather simple and predictable: from any initial state, the system converges to the sole stable state. If (α, γ)$\in H$, the limit behavior of cHGN is in no way simple. However, to determine the behavior, it is sufficient to study the corresponding model for only one set of parameters, as the behavior of system will be qualitatively similar to the mode studied for all the rest values belonging to the region H.

When values of the parameters α and γ are intermediate, the operation modes are determined not only by the graph's structure, but also by particular values of the parameters. Let us exemplify the situation by the class 3 cHGN with the structural graph $G_{21,5}$. At small values of α and γ (for example, $\tau = 0$, $\gamma = 1$, and $\alpha = 10$), this model has only one stable stationary point, whereas at rather high values (for example, $\tau = 0$, $\gamma = 10$, and $\alpha = 10$), it has only one stable limit cycle.

However, the phase-plane portrait of the model is richer in the case of intermediate values (for example, $\tau = 0$, $\gamma = 10$, and $\alpha = 5$): three stable oscillatory modes of operation are present; of them, two are cycles and the third has no period. Diagram of aperiodic behavior discovered in the model $M_3(21,5)$ is shown in Figure 3. Figure 3*a* demonstrates the time curves of the first variable calculated according to model (3) constructed using structural graph $G_{21,5}$ at $\tau = 0$, $\gamma = 10$, and $\alpha = 5$ for the initial data $x_1(0) = 1$, $x_i(0) = 0$, $i = 2, \ldots, 21$ (note that the system reaches cyclic operation mode when the following variants of initial data set are specified: $x_1(0) = 10^{-8}$, $x_i(0) = 0$, $i = 2, \ldots, 21$ or $x_1(0) = 10^{+8}$, $x_i(0) = 0$, $i = 2, \ldots, 21$).

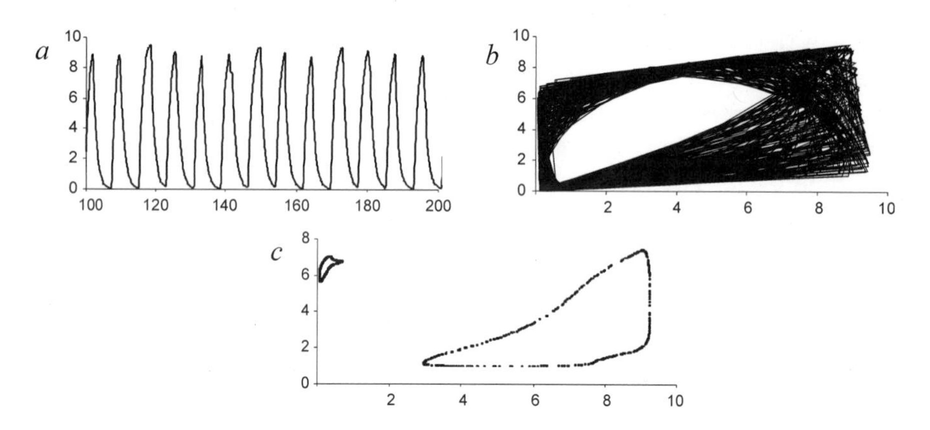

Figure 3. An example of aperiodic behavior in the symmetric hypothetical gene network of class 3 constructed using the graph $G_{21,5}$.

Aperiodicity of the 21-dimensional curve is well evident on the phase curve shown in Figure 3*b*. Analysis of the trajectory projection onto the Poincare plane slicing the 21-dimensional α-cube $[0{:}\alpha, \ldots, 0{:}\alpha]$ by the leading diagonal (Figure 3*c*) suggests that the curve in question is quasi-periodic. This example is also noteworthy, as it demonstrates an aperiodic behavior of a completely symmetrical gene network.

As was noted above, we believe that the cHGN properties in the region *H* are formed exclusively by the effect of structural graph. Therefore, it is interesting to discover the patterns of graphs that would allow the possible gene network behavior to be predicted. Analysis of the properties of structural graph in comparison with the dynamic properties of the corresponding cHGNs allowed us to discover a number of such patterns. We assume that they hold for all the cHGN constructions, not only for those tested in numerical experiments. In particular, we discovered the criterion of the number of stationaries for cHGNs of classes 1–4. Further description of the results requires certain concepts of the graph theory to be introduced. We will mainly use the terminology according to the monograph of Harary (1973). Let a directed graph G = G(V, W), where V is the number of nodes and W, number of arcs, be specified. Let $u, v \in$ V, $(u, v) \in$ W. Let us consider that the node u is adjacent to v and v, adjacent to u. The in-degree (out-degree) of the node is the number of nodes adjacent from (to) v. Let us name as 1-base the minimal set U of such pairwise nonadjacent nodes so that any node of the directed graph G either belongs to U or is adjacent from a node belonging to U. Let us introduce the concept of 1m-base. We will call the set U of directed graph nodes the 1m-base if and only if: (i) all the nodes from the set V\U are pairwise nonadjacent; (ii) all the nodes with zero in-degrees belong to U; and (iii) any node with a nonzero in-degree belonging to U is adjacent to at least one node belonging to V\U.

Let us name directed graph symmetric if and only if all its nodes have equal in-degrees and there exists such arrangements of nodes on the circle that all the nodes accessible from u are located directly after it in a clockwise order. Examples of such directed graphs are shown in Figure 1*a*, *b*, *c*, *e*. As the number of nodes n and out-degree $(k-1)$ characterize symmetric directed graphs completely, let us designate these directed graphs as $G_{n,k}$. Evidently, for a fixed n, exactly $n-1$ various $G_{n,k}$ exist: $k = 2, \ldots, n$. Let us consider the directed graphs $G(V_1, W_1), \ldots, G(V_m, W_m)$, $m > 0$. Let us name the directed graph G(V, W) meeting the conditions

(i) $V = \bigcup_{i=1}^{m} V_i$,

(ii) $W = \bigcup_{i=1}^{m} \left(W_i \cup \{(u,v) \mid u \in V_i, v \in V \setminus V_i\}\right)$

the strong union of directed graphs $G(V_i, W_i)$, $i = 1, \ldots, m$.

Criterion of the number of stationaries:

Let an oriented graph G be given. Let us construct cHGNs of classes 1–4 *using this graph. Then, for any prespecified* τ, *there exists such* $\gamma_0(\tau)$ *that for any* $\gamma > \gamma_0$ *there exists such* $\alpha_0(\gamma, \tau)$ *that for any* $\alpha > \alpha_0$, *the number of stable stationaries of class* 1–4 *cHGNs is equal to* 1-(1*m*-)*bases of the directed graph G(V, W).* It follows necessarily from this criterion that if a structural graph has no 1-(1m-)bases, the corresponding cHGN for any τ and sufficiently high γ and α, will have no stationaries and, consequently, will have at least one oscillatory mode of operation. We yet failed to relate the oscillatory properties of cHGNs to any specific organization characteristics of structural graphs of the general form. In the case of symmetric structural graph, it is possible not only to simplify considerably the initial formulation of this criterion, but also to describe cHGN cyclic properties.

Let the directed graph $G_{n,k}$ be symmetric. Let d designate the greatest common divisor of n and k. Then, the following formulation gives a complete description of 1-bases and 1m-bases of the graph $G_{n,k}$. If $d = k$, then $G_{n,k}$ has exactly k 1-bases and k 1m-bases. The sets of nodes $U_i = \{u_{i+dj}, j = 0, \ldots, n/d - 1\}$ are 1-bases; the sets $V \backslash U_i = \{u_{i+dj}, j \neq 0, \ldots, n/d - 1\}$, $i = 1, \ldots, k$ are 1m-bases. For $d \neq k$, not a single 1-(1m)base exists.

Let us construct models (2)–(5) using the graph $G_{n,k}$ and designate them for the sake of definiteness as $M_i(n, k)$, where $i = 1, \ldots, 4$ indicates the number of class. Then, the criterion of the number of stationaries may be formulated in the following equivalent form:

(n, k)-criterion: *if $d = k$, then for any prespecified τ, there exists such $\gamma_0(\tau)$ that for any $\gamma > \gamma_0$ there exists such $\alpha_0(\gamma, \tau)$ that for any $\alpha > \alpha_0$, the model $M_i(n, k)$, $i = 1$–4 has exactly k stable stationaries.*

The next formulation allows the number of oscillatory modes of cHGNs of classes 1, 2, and 4 constructed using the graph $G_{n,k}$ to be calculated:

(n, k)-criterion of the number of cycles: *if $d \neq k$, then for any prespecified τ, there exists such $\gamma_0(\tau)$ that for any $\gamma > \gamma_0$ there exists such $\alpha_0(\gamma, \tau)$ that for any $\alpha > \alpha_0$, the model $M_i(n, k)$, $i = 1, 2, 4$ has d stable limit cycles.*

As for the cHGNs of class 3, the above (n, k)-criterion of the number of cycles fails to hold. For example, only one stable limit cycle is found numerically in the model $M_3(6,4)$ for rather high α and γ (Figure 1*e*; calculations not shown). Numerical experiments suggest that the (n, k)-criterion holds for the cHGNs of class 3 in the following formulation: *if $d \neq k$, then for any prespecified τ, there exists such $\gamma_0(\tau)$ that for any $\gamma > \gamma_0$ there exists such $\alpha_0(\gamma, \tau)$ that for any $\alpha > \alpha_0$, the model $M_3(n, k)$, has only one stable limit cycle.*

4. CONCLUSION

Transition to analysis of behavior of gene networks regulated by combinations of the types of mechanisms considered multiply complicates search for the characteristics of structural graphs appropriate for predicting stationaries and oscillatory modes of operation.

Data on investigation of Edwards and Glass class of Boolean (threshold) gene network models (Edwards and Glass, 2000), which represent the limit case of models (1), illustrate the complexity of this problem. According to authors' estimate, five elements are enough to construct over 10^{20} hypothetical gene networks with different types of limit operation modes.

However, despite the fact that the results reported describe the properties of a limited diversity of genetic objects, they form certain theoretical background for solving particular problems of gene network construction. For example, these results demonstrate the principal possibility of construction of the gene network with any specified number of stationaries or limit cycles. The following simple algorithm, which we will give without substantiation, solves the problem of constructing hypothetical gene networks with any prespecified number of stationaries and cycles.

Let us assume that we need to construct the gene network with *k* stationaries and *d* cycles. Let us take the directed graphs $G_{n_1k,k}$ and G_{n_2d,n_3d}, where n_1, n_2, $n_3 > 0$, n_2 and n_3 are coprime numbers. Then, the cHGN of class 1 constructed using their strong union possesses the properties sought for.

Note also that the necessary conditions for existence of gene network stationary and/or cyclic operation modes, predicted from analyzing the properties of structural graph, are the presence of certain nonlinearity of the processes involved in activity regulation of gene network genetic elements (a high value of parameter γ) and high enough efficiency of gene expression in the absence of repressors (a high value of parameter α). When constructing real gene networks, an adequate multimerization degree of repressor proteins should provide this necessary nonlinearity, while use of "strong" promoters and/or efficient translation initiation sites, the efficiency of expression (Elowitz and Leibler, 2000; Gardner *et al.,* 2000). Thus, the results obtained may find application in the fields of biotechnology and pharmacogenetics, in particular, for constructing gene networks with prespecified properties, searching for optimal strategies for control of gene networks, and constructing "intelligent" genetic systems capable of providing a maximally balanced correction of the parameters of the body functions (impaired, for example, due to a genetic defect) to bring them back to an individual norm.

ACKNOWLEDGMENTS

This work was supported in part by the Russian Foundation for Basic Research (grants Nos. 02-04-48802, 03-07-96000, 03-04-48829, 01-07-90376-в, 02-07-90359, 03-07-96833-p2003, 03-01-00328, 03-04-48506-a, 03-07-06077-мас, 03-01-00328); Russian Ministry of Industry, Science, and Technologies (grant No. 43.073.1.1.1501); grant PCB RAS (No. 10.4); the Siberian Branch of the Russian Academy of Sciences (integration project No. 119); NATO (grant No. LST.CLG.979816).

STATISTICAL ANALYSIS OF MICROARRAY DATA: IDENTIFICATION AND CLASSIFICATION OF YEAST CELL CYCLE GENES

Yu.V. KONDRAKHIN[1,2], O.A. PODKOLODNAYA[2,3]*, A.V. KOCHETOV[2,3], G.N. EROKHIN[1], N.A. KOLCHANOV[2,3]
[1]Ugra Research Institute of Information Technologies, ul. Mira 151, Khanty-Mansyisk, 628011 Russia; [2]Institute of Cytology & Genetics, Siberian Branch of the Russian Academy of Sciences, prosp. Lavrentieva 10, Novosibirsk, 630090 Russia, e-mail: opodkol@bionet.nsc.ru; [3]Novosibirsk State University, ul. Pirogova 2, Novosibirsk, 630090 Russia
* *Corresponding author*

Abstract We are proposing two methods, based on regression approach and estimation–maximization technique, for an in-depth statistical analysis of microarray profiles. The cell cycle-regulated genes were identified and classified basing on statistical analysis of two characteristics—the cell cycle period and the time point of maximal gene expression. Application of the methods proposed to experimental data of Spellman *et al.* (1998) and Cho *et al.* (1998) allowed us to find 1628 genes involved in the yeast cell cycle and 171 genes associated with the cell cycle. Thus, we succeeded in increasing the number of genes identified as regulated during the cell cycle more than twofold compared with the analogous data published so far. We also demonstrated that various techniques for synchronizing cell cultures might influence specific features of the cell cycle, in particular, change essentially the activation periodicity of certain genes. In this statistical analysis, an increased attention was paid to a high robustness and reliability of the results obtained, which were controlled using variance characteristics.

Key words: microarray data analysis, periodicity, cell cycle-regulated genes, estimation–maximization algorithm

1. INTRODUCTION

Microarray provides researchers with a unique opportunity for investigation of gene expression patterns through differential mRNA quantification. It allows the basic processes, such as mitosis, meiosis, cell differentiation, etc., to be reconstructed. However, although microarray data might be a valuable information source, their interpretation is a complex process that demands development of special mathematical approaches. Only a perfect combination of experimentally measured mRNA levels with appropriate bioinformatics tools will allow detailed components of gene networks regulating all the cellular processes to be revealed.

A vast majority of the methods used now for analysis of microarray data are based essentially on various approaches of applied statistics. Two groups of approaches are most intensely used—the methods involving decrease in dimensionalities, on the one hand, and the methods based on clustering and classification, on the other. The main principle of the first group is to reveal a low-dimensional projection of the initial high-dimensional data. Principle component analysis (Chu *et al.*, 1998), multidimensional scaling (DeRisi *et al.*, 1997), correspondence analysis (Fellenberg *et al.*, 2001), and singular value decomposition method (Alter *et al.*, 2000) are most typical representatives of the first group.

The second group of methods is oriented to creation and analysis of various subsets of genes through analyzing the exhaustive set of microarray profiles and is exemplified by hierarchical cluster analysis (Eisen *et al.*, 1998), *k*-means cluster analysis (Tavazoie *et al.*, 1999), self-organizing map approach (Tamayo *et al.*, 1999), and classification methods (Dudoit *et al.*, 2000).

Note that the majority of these methods, as a rule, are not designed capable of taking into account meticulously the specific features of microarray data, such as, for example, time dependence or periodicity. However, development of the methods able to take into account such specific features is in high demand, as periodical processes are abundant in dynamical genetic systems.

The goal of our work was to identify and classify the genes whose RNA levels changed periodically during the cell cycle using a new statistical approach. We have analyzed the microarray data obtained by Spellman *et al.* (1998) and Cho *et al.* (1998) from synchronized cells as an example of the approach utility. Spellman *et al.* (1998) also analyzed these data by a method combining use of Fourier transform and correlation function and identified 800 genes related to the cell cycle. We have developed a more comprehensive method for analyzing periodicity and succeeded in identifying over 1600 genes related to the cell cycle.

2. MATERIAL AND METHODS

2.1 Data

Microarray data and supplemental files containing list of well-known cell cycle-regulated genes and gene classification obtained by Spellman *et al.* (1998) were extracted from the Stanford Microarray Database (Sherlock *et al.*, 2001) publicly available at http://genome-www5.stanford.edu/MicroArray/SMD/helpindex.html.

2.2 HRAMP algorithm

We have developed two algorithms for mathematical processing of microarray profiles. The first algorithm, HRAMP (Harmonic Regression Analysis of Microarray Profiles), is designed for processing individual microarray profiles to determine the period of cell cycle and the time point of maximal expression. Assuming a stability of periodicity during several successive cell cycles, we are describing using the general cosine function

$$Y(t) = \mu + A \times \cos(\omega \times t + \varphi) + \varepsilon_t\,, \tag{1}$$

each individual profile PROFILE = $\{(t_1, Y_1), \ldots, (t_n, Y_n)\}$, comprising n expression levels of Y_i measured at given time points t_i. In equation (1), μ is the mean value; ω, angular frequency; A, amplitude; φ, phase; and ε_t is the general error comprising the experimental errors of microarray measurements and deviation of the profile approximation by cosine function from the actual unknown curve. Trigonometric algebra reduces the function $Y(t)$ to the following transformed form:

$$Y(t) = \mu + A_1 \times \cos(\omega \times t) + A_2 \times \sin(\omega \times t) + \varepsilon_t\,, \tag{2}$$

where $A_1 = \cos\varphi$ and $A_2 = -\sin\varphi$.

If the angular frequency ω is fixed, then the transformed function $Y(t)$ depends linearly on the parameters μ, A_1, and A_2. Consequently, it is not difficult to obtain simultaneously the estimations of these three parameters by applying a standard multiple regression algorithm to individual microarray profile. The initial parameters A and φ are calculated by the following equations:

$$A = (A_1^{\ 2} + A_2^{\ 2})^{-1/2},\ \varphi = \mathrm{arctg}\,(-A_2 / A_1).$$

The HRAMP algorithm described above implies occurrence of a specified value of the angular frequency ω. The optimal value of this parameter is computed by applying the HRAMP algorithm to different values of ω from a predefined range (ω_1, ω_2). We used the principle of minimization of residual variation *Var* as the optimality criterion, where *Var* is defined as

$$Var = (\Sigma_{i=1,\ldots,n} (Y_i - Y_i^*)^2)/(n-3)$$

and Y_i^* is a regression estimation of the profile point Y_i calculated for a fixed angular frequency ω. In other words,

$$\omega_{\text{opt}} = \underset{\omega \in (\omega_1, \omega_2)}{\operatorname{argmin}} Var = \underset{\omega \in (\omega_1, \omega_2).}{\operatorname{argmin}} (\Sigma_{i=1,\ldots,n} (Y_i - Y_i^*)^2)/(n-3),$$

Noteworthy, the complete cell cycle period T and the time point of the maximal expression level $t_{\max}$, defined as

$$T = 2\pi/\ \omega_{\text{opt}},\ t_{\max} = \operatorname{argmax}(\mu + A \times \cos(\omega_{\text{opt}} \times t + \varphi)),$$

appeared more natural and convenient for analyzing the microarray profiles than the estimated parameters ω_{opt}, μ, A, and φ. The quality of regression was tested through the variation ratio R,

$$R = Var\ /\ Var_0,$$

where $Var_0 = \Sigma_{i=1,\ldots,n} (Y_i - y_0)^2/(n-1)$, $y_0 = \Sigma_{i=1,\ldots,n} Y_i/n$.

The smaller is the value of R, the more significant is the regression, whereas the R values close to unity indicate that a profile is virtually indescribable by a cosine-shaped curve, suggesting, in turn, the lack of periodicity.

2.3 PEHEM algorithm

The second algorithm, PEHEM (determination of the PEak of Histogram using an Estimation–Maximization technique), was developed to assess the sample properties of period T and the time point of maximal expression level $t_{\max}$. Let us describe the algorithm in terms of the period T.

Now, we assume that our period sample $X = \{T_1, \ldots, T_N\}$ consists of a given number of periods T_i calculated by the HRAMP algorithm involving a predefined set of N microarray profiles. We also assume that each period T_i

falls into the interval $[T_1^*, T_2^*]$. To identify a maximal value of the period histogram, it is necessary to estimate the shape of sample distribution. Let us assume that the period distribution is a mixture of two components, one uniform and the other normal, then the density $f(T)$ of the period is

$$f(T) = p \times f_1(T) + (1 - p) \times f_2(T),\ 0 < p < 1,\ T_1^* \leq T \leq T_2^*.$$

The function $f_1(T)$ is the density of the normal distribution with the center parameter a and the scale parameter b:

$$f_1(T) = \exp(-(T - a)^2/2b^2) / ((2\pi)^{1/2} b).$$

The second function $f_2(T)$ represents the density of the uniform distribution in the interval $[T_1^*, T_2^*]$:

$$f_2(T) = 1/(T_2^* - T_1^*).$$

Thus, this representation of the density $f(T)$ contains three unknown parameters p, a, and b, where p is the mixture parameter that defines the proportion of the normal component. All the three parameters are estimated by the maximum likelihood approach using the estimation-maximization algorithm basing on the observed period sample $X = \{T_1, \ldots, T_N\}$. This algorithm is implemented as an iterative procedure, where each iteration consists of two steps—estimation and maximization.

Estimation. The probabilities g_{ij} are estimated under the condition that the values of parameters p, a, and b are known:

$$g_{i1} = p \times f_1(T_i)/(p \times f_1(T_i) + (1 - p)/(T_2^* - T_1^*)),\ g_{i2} = 1 - g_{i1},\ i = 1, \ldots, N,$$

where g_{i1} is regarded as a posterior probability that the ith period T_i is normally distributed, while another value g_{i2} is the probability that the distribution of T_i is uniform, $g_{i1} + g_{i2} = 1$.

Maximization. The parameter values p, a, and b are calculated by maximizing the log-likelihood function $L(p, a, b)$ under the condition that the probabilities g_{ij} are known. $L(p, a, b)$ is first reduced to the form

$$\begin{aligned} L(p, a, b) &= \ln(\Pi_{i=1,\ldots,N} f(T_i)) = [\ln(p)\, \Sigma_{i=1,\ldots,N}\, g_{i1} + \\ &+ \ln(1 - p)\, \Sigma_{i=1,\ldots,n}\, g_{i2}] + [\Sigma_{i=1,\ldots,N}\, g_{i1} \ln(f_1(T_i)) + \\ &+ \Sigma_{i=1,\ldots,N}\, g_{i2} \ln(f_2(T_i))] - [\Sigma_{i=1,\ldots,N}\, g_{i1} \ln(g_{i1}) + \\ &+ \Sigma_{i=1,\ldots,N}\, g_{i2} \ln(g_{i2})], \end{aligned}$$

from which p, a, and b can be derived:

$$p = \Sigma_{i=1,\ldots,N}\, g_{i1}/N,\ a = \Sigma_{i=1,\ldots,N}\, g_{i1}T_i \,/\, \Sigma_{i=1,\ldots,N}\, g_{i1},$$

$$b^2 = \Sigma_{i=1,\ldots,N}\, g_{i1}(T_i - a)^2 \,/\, \Sigma_{i=1,\ldots,N}\, g_{i1}.$$

3. RESULTS AND DISCUSSION

3.1 Identification of cell cycle-regulated genes

Cell cycle-regulated genes were identified and classified basing on statistical analysis of the three subsets of microarray data obtained using yeast cultures synchronized by three independent methods: α-factor arrest, arrest of cdc15 temperature-sensitive mutant, and cdc28 experiment. Each subset of experimental data was processed independently. Initially, we suggested that the period T of cell cycle was unknown.

We used the following strategy to identify a gene as the cell cycle–regulated. For each microarray profile, we applied the HRAMP algorithm 10 000 times changing the angular frequency in a range of $(\omega_1, \omega_2) =$ $= (2\pi/900, 2\pi/30)$ using the equation $\omega_{(i)} = \omega_1 + i \times (\omega_2 - \omega_1)/10\,000$, $i = 1, \ldots, 10\,000$. Finally, the optimal frequency ω_{opt} was determined that corresponded to the least residual variation *Var*. To calculate the generalized cycle period T, we constructed a training sample $X = \{T_1, \ldots, T_N\}$ of individual periods T_i. The condition $R < 0.3$ was used as a criterion for selection. Fulfillment of this inequality virtually guarantees that only the most significant periods T_i fall into the sample X, thereby providing a high robustness in determining T.

The generalized period T was calculated through applying the PEHEM algorithm to the selected sample X. The sample arithmetic mean was used as an initial approximation of the period, necessary for the iteration estimation–maximization procedure. The histogram of the period distribution constructed using the sample X for the case of cdc15-based synchronization is shown in Figure 1.

The analysis performed suggested us that the cultures synchronized by the three methods displayed different cell cycle periods: α-factor, $T = 62.26$; cdc15, $T = 115.37$; and cdc28, $T = 85.49$. Consequently, we infer that different synchronization techniques may cause considerable distinctions in the cell cycle process.

These values of cell cycle periods played an essential role for a direct identification of the cell cycle-regulated gene. We estimated regression curve (2) for each gene, applying the HRAMP algorithm to three microarray profiles obtained by three types of cell synchronization. In doing so, the following fixed values of angular frequency were used for each synchronization type:

$$\omega = 2\pi/T = 2\pi/62.26, \quad \omega = 2\pi/115.37, \quad \text{and } \omega = 2\pi/85.49.$$

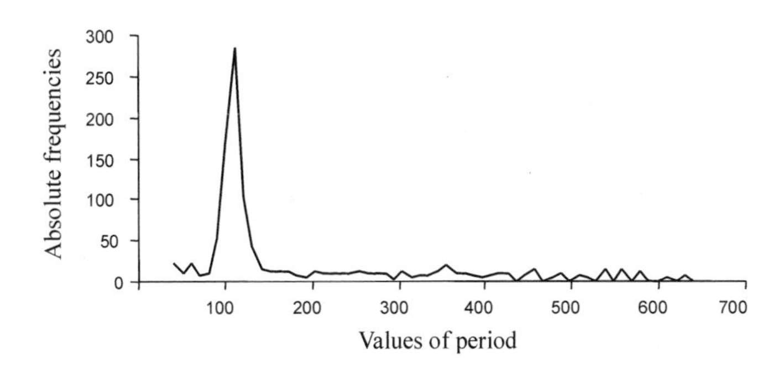

Figure 1. Histogram of the period distribution for the case of cdc15-based synchronization.

In general, we identified a gene as a cell cycle-regulated gene, if essential significance of regression was observed at least in one microarray profile of the three. In turn, the significance of regression was tested using the two following inequalities:

$$R < 0.7, \; Var_0 > 0.04. \tag{3}$$

Using this decision rule, we identified 1628 yeast genes as the cell cycle–regulated genes. The complete list of the genes predicted, LIST1, is available at http://wwwmgs.bionet.nsc.ru/mgs/info/kondrakhin/list1.txt.

Comparison of this list with the 800 genes identified by Spellman *et al.* (1998) is shown schematically in Figure 2. We also identified the majority of the latter genes (namely, 686 of 800) as the cell cycle–regulated genes. However, we supplemented this set with additional 942 genes related to the cell cycle, thereby expanding the list of cell cycle–regulated genes more than twofold.

A cell cycle dependence of expression of additional genes predicted in our analysis but not identified previously by Spellman *et al.* (1998) is in good agreement with known experimental data. For example, this group includes *PPS1* gene, coding for a protein phosphatase (Ernsting and Dixon, 1997), and *DBF4* gene, coding for a regulatory protein (Ferreira *et al.*, 2000) whose transcription increases at the S-phase of cell cycle.

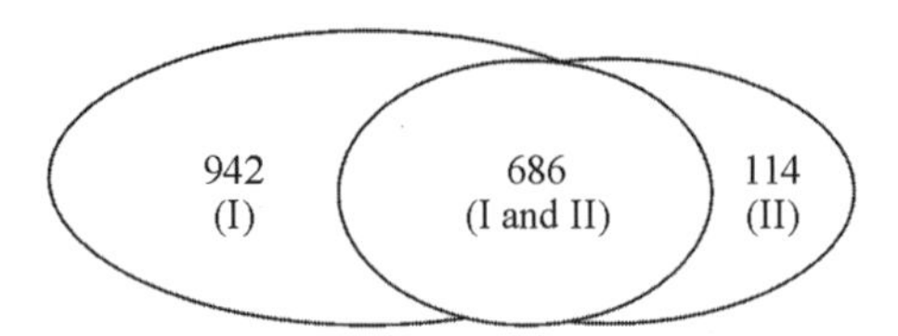

Figure 2. A schematic ratio of (I) the genes we identified as cell cycle-regulated and (II) the genes identified by Spellman *et al.* (1998).

Note that we detected another interesting subsample of genes, whose profiles were definitely periodic; however, according to our criteria (3), we did not ascribe them to the cell cycle–regulated genes because the periods of their microarray profiles were severalfold longer than the cell cycle period. Let us illustrate this effect by an example of the gene *YCR098C*. When cdc15 microarray profile was processed individually, the optimal angular frequency appeared equal to $\omega_{opt} = 2\pi/462.96$, that is, the profile in question had a period of 462.96, exceeding the cell cycle period $T = 115.37$ approximately fourfold. In addition, according to the variance characteristics

$$Var_0 = 1.127594,\ Var = 0.091841, \text{ and } R = Var\,/\,Var_0 = 0.081,$$

the regression quality corresponding to the period of 462.96 is characterized as a very high. Figure 3*a*, showing the theoretically calculated curve and the experimental profile of gene *YCR098C* during cdc15-based cell synchronization, demonstrates this visually.

The cdc28 profile (Figure 3*b*) displaying the following characteristics

$$\omega_{opt} = 2\pi/145.35,\ Var_0 = 1.105831,\ Var = 0.572945, \text{ and } R = 0.518,$$

suggests the same inference.

In particular, the observed period of the profile amounts to 145.35, being approximately twofold longer than the cell cycle period, $T = 85.49$. However, when estimating regression for conventional cell cycle periods, that is, using the values $\omega = 2\pi/115.37$ (cdc15) and $\omega = 2\pi/85.49$ (cdc28), we observed the following variation characteristics:

cdc15, $Var_0 = 1.127594$, $Var = 1.204566$, and $R = 1.068$,
cdc28, $Var_0 = 1.105831$, $Var = 1.174824$, and $R = 1.062$.

The values of variation ratio R close to unity indicate evidently a complete absence of the periods $T = 115.37$ and $T = 85.49$, typical of cdc15- and cdc28-based synchronizations.

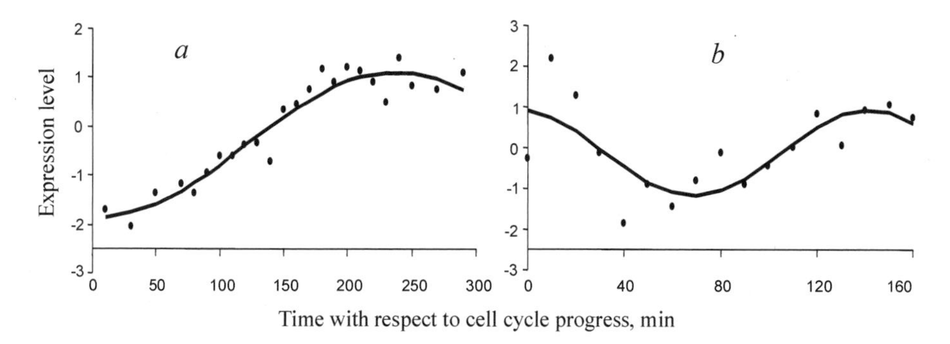

Figure 3. Experimental profiles (dots) and theoretically calculated profiles (firm curves) of the gene *YCR098C* for (*a*) cdc15- and (*b*) cdc28-based synchronizations.

In addition to the set LIST1, we constructed a new set, LIST2, containing 171 genes displaying highly untypical periods. This list of genes associated with the cell cycle is available at http://wwwmgs.bionet.nsc.ru/mgs/info/kondrakhin/list2.txt. Several genes also exhibiting unusual periods in individual profiles were nonetheless identified according to other profiles and included into the main set LIST1. Again, this effect allowed us to infer that different methods of cell synchronization might result in an unusual behavior of certain genes, namely, periods of cyclic behavior of such genes according to microarray profiles exceeded severalfold the conventional cell cycle period. This inference complies well with the fact that different cell cycle periods were identified while analyzing the three synchronization types in question. Identification of the gene set LIST2 suggests us that there may exist a phenomenon of cell memory. This means that expression of certain genes is beyond the frames of a single cell cycle. It seems likely that the process of transcription regulation of such genes is oriented to maintain a certain functional relation between individual independent cell cycles. However, a detailed description of this phenomenon requires further studies.

3.2 Classification of identified genes

Although the period played the major role when identifying the cell cycle-regulated genes, their classification was based on the other characteristic—the time point of maximal expression level t_{max}. The classification procedure was applied only to the genes identified as cell cycle–regulated genes according to our criterion (3). We used a set of 104 well-known genes (Spellman *et al.*, 1998) as a training sample to determine the most typical values of t_{max} for the genes involved in a specified cell cycle phase. This set of genes was divided into five subsets. Each gene belongs to one and only one of the subsets depending on the cell cycle phase (M/G1, G1, S, S/G2, or G2/M) when the expression of this gene peaks. This means that a separate sample X of t_{max} values was constructed for each of the five phases and processed by the PEHEM algorithm. The values of t_{max} calculated for each cell cycle phase are listed in Table 1. While classifying a gene, the cell cycle phase when the least difference between the studied t_{max} value and one of the values listed in Table 1 was observed was ascribed to this gene.

Table 1. Values of the time point of maximal expression level t_{max} calculated for each cell cycle phase by the PEHEM algorithm (the last row, sizes of the samples processed by the algorithm)

Experiment type	Cycle phase				
	M/G1	G1	S	S/G2	G2/M
α-factor	8.63	22.5	35.2	40.5	53.6
cdc15	21.2	45.2	60.6	72.0	102.3
cdc28	9.6	26.8	37.8	49.1	66.9
Sample size	18	54	9	8	15

The main sample LIST1 contains the results of classification of each gene identified; the initial fragment of LIST1 is shown below.

	α-factor	cdc15	cdc28
YAL007C	S	S	–
YAL012W	G2/M	–	–
YAL022C	–	G2/M	G2/M
YAL023C	S	–	–
YAR008W	–	M/G1	G1
YBL002W	S	S	S
YBL003C	S	S	S

Figure 4. Initial fragment of the main sample LIST1 containing the results of classification of the identified genes.

As is evident, the genes *YAL012W* and *YAL023C* were identified only in one type of cell synchronization (namely, alpha experiment). In the rest cases listed here, classification involved two or three synchronization types. Although the data of these three experiments were processed independently, the majority of genes in question displayed a stable classification results, that is, the same cell cycle phase was determined by independent analysis of the same gene (for example, genes *YAL007C, YAL022C, YBL002W*, and *YBL003C*). Even when different cell cycle phases were ascribed to a gene while analyzing different types of synchronization (for example, *YAR008W*), the phases determined were neighboring. This confirms additionally the robustness of our results and the absence of serious errors in classification of the majority of genes.

4. CONCLUSION

The main goal of our work was to perform an in-depth statistical analysis of the microarray data obtained by Spellman *et al.* (1998) and Cho *et al.* (1998) aiming to identify and classify the yeast cell cycle–regulated genes. For this purpose, we used the specially developed statistical methods and succeeded in identifying over 1600 genes, which exceeds twofold the number of genes identified by Spellman *et al.* (1998). We also discovered that different methods of cell culture synchronization might result in essential differences in the cell cycle duration. For example, the cell cycle period in the case of cdc28-based synchronization appeared almost twofold longer compared with the α-factor arrest–based synchronization.

Finally, we identified about two hundred genes whose expression evidently changed in a periodical manner, although the values of their microarray profile periods exceeded essentially the cell cycle duration. Occurrence of such genes suggests the phenomenon of cell memory. Presumably, certain specific processes were involved in regulation of expression levels of these genes; and these processes were determined not by individual successive cell cycles, but rather oriented to maintain a certain functional relation between individual cell cycles.

For a comprehensive analysis of periodicity, we used the general cosine function-based approximation of individual microarray profiles. On the one hand, the cell cycle period played the major role in identification of the genes; on the other, the problem of classification was solved using another important characteristic—the time point of maximal expression level of each gene. The quality of regression approximations we obtained was controlled using variance characteristics; this provided for a certain degree of reliability and robustness of the results obtained.

ACKNOWLEDGMENTS

This work was supported by the Russian Foundation for Basic Research (grants Nos. 01-07-90376-в, 02-04-48508, 02-07-90359, 03-07-96833-р2003, 03-07-96833, 03-04-48506-а, 03-07-06077-мас, 03-01-00328); grant PCB RAS (No. 10.4); Russian Ministry of Industry, Science, and Technologies (grants Nos. 43.073.1.1.1501, 43.106.11.0011 and Sc.Sh.-2275.2003.4); the Siberian Branch of the Russian Academy of Sciences (integration project No. 119); NATO (grant No. LST.CLG.979816).

References

Aalfs J.D., Kingston R.E. What does 'chromatin remodeling' mean? Trends Biochem Sci 2000; 25:548–55.

Adams M.D., Celniker S.E., Holt R.A. *et al.* The genome sequence of *Drosophila melanogaster*. Science 2000; 287:2185–95.

Afonnikov D.A., Oshchepkov D.Yu., Kolchanov N.A. Detection of conserved physico-chemical characteristics of proteins by analyzing clusters of positions with co-ordinated substitutions. Bioinformatics 2001; 17:1035–46.

Agresti, A. *An Introduction to Categorical Data Analysis*. John Wiley & Sons, 2nd edition, 1996.

Alter O., Brown P.O., Botstain D. Singular value decomposition for genome-wide expression data processing and modeling. Proc Natl Acad Sci USA 2000; 97:10101–6.

Altschul S.F., Koonin E.V. Iterated profile searches with PSI-BLAST—a tool for discovery in protein databases. TIBS 1998; 23:444–7.

Ananko E.A., Bazhan E.A., Belova O.E., Kel A.E. Mechanisms of transcription of the interferon-induced genes: a description in the IIG-TRRD information system. Mol Biol (Mosk) 1997; 31:592–605.

Ananko E.A., Naumochkin A.N., Fokin O.N., Frolov A.S. Programs for data input to the Transcription Regulatory Regions Database. Proceedings of the First International Conference on Bioinformatics of Genome Regulation and Structure; 1998 August 24–31; Novosibirsk: ICG, 1998; 1.

Ananko E.A., Oshchepkov D.Yu., Levitskii V.G., Pozdnyakov M.A. Analysis of the regulatory regions of genes involved in the immune system operation. Proceedings of the Third International Conference on Bioinformatics of Genome Regulation and Structure; 2002 July 14–20; Novosibirsk: ICG, 2002; 1.

Ananko E.A., Podkolodny N.L., Stepanenko I.L., Ignatieva E.V., Podkolodnaya O.A., Kolchanov N.A. GeneNet: a database on structure and functional organisation of gene networks. Nucleic Acids Res 2002; 30:398–401.

Anderson, T.W. *An Introduction to Multivariate Statistical Analysis*. John Wiley & Sons Inc. New York, 1958.

Andersson S.G. *et al.* The genome sequence of *Rickettsia prowazekii* and the origin of mitochondria. Nature 1998; 396:133–40.

Andersson S.G., Sharp P.M. Codon usage and base composition in *Rickettsia prowazekii*. J Mol Evol 1996; 42(5):525–36.

Andrade M.A., Ponting C.P., Gibson T.J., Bork P. Homology-based method for identification of protein repeats using statistical significance estimates. J Mol Biol 2000; 298:521–37.

Antonarakis, S.E., Krawczak, M., Cooper, D.N. "The nature and mechanisms of human gene mutation". In *The Metabolic & Molecular Bases of Inherited Disease*, C.R. Scriver, A.L. Beaudet, W.S. Sly, D. Valle, eds. 8th ed., New York: McGraw-Hill, 2001.

Apostolico A., Bock M.E., Lonardi S., Xu X. Efficient detection of unusual words. J Comput Biol 2000; 7(1/2):71–94.

Aravind L., Koonin E.V. Gleaning non-trivial structural, functional and evolutionary information about proteins by iterative database searches. J Mol Biol 1999; 287:1023–40.

Bader G.D., Donaldson I., Wolting C., Ouellette B.F., Pawson T., Hogue C.W. BIND—The Biomolecular Interaction Network Database. Nucleic Acids Res 2001; 29:242–5.

Baker P.J., Britton K.L., Rice D.W., Rob A., Stillman T.J. Structural consequences of sequence patterns in the fingerprint region of the nucleotide binding fold. J Mol Biol 1992; 228:662–71.

Baldi P., Brunak S., Frasconi P., Soda G., Pollastri G. Exploiting the past and the future in protein secondary structure prediction. Bioinformatics 1999; 15:937–46.

Bartel B., Bartel D.P. MicroRNAs: At the root of plant development? Plant Physiol 2003; 132:709–17.

Bashford D., Case D.A. Generalized Born models of macromolecular solvation effects. Annu Rev Phys Chem 2000; 51:129–52.

Batagelj V., Mrvar A. Pajek—program for large network analysis. Connections 1998; 21:47–57.

Bateman E., Iida C.T., Kownin P., Paule M.R. Footprinting of ribosomal RNA genes by transcription initiation factor and RNA polymerase I. Proc Natl Acad Sci USA 1985; 82(23):8004–8.

Baxevanis A.D. The Molecular Biology Database Collection: an update compilation of biological database resources. Nucleic Acids Res 2001; 29:1–10.

Beaudoing E., Freier S., Wyatt J., Claverie J.M., Gautheret D. Patterns of variant polyadenylation signal usage in human genes. Genome Res 2000; 10:1001–10.

Becker O.M., Karplus M. The topology of multidimensional potential energy surfaces: theory and application to peptide structure and kinetics. J Chem Phys 1997; 106:1495–517.

Becker P.B. Nucleosome sliding: facts and fiction. EMBO J 2002; 21:4749–53.

Beg Z.H., Reznikov D.C., Avigan J. Regulation of 3-hydroxy-3-methylglutaryl coenzyme A reductase activity in human fibroblasts by reversible phosphorylation: modulation of enzymatic activity by low density lipoprotein, sterols, and mevalonolactone. Arch Biochem Biophys 1986; 244:310–22.

Beg Z.H., Stonik J.A., Brewer H.B. jr. Human hepatic 3-hydroxy-3-methylglutaryl coenzyme A reductase: evidence for the regulation of enzymic activity by a bicyclic phosphorylation cascade. Biochem Biophys Res Commun 1984; 119:488–98.

Belyaev D.K. Problems and prospects of research into animal genetics and breeding. Genetika 1987; 23:937–946.

Bender E.A., Kochman F. The distribution of subwords counts is usually normal. Eur J Combinatorics 1993; 14:265–275.

Ben-Naim A., Marcus Y. J Chem Phys 1984; 81:2016–27.

Berg C. van den, Willemsen V., Hage W., Weisbeck P., Scheres B. Cell fate in the Arabidopsis root meristem determined by directional signalling. Nature 1995; 378:62–5.

Berg O.G., Hippel P.H. von. Selection of DNA binding sites by regulatory proteins II. The binding specificity of cyclic AMP receptor protein to recognition sites. J Mol Biol 1988; 193:723–50.

Bernstein F.C., Koetzle T.F., Williams G.J.B., Meyer E.F., Brice M.D., Rodgers J.R., Kennard O., Shimanouchi T., Tasumi M. The Protein Data Bank: a computer-based archival file for macromolecular structures. J Mol Biol 1977; 112:535–42.

Berry R.S., Breitengraser-Kunz R. Topography and dynamics of multidimensional interatomic potential surfaces. Phys Rev Lett 1995; 74:3951–4.

Blanchette M., Sinha S. Separating real motifs from their artifacts. Bioinformatics (ISMB special issue) 2001; 817:30–8.

Blank T.A., Becker P.B. The effect of nucleosome phasing sequences and DNA topology on nucleosome spacing. J Mol Biol 1996; 260:1–8.

Blattner F.R. *et al.* The complete genome sequence of *Escherichia coli* K-12. Science 1997; 277:1453–62.

Blażewicz J., Hammer P.L., Lukasiak P. Prediction of protein secondary structure using Logical Analysis of Data algorithm. Comput Meth Sci Tech 2001a; 7(1):7–25.

Blażewicz, J., Hammer, P.L., Lukasiak, P. "Protein Secondary Structure Prediction using logical analysis of data". In *Mathematics and Simulation with Biological, Economical and Musicoacoustical Applications*. WSES Press, 2001b.

Böckmann R., Grubmüller H. Nanoseconds molecular dynamics simulation of primary mechanical energy transfer steps in F1-ATP synthase. Nature Struct Biol 2002; 9:198–202.

Boeckmann B., Bairoch A., Apweiler R., Blatter M.-C., Estreicher A., Gasteiger E., Martin M.J., Michoud K., O'Donovan C., Phan I., Pilbout S., Schneider M. The Swiss-Prot protein knowledgebase and its supplement TrEMBL in 2003. Nucleic Acids Res 2003; 31:365–70.

Booch, G. *Object Oriented Design with Applications*. The Benjamin/Cummings Publishing Co., Inc., 1991.

Bork P., Koonin E.V. Protein sequence motifs. Curr Opin Struct Biol 1996; 6:366–76.

Bornberg-Bauer E., Rivals E., Vingron M. Computational approaches to identify leucine zippers. Nucleic Acids Res 1998, 26:2740–6.

Boros E., Hammer P.L., Ibaraki T., Kogan A. Logical analysis of numerical data. Rutcor Research Report 1997; 04–97.

Boros E., Hammer P.L., Ibaraki T., Kogan A., Mayoraz E., Muchnik I. An implementation of logical analysis of data. Rutcor Research Report 1996; 22–96.

Boros E., Hammer P.L., Kogan A., Mayoraz E., Muchnik I. Logical analysis of data—overview. Rutcor Research Report 1994; 1–94.

Bowen N.J., McDonald J.F. *Drosophila* euchromatic LTR retrotransposons are much younger than the host species in which they reside. Genome Res 2001; 11(9):1527–40.

Bowman J.L., Eshed Y. Formation and maintenance of the shoot apical meristem. Trends in Plant Sci 2000; 5(3):110–15.

Brand U., Fletcher J.C., Hobe M., Meyerowitz E.M., Simon R. Dependence of stem cell fate in Arabidopsis on a feedback loop regulated by CLV3 Activity. Science 2000; 289:617–19.

Branden, C., Tooze, J. *Introduction to Protein Structure*. New York, London: Garland Publishing, 1991.

Brennecke J., Hipfner D.R., Stark A., Russell R.B., Cohen S.M. *bantam* encodes a developmentally regulated microRNA that controls cell proliferation and regulates the proapoptotic gene *hid* in *Drosophila*. Cell 2003; 113:25–36.

Brody T. The Interactive Fly: gene networks, development and the Internet. Trends Genet 1999; 15:333–4.

Brooks B.R., Brucceroli R.E., Olafson B.D., States D.J., Swaminathan S., Karplus M. CHARMM: A program for macromolecular energy, minimization, and dynamics calculations. J Comput Chem 1983; 4:187–217.

Brow D.A., Guthrie C. Transcription of a yeast U6 snRNA gene requires a polymerase III promoter element in a novel position. Genes Dev 1990; 4(8):1345–56.

Brown A.J., Sun L., Feramisco J.D., Brown M.S., Goldstein J.L. Cholesterol addition to ER membranes alters conformation of SCAP, the SREBP escort protein that regulates cholesterol metabolism. Molecular Cell 2002; 10:237–45.

Brown M.S., Goldstein J.L. A proteolytic pathway that controls the cholesterol content of membranes, cells and blood. Proc Natl Acad Sci USA 1999; 96:11041–8.

Brown M.S., Goldstein J.L. Receptor-mediated endocytosis: insights from the lipoprotein receptor system (review). Proc Natl Acad Sci USA 1979; 76:3330–7.

Brunk B., Crabtree J., Diskin S., Mazzarelli J., Zigouras N., Alkalaeva E., Bogdanova V., Trifonoff V., Vorobjeva N., Katokhin A., Kolchanov N., Stoeckert C. Manual annotation of the human and mouse gene index: www.allgenes.org. Proceedings of the Thrird International Conference on Bioinformatics of Genome Regulation and Structure; 2002 July 14–20; Novosibirsk: ICG, 2002; 1.

Buhler J., Tompa M. Finding Motifs Using Random Projections. Proceedings RECOMB'01, Montréal. ACM, 2001.

Butler J.E.F., Kadonaga J.T. The RNA polymerase II core promoter: a key component in the regulation of gene expression. Genes Dev 2002; 16:2583–92.

Caflisch A., Fischer S., Karplus M. Docking by Monte-Carlo minimization with a solvation correction: Application to an FKBP-substrate complex. J Comput Chem 1997; 18:723–43.

Calimet N., Schaefer M., Simonson T. Protein molecular dynamics with the Generalized Born/ACE solvent model. Proteins 2001; 45:144–58.

Cao H., Widlund H.R., Simonsson T., Kubista M. TGGA repeats impair nucleosome formation. J Mol Biol 1998; 281:253–60.

Chaley M.B., Korotkov E.V., Kudryashov N.A. Latent periodicity of 21 bases typical for MCP II gene is widely present in various bacterial genes. DNA Res 2003; 14:33–52.

Chaley M.B., Korotkov E.V., Skryabin K.G. Method revealing latent periodicity of the nucleotide sequences for a case of small samples. DNA Res 1999; 6:153–63.

Chalker D.L., Sandmeyer S.B. Sites of RNA polymerase III transcription initiation and Ty3 integration at the U6 gene are positioned by the TATA box. Proc Natl Acad Sci USA 1993; 90(11):4927–31.

Chekmarev, S. "Confinement Technique for Simulating Finite Many-Body Systems". In *Atomic Clusters and Nanoparticles*, C. Guet, P. Hobza, F. Spiegelmann, F. David, eds. Les Ulis: Springer-Verlag and EDP Sciences, 2001.

Chekmarev S.F., Krivov S.V. Confinement of the molecular dynamics trajectory to a specified catchment area on the potential surface. Chem Phys Lett 1998; 287:719–24.

Chen F., Kuhn D.C., Sun S.C., Gaydos L.J., Demers L.M. Dependence and reversal of nitric oxide production on NF-kappa B in silica and lipopolysaccharide-induced macrophages. Biochem Biophys Res Commun 1995; 214:839–46.

Chen J., Seeman N.C. Synthesis from DNA of a molecule with the connectivity of a cube. Nature 1991; 350:631–3.

Chen Y., Keller J.M. Transcriptional state and chromatin structure of the murine entactin and laminin gamma1 genes. J Cell Biochem 2001; 82:225–33.

Cheng M., Sexl V., Sherr C.J., Roussel. M.F. Assembly of cyclin D dependent kinase and titration of p27Kip1 regulated by mitogen-activated protein kinase kinase (MEK1). Proc Natl Acad Sci USA 1998; 95:1091–6.

Cho R.J., Campbell M.J., Winzeler E.A., Steinmetz L., Conway A., Wodicka L. A genome-wide transcriptional analysis of the mitotic cell cycle. Mol Cell 1998; 2:65–73.

Cho Y., Gorina S., Jeffrey P.D., Pavletich N.P. Crystal structure of a p53 tumor suppressor-DNA complex: understanding tumorigenic mutations. Science 1994; 265:346–55.

Chu S., DeRisi J., Eisen M., Mulholland J., Botstein D., Brown P.O., Herskowitz I. The transcriptional program of sporulation in budding yeast. Science 1998; 282:699–705.

Chuzhanova N.A., Anassis E.J., Ball E., Krawczak M., Cooper D.N. Meta-analysis of indels causing human genetic disease: mechanisms of mutagenesis and the role of local DNA sequence complexity. Hum Mutation 2003; 21:28–44.

Chytil M., Verdine G.L. The Rel family of eukaryotic transcription factors. Curr Opin Struct Biol 1996; 6:91–100.

Clayton D.A. Transcription of the mammalian mitochondrial genome. Annu Rev Biochem 1984; 53:573–94.

Clement J., Dolley E., Regnier M. Approximate Words, 2003 (submitted).

Cocks B.G., Chang C.C., Carballido J.M., Yssel H., Vries J.E. de, Aversa G. A novel receptor involved in T-cell activation. Nature 1995; 376:260–3.

Conte C., Calco V., Desset S., Leblanc P., Dastugue B., Vaury C. Impact of multiple insertions of two retroelements, ZAM and Idefix at an euchromatic locus. Genetica 2000; 109:53–9.

Cook S.J., Aziz N., McMahon M. The repertoire of *fos* and *jun* proteins expressed during the G1 phase of the cell cycle is determined by the duration of mitogen-activated protein kinase activation. Mol Cell Biol 1999; 19:330–41.

Cooper, D.N., Krawczak, M. *Human Gene Mutation*. Oxford: BIOS Scientific, 1993.

Cornell W.D., Cieplak P., Bayly C.I., Gould I.R., Merz K.M., Ferguson D.M., Spellmeyer D.C., Fox T., Caldwell J.W., Kollman P.A. A second generation force field for the simulation of proteins, nucleic acids, and organic molecules. J Am Chem Soc 1995; 117:5179–97.

Crama Y., Hammer P.L., Ibaraki T. Cause-effect relationships and partally defined Boolean functions. Annals of Operations Res 1998; 16:299–326.

Cuff J. *et al.* Jpred: A consensus secondary structure prediction server. Bioinformatics 1998; 14:892–3.

Cuff J., Barton G. Application of multiple sequence alignment profiles to improve protein secondary structure prediction. Proteins 2000; 40:502–11.

Cuff J.A., Barton G.J. Evaluation and improvement of multiple sequence methods for protein secondary structure prediction. Proteins: Struct Funct Genet 1999; 34:508–19.

Dalphin M.E., Brown C.M., Stockwell P.A., Tate W.P. The translational signal database, TransTerm: more organisms, complete genomes. Nucleic Acids Res 1997; 25:246–7.

Dassow G. von, Meir E., Munro E.M., Odell G.M. The segment polarity network is a robust developmental module. Nature 2000; 406:188–92.

Davuluri R.V., Suzuki Y., Sugano S., Zhang M.Q. CART classification of human 5' UTR sequences. Genome Res 2000; 10:1807–16.

DeBoer J.G., Ripley L.S. Demonstration of the production of frameshift and base-substitution mutations by quasi-palindromic sequences. Proc Natl Acad Sci USA 1984; 81:5528–31.

DeGregori J., Kowalik T., Nevins J.R. Cellular targets for activation by the E2F1 transcription factor include DNA synthesis- and G1/S-regulatory genes. Mol Cell Biol 1995; 15:4215–24.

Dej K.J., Gerasimova T., Corces V.G., Boeke J.D. A hotspot for the *Drosophila gypsy* retroelement in the ovo locus. Nucleic Acids Res 1998; 26:4019–24.

Dembo, A., Zeitouni, O. *Large Deviations Techniques*. Boston: Jones and Bartlett, 1992.

Denise A., Régnier M., Vandenbogaert M. Assessing statistical significance of overrepresented oligonucleotides. WABI'01 Proceedings of the First International Workshop on Algorithms in Bioinformatics, Aarhus, Denmark, August 2001; preliminary version as INRIA research report 4132. Springer-Verlag, 2001.

DeRisi J.L., Iyer V.R., Brown P.O. Exploring the metabolic and genetic control of gene expression on a genomic scale. Science 1997; 278:680–6.

Dickerson T.D., Drew H.R. Structure of B-DNA dodecamer. II. Influence of base sequence on helix structure. J Mol Biol 1981; 149:761–86.

Diehl J.A., Cheng M., Roussel M.F., Sherr C.J. Glycogen synthase kinase-3b regulates cyclin D1 proteolysis and subcellular localization. Genes Dev 1998; 12:3499–511.

Dornberger U., Leijon M., Fritzsche H. Solution structure and base pair opening rates of GGCC containing oligonucleotides. J Biomol Struct Dynamics 1999; 16:1251–2.

Draper, N., Smith, H. *Applied Regression Analysis*. New York: John Wiley & Sons, 1981.

Driever W., Nusslein-Volhard C. A gradient of bicoid protein in *Drosophila* embryos. Cell 1988; 54:83–93.

Dudoit S., Fridlyand J., Speed T.P. Comparison of discrimination methods for the classification of tumors using gene expression data. Technical report 576, Dept. of Statistics, University of California, Berkeley, CA, USA, 2000.

Dyson N. The regulation of E2F by pRB-family proteins. Genes Dev 1998; 12:2245–62.

Dzhumagaliev E.B., Bayev A.A., Il'in Iu.V. Primary structure of long terminal repeats and adjacent genome regions for mobile dispersed gene mdg4 in *Drosophila melanogaster*. Dokl Akad Nauk SSSR 1983; 273:214–8.

EcoCyc, http://ecocyc.org/

Edwards R., Glass L. Combinatorial explosion in model gene networks. CHAOS 2000; 10:691–704.

Efstratiadis A., Posakony J.W., Maniatis T., Lawn R.M., O'Connell C., Spritz R.A., DeRiel J.K., Forget B.G., Weissman S.M., Slightom J.L., Blechl A.E., Smithies O., Baralle F.E., Shoulders C.C., Proudfoot N.J. The structure and evolution of the human β-globin gene family. Cell 1980; 21:653–68.

Eidhammer I., Jonassen I., Taylor W.R. Structure comparison and structure patterns. J Comput Biol 2000; 7:685–716.

Eisen M.B., Spellman P.T., Brown P.O., Botstain D. Cluster analysis and display of genome-wide expression patterns. Proc Natl Acad Sci USA 1998; 95:14863–8.

Eisenhaber B., Bork P., Eisenhaber F. Sequence properties of GPI-anchored proteins near the omega-site: constraints for the polypeptide binding site of the putative transamidase. Protein Eng 1998; 11:1155–61.

Ekin O., Hammer P.L., Kogan A. Convexity and logical analysis of data. Rutcor Research Report 1998; 5–98.

Elowitz M.B., Leibler S. A synthetic oscillatory network of transcriptional regulators. Nature 2000; 403:335–8.

Ernsting B.R., Dixon J.E. The PPS1 gene of *Saccharomyces cerevisiae* codes for a dual specificity protein phosphatase with a role in the DNA synthesis phase of the cell cycle. J Biol Chem 1997; 272:9332–43.

Eschenlauer J.B., Kaiser M.W., Gerlach V.L., Brow D.A. Architecture of a yeast U6 RNA gene promoter. Mol Cell Biol 1993; 13(5):3015–26.

Eskin E., Pevzner P. Finding composite regulatory patterns in DNA sequences. Bioinformatics 2002; 1(1):1–9.

Espenshade P.J., Li W.P., Yabe D. Sterols block binding of COPII proteins to SCAP, thereby controlling SCAP sorting in ER. Proc Natl Acad Sci USA 2002; 99:11694–9.

Etzold T., Ulyanow A., Argos P. SRS: information retrieval system for molecular biology data banks. Methods in Enzymology 1996; 266:114–28.

Evans D.A., Wales D.J. Free energy landscapes of model peptides and proteins. J Chem Phys 2003; 118:3891–97.

Eyring, H., Lin, S.H., Lin, S.M. *Basic Chemical Kinetics*. New York: Wiley, 1980.

Falquet L., Pagni M., Bucher P., Hulo N., Sigrist C.J.A., Hofmann K., Bairoch A. The PROSITE database, its status in 2002. Nucleic Acids Res 2002; 30:235–8.

Featherstone M. Coactivators in transcription initiation: here are your orders. Curr Opin Genet Devel 2002; 12:149–55.

Fellenberg K., Hauser N.C., Brors B., Neutzner A., Hoheisel J.D., Vingron M. Correspondence analysis applied to microarray data. Proc Natl Acad Sci USA 2001; 98:10781–6.

Felsenstein J. Phylogenies and the comparative method. Am Nat 1985; 125:1–15.

Ferguson J.C. Multivariable curve interpolation. J ACM 1964; 2:221–8.

Ferreira M.G., Santocanale C., Drury L.S., Diffley J.F.X. Dbf4p, an essential S phase-promoting factor, is targeted for degradation by the anaphase-promoting complex. Mol Cell Biol 2000; 20:242–8.

Festenstein R., Kioussis D. Locus control regions and epigenetic chromatin modifiers. Curr Opin Genet Devel 2000; 10:199–203.

Fickett J.W., Hatzigeorgiou A.C. Eukaryotic promoter recognition. Genome Res 1997; 7:861–78.

Finnegan E.J., Peacock W.J., Dennis E.S. DNA methylation, a key regulator of plant development and other processes. Curr Opin Genet Devel 2000; 10:217–23.

Fischer S., Karplus M. Conjugate peak refinement: an algorithm for finding reaction paths and accurate transition states in systems with many degrees of freedom. Chem Phys Lett 1992; 194:252–61.

Fischer S., Michnick S., Karplus M. A mechanism for rotamase catalysis by the FK506 binding protein (FKBP). Biochemistry 1993; 32:13830–7.

Fischetti V., Landau G., Schmidt J., Sellers, P. Identifying periodic occurrences of a template with applications to a protein structure. Proceedings of the Third Annual Symposium on Combinatorial Pattern Matching. Lecture Notes in Computer Science; Springer-Verlag 1992; 644.

Fishburn, P.C. *Utility Theory for Decision Making*. New York: John Wiley and Sons, 1970.

Fisher A.J., Smith C.A., Thoden J.B., Smith R., Sutoh K., Holden H.M., Rayment I. X-Ray structures of the myosin motor domain of *Dictyostelium discoidum* complexed with Mg.ADP.BeF_x and Mg.ADP.AlF_4. Biochemistry 1995; 28:8960–72.

Fitzgerald D.J., Anderson J.N. DNA structural and sequence determinants for nucleosome positioning. Gene Ther Mol Biol 1999; 4:349–62.

Flajolet, Ph., Sedgewick, R. *Analysis of Algorithms*. Addison-Wesley, 1995.

Fletcher J.C., Brand U., Running M.P., Simon R., Meyerowitz E.M. Signaling of cell fate decisions by clavata3 in *Arabidopsis* shoot meristems. Science 1999; 283:1911–4.

Florenes V.A., Bhattacharya N., Bani M.R., Ben-David Y., Kerbel R.S., Slingerland J.M. TGF-beta mediated G1 arrest in a human melanoma cell line lacking p15INK4B: evidence for cooperation between p21Cip1/WAF1 and p27Kip1. Oncogene 1996; 13:2447–57.

Forst, Wendell. *Theory of Unimolecular Reactions*. New York: Acad. Press, 1973.

Fowlkes D.M., Shenk T. Transcriptional control regions of the adenovirus VAI RNA gene. Cell 1980; 22(2 Pt 2):405–13.

Frank A.C., Lobry J.R. Asymmetric substitution patterns: a review of possible underlying mutational or selective mechanisms. Gene 1999; 238:65–77.

Frank D.E., Saecker R.M., Bond J.P., Capp M.W., Tsodikov O.V., Melcher S.E., Levandoski M.M., Record M.T., Jr. Thermodynamics of the interactions of Lac repressor with variants of the symmetric Lac operator: effects of converting a consensus site to a non-specific site. J Mol Biol 1997; 267:1186–206.

Fraser C.C., Howie D., Morra M., Qiu Y., Murphy C., Shen Q., Gutierrez-Ramos J.C., Coyle A., Kingsbury G.A., Terhorst C. Identification and characterization of SF2000 and SF2001, two new members of the immune receptor SLAM/CD2 family. Immunogenetics 2002; 53:843–50.

Fraser C.M. *et al.* Complete genome sequence of *Treponema pallidum*, the syphilis spirochete. Science 1998; 281:375–88.

Freier A., Hofestädt R., Lange M., Scholz U. BioDataServer: A SQL-based service for the online integration of life science data. In Silico Biology 2002; 2.

Freund R., Meselson M. Long terminal repeat nucleotide sequence and specific insertion of the *gypsy* transposon. Proc Natl Acad Sci USA 1984; 81:4462–4.

Frishman D., Argos P. Knowledge based secondary structure assignment. Proteins: Struct Funct Genet 1995; 23:566–579.

Frishman D., Argos P. Seventy-five percent accuracy in protein secondary structure prediction. Proteins 1997; 329–35.

Furman D.P., Katokhin A.V., Oshchepkov D.Yu., Stepanenko I.L. Do Drosophila retrotransposon LTRs contain functional sites contain functional sites capable of providing heat-shock-inducible transposition? Proceedings of the Third International Conference on Bioinformatics of Genome Regulation and Structure; 2002 July 14–20; Novosibirsk: ICG, 2002; 1.

Futcher B., Latter G.I., Monardo P., McLaughlin C.S., Garrels J.I. A sampling of the yeast proteome. Mol Cell Biol 1999; 19:7357–68.

Gabrielsen O.S., Sentenac A. RNA polymerase III (C) and its transcription factors. Trends Biochem Sci 1991; 16:412–16.

Galimzyanov A.V. Software automated package for analyzing the dynamics of control gene networks. Proceedings of the Second International Conference on Bioinformatics of Genome Regulation and Structure; 2000 August 7–11; Novosibirsk: ICG, 2000; 1.

Galimzyanov A.V., Tchuraev R.N. Mathematical model of the *Drosophila melanogaster* early ontogenesis control subsystem (construction and analysis of the dynamics). Preprint. Ufa: Gilem, 2002.

Gallie D.R. Translational control of cellular and viral mRNAs. Plant Mol Biol 1996; 32:145–58.

Gardner T.S., Cantor C.R., Collins J.J. Construction of a genetic toggle switch in *Escherichia coli.* Nature 2000; 403:339–42.

Gardner T.S., Dolnik M., Collins J.J. A theory for controlling cell cycle dynamics using a reversibly binding inhibitor. Proc Natl Acad Sci USA 1998; 95:14190–14195.

Garnier J., Osguthorpe D., Robson B. Analysis of the accuracy and implications of simple methods for predicting the secondary structure of globular proteins. J Mol Biol 1978; 120:97–120.

Gartenberg M.R., Crothers D.M. DNA sequence determinants of CAP-induced bending and protein binding affinity. Nature 1988; 333:824–829.

Gautier C. Compositional bias in DNA. Curr Opin Genet Dev 2000; 10:656–61.

Geiduschek E.P., Tocchini-Valentini G.P. Transcription by RNA polymerase III. Ann Rev Biochem 1988; 57:873–914.

Gerstein M., Sonnhammer E.L.L., Chothia C. Volume changes in protein evolution. J Mol Biol 1994; 236:1076–8.

Ghosh D. Object-oriented transcription factors database (ooTFD). Nucleic Acids Res 2000; 28(1):308–10.

Ghosh G., Duyne G.V., Ghosh S., Sigler P.B. Structure of NF-κB p50 homodimer bound to a kB site. Nature 1995; 373:303–10.

Gibrat J.F., Madej T., Bryant S.H. Surprising similarities in structure comparison. Curr Opin Struct Biol 1996; 6:377–85.

Gierlik A., Kowalczuk M., Mackiewicz P., Dudek M.R., Cebrat S. Is there replication-associated mutational pressure in the *Saccharomyces cerevisiae* genome? J Theor Biol 2000; 202:305–14.

Gilbert W. The RNA world. Nature 1986; 319:618.
Gimpl G., Burger K., Fahrenholz F. A closer look at the cholesterol sensor. Trends Biochem Sci 2002; 27:596–9.
Glass A., Gierl L. A system architecture for genomic data analysis. In Silico Biology. Special. Issue: GCB'01 2002.
Glickman B.W., Ripley L.S. Structural intermediates of deletion mutagenesis: a role for palindromic DNA. Proc Natl Acad Sci USA 1984; 81:512–6.
Glukhov I.L., Il'in Iu.V., Ivanov V.A. Specific endonuclease activity of integrase encoded by MDG4 (*gypsy*) retrotransposon. Mol Biol (Mosk) 2000; 34:277–84.
Goesmann A., Meyer F., Kalinowski J., Giegerich R. PathFinder: reconstruction and dynamic visualization of metabolic pathways. Bioinformatics 2002; 18:124–9.
Goffeau A. *et al.* Life with 6000 genes. Science 1996; 274:546, 563–7.
Goldberg, D. *Genetic algorithms in search, optimization, and machine learning*. Addison-Wesley, San Mateo, CA, 1989.
Goldbeter A. A minimal cascade model for the mitotic oscillator involving cycline and cdc2 kinase. Proc Natl Acad Sci USA 1991; 88:9107–9111.
Goldstein J.L., Brown M.S. The low-density lipoprotein pathway and its relation to atherosclerosis (review). Annu Rev Biochem 1977; 46:897–930.
Goldstein J.L., Dana S.E., Faust J.R., Beaudet A.L., Brown M.S. Role of lysosomal acid lipase in the metabolism of plasma low density lipoprotein. Observations in cultured fibroblasts from a patient with cholesteryl ester storage disease. J Biol Chem 1975; 250:8487–95.
Goldstein J.L., Rawson R.B., Brown M.S. Mutant mammalian cells as tools to delineate the sterol regulatory element-binding protein pathway for feedback regulation of lipid synthesis. Arch Biochem Biophys 2002; 397:139–48.
Golygina V.V., Kiknadze I.I., Fedotov A.M., Kolchanov N.A. The database "Chironomidae: species, populations, genetic variability". Abstracts of the X International Balbiani Ring Workshop, Varna, 2001; 21.
Gough P.J., Greaves D.R., Suzuki H., Hakkinen T., Hiltunen M.O., Turunen M., Herttuala S.Y., Kodama T., Gordon S. Analysis of macrophage scavenger receptor (SR-A) expression in human aortic atherosclerotic lesions. Arterioscler Thromb Vasc Biol 1999; 19:461–71.
Gregory D.S., Martin A.C., Cheetham J.C., Rees A.R. The prediction and characterization of metal binding sites in proteins. Protein Eng 1993; 6:29–35.
Grigoriev A. Analyzing genomes with cumulative skew diagrams. Nucleic Acids Res 1998; 26:2286–90.
Guerrier-Takada C., Gardiner K., Marsh T. *et al.* The RNA moiety of ribonucleases P is the catalytic subunit of the enzyme. Cell 1983; 35:849–857.
Guibas L., Odlyzko A.M. String overlaps, pattern matching and nontransitive games. J Combinatorial Theory Series A 1981; 30:183–208.
Gusev V.D., Nemytikova L.A., Chuzhanova N.A. A fast method for identification of interconnections between functionally and/or evolutionarily related genetic sequences. Mol Biol (Mosk) 2001; 35:867–73.
Gusev V.D., Nemytikova L.A., Chuzhanova N.A. On the complexity measures of genetic sequences. Bioinformatics 1999; 15:994–9.
Haas L.M., Schwarz P.M., Kodali P., Kotlar E., Rice J.E., Swope W.C. DiscoveryLink: a system for integrated access to life sciences data sources. IBM Systems J 2001; 40:489–511.
Hambleton J., Weinstein S.L., Lem L., DeFranco A.L. Activation of c-Jun *N*-terminal kinase in bacterial lipopolysaccharide-stimulated macrophages. Proc Natl Acad Sci USA 1996; 93:2774–8.

Hammer P.L. Partially defined boolean functions and cause-effect relationships. International Conference on Multi-Attribute Decision Making Via OR-Based Expert Systems; April 1986; University of Passau, Germany.

Hamspon S., Kibler D., Baldi P. Distribution patterns of over-represented k-mers in non-coding yeast DNA. Bioinformatics 2002; 18:513–28.

Han, J. "Data Mining". In *Encyclopedia of Distributed Computing*, J. Urban, P. Dasgupta, eds. Kluwer Academic Publishers, 1999.

Hanks S., Quinn A.M. Protein kinase catalytic domain sequence database: Identification of conserved features of primary structure and classification of family members. Meth Enzymol 1991; 200:38–62.

Hansen, J.P., McDonald, I.R. *Theory of Simple Liquids*. New York: Acad. Press 1986.

Harary, F. *Graph Theory*. Moscow: Mir, 1973.

Hart R., Royyuru A.K., Stolovitzky G., Califano A. Systematic and automated discovery of patterns in PROSITE families. Proceedings RECOMB'00, Tokyo. ACM, 2000; 147–54.

Hartwell L.H., Weinert T.A. Checkpoints: controls that ensure the order of cell cycle events. Science 1989; 246:629–34.

Heinemeyer T., Chen X., Karas H., Kel A.E., Kel O.V., Liebich I., Meinhardt T., Reuter I., Schacherer F., Wingender E. Expanding the TRANSFAC database towards an expert system of regulatory molecular mechanisms. Nucleic Acids Res 1999; 27(1):318–22.

Heinemeyer T., Wingender E., Reuter I., Hermjakob H., Kel A.E., Kel O.V., Ignatieva E.V., Ananko E.A., Podkolodnaya O.A., Kolpakov F.A., Podkolodny N.L., Kolchanov N.A. Databases on transcriptional regulation: TRANSFAC, TRRD and COMPEL. Nucleic Acids Res 1998; 26:362–7.

Helden J. van, Andre B., Collado-Vides J. Extracting regulatory sites from the upstream region of yeast genes by computational analysis of oligonucleotide frequencies. J Mol Biol 1998; 281:827–42.

Hermann R.B. Theory of hydrophobic bonding. J Phys Chem 1972; 76:2754–9.

Herrero C., Marques L., Lloberas J., Celada A. IFN-gamma-dependent transcription of MHC class II IA is impaired in macrophages from aged mice. J Clin Invest 2001; 107:485–93.

Hertz G.Z., Stormo G.D. Identifying DNA and protein patterns with statistically significant alignments of multiple sequences. Bioinformatics 1999; 15(7–8):563–77.

Hof P., Pluskey S., Dhe-Paganon S., Eck M.J., Shoelson S.E. Crystal structure of the tyrosine phosphatase SHP-2. Cell 1998; 92:441–50.

Hofacker I. L., Fontana W., Stadler P. F., Bonhoeffer L. S., Tacker M., Schuster P. Fast Folding and Comparison of RNA Secondary Structures. Monatshefte f. Chemie 1994; 125:167–88.

Hofmann F., Martelli F., Livingston D.M., Wang Z. The retinoblastoma gene product protects E2F-1 from degradation by the ubiquitin-proteasome pathway. Genes Dev 1996; 10:2949–59.

Honig B., Nicholls A. Classical electrostatics in biology and chemistry. Science 1995; 268:1144–9.

Howie D., Simarro M., Sayos J., Guirado M., Sancho J., Terhorst C. Molecular dissection of the signaling and costimulatory functions of CD150 (SLAM): CD150/SAP binding and CD150-mediated costimulation. Blood 2002; 99:957–65.

Hsu C.W., Lin C.J. IEEE Transaction on Neural Networks, 2002; 13:415–25.

Hu P.P., Shen X., Huang D., Liu Y., Counter C., Wang X.-F. The MEK pathway is required for stimulation of p21(WAF1/CIP1) by transforming growth factor-beta. J Biol Chem 1999; 274: 35381–7.

Hua S.J., Sun Z.R. A novel method of protein secondary structure prediction with high overlap measure: support vector machine approach. J Mol Biol 2001; 308:397–407.

Huang C.H., Chen C.Y., Tsai H.H., Chen C., Lin Y.S., Chen C.W. Linear plasmid SLP2 of *Streptomyces lividans* is a composite replicon. Mol Microbiol 2003; 47:1563–76.

Hutchinson G.B. The prediction of vertebrate promoter regions using differential hexamer frequency analysis. Comput Appl Biosci 1996; 12:391–8.

Hutvagner G., Zamore P.D. A microRNA in a multiple-turnover RNAi enzyme complex. Science 2002; 297:2056–60.

Hwang P.M., Li C., Morra M., Lillywhite J., Muhandiram D.R., Gertler F., Terhorst C., Kay L.E., Pawson T., Forman-Kay J.D., Li S.C. A "three-pronged" binding mechanism for the SAP/SH2D1A SH2 domain: structural basis and relevance to the XLP syndrome. Embo J 2002; 21:314–23.

Ignatieva E.V., Merkulova T.I., Vishnevskii O.V., Kel A.E. Transcription regulation of lipid metabolism genes as described in the TRRD database. Mol Biol (Mosk) 1997; 31:684–700.

Ikemura T. Codon usage and tRNA content in unicellular and multicellular organisms. Mol Biol Evol 1985; 2:13–24.

Im W., Beglov D., Roux B. Continuum solvation model: Electrostatic forces from numerical solutions to the Poisson–Boltzmann equation. Comp Phys Comm 1998; 111:59–75.

Ioshikhes I., Trifonov E.N. Nucleosomal DNA sequence database. Nucleic Acids Res 1993; 21:4857–9.

Ivanisenko V.A., Grigorovich D.A., Kolchanov N.A. PDBSite: a database on protein active sites and their environment. Proceedings of the Third International Conference on Bioinformatics of Genome Regulation and Structure; 2002 July 14–20; Novosibirsk: ICG, 2002.

Jegla D.E., Sussex I.M. Cell lineage patterns in the shoot meristem of the sunflower embryo in the dry seed. Dev Biol 1989; 131(1): 215–25.

Joazeiro C.A.P., Kassavetis G.A., Geiduschek E.P. Alternative outcomes in assembly of promoter complexes: the roles of TBP and a flexible linker in placing TFIIIB on tRNA genes. Genes Dev 1996; 10:725–39.

Jones D. Protein secondary structure prediction based on position-specific scoring matrices. J Mol Biol 1999; 292:195–202.

Jönsson H., Shapiro B.E., Meyerowitz E.M., Mjolsness E. Resources and signaling in multicellular models of plant development. Proceedings of the Third International Conference on Systems Biology (ICSB2002); 2002 December 11–15; Stockholm, Sweden.

Jönsson, H., Shapiro, B.E., Meyerowitz, E.M., Mjolsness, E. "Signalling in multicellular models of plant development." In *On growth, form, and computers.* S. Kumar, P.J. Bentley, eds. London: Academic Press, 2003 (to appear).

Jordan I.K., Matyunina L.V., McDonald J.F. Evidence for the recent horizontal transfer of long terminal repeat retrotransposon. Proc Natl Acad Sci USA. 1999; 96:12621–25.

Jordan K., Haas A., Logan T., Hall D. Detailed analysis of the basic domain of the E2F1 transcription factor indicate that it is unique among bHLH proteins. Oncogene 1994; 9:1177–1185.

Joyce, G.F., Orgel, L.E. "Prospects for understanding the origin of the RNA world". In *The RNA World,* R.F. Gesteland, J.F. Atkins, eds. New York: Cold Spring Harbor Laboratory Press, 1993; 1–25.

Kabsch W. A solution for the best rotation to relate two sets of vectors. Acta Cryst 1976; A32:922–3.

Kabsch W., Sander C. Dictionary of protein secondary structure: pattern recognition of hydrogen bonded and geometrical features. Biopolymers 1983; 22:2577–637.

Kakuda D.K., Sweet M.J., MacLeod C.L., Hume D.A., Markovich D. CAT2-mediated L-arginine transport and nitric oxide production in activated macrophages. Biochem J 1999; 340:549–53.

Kaminker J.S., Bergman C. M., Kronmiller B., Carlson J., Svirskas R., Patel S., Frise E., Wheeler D.A., Lewis S.E., Rubin G.M., Ashburner M., Celniker S.E. The transposable elements of the

Drosophila melanogaster euchromatin: a genomics perspective. Genome Biology 2002; 3:research/0084.1–0084.20.

Kananyan G.Kh., Ratner V.A., Tchuraev R.N. Enlarged model of lambda phage ontogenesis. J Theor Biol 1981; 88:393–407.

Kanehisa, M., Goto, S. "KEGG for Computational Genomics". In *Current Topics in Computational Molecular Biology*, T. Jiang, Y. Xu, M.Q. Zhang, eds. Cambridge, MA: MIT Press, 2002; 301–15.

Kantakamalakul W., Politis A.D., Marecki S., Sullivan T., Ozato K., Fenton M.J., Vogel S.N. Regulation of IFN consensus sequence binding protein expression in murine macrophages. J Immunol 1999; 162:7417–25.

Karp P.D., Riley M., Saier M., Paulsen I.T., Collado-Vides J., Paley S.M., Pellegrini-Toole A., Bonavides C., Gama-Castro S. The EcoCyc Database. Nucleic Acids Res 2002; 30:56–8.

Kassavetis G.A., Riggs D.L., Negri R., Nguyen L.H., Geiduschek E.P. Transcription factor IIIB generates extended DNA interactions in RNA polymerase III transcription complexes on tRNA genes. Mol Cell Biol 1989; 9(6):2551–66.

Kasschau K.D., Xie Z., Allen E., Llave C., Chapman E.J., Krizan K.A., Carrington J.C. P1/HC-Pro, a viral suppressor of RNA silencing, interferes with *Arabidopsis* development and miRNA function. Dev Cell 2003; 4:205–17.

Katti M.V., Sami-Subbu R., Ranjekar P.K., Gupta V.S. Amino acid repeat patterns in protein sequences: their diversity and structural–functional implications. Protein Sci 2000; 9:1203–9.

Kawashima S., Ogata H., Kanehisa M. AAindex: amino acid index database. Nucleic Acids Res 1999; 27:368–9.

Kel A.E., Kel-Margoulis O.V., Farnham P.J., Bartley S.M., Wingender E., Zhang M.Q. Computer-assisted identification of cell cycle-related genes: new targets for E2F transcription factors. J Mol Biol 2001; 309:99–120.

Kel O.V., Romashchenko A.G., Kel A.E., Wingender E., Kolchanov N.A. A compilation of composite regulatory elements affecting gene transcription in vertebrates. Nucleic Acids Res 1995; 23(20):4097–103.

Kel-Margoulis O.V., Kel A.E., Reuter I., Deineko I.V., Wingender E. TRANSCompel: a database on composite regulatory elements in eukaryotic genes. Nucleic Acids Res 2002; 30:332–4.

Ketting R.F., Fischer S.E., Bernstein E., Sijen T., Hannon G.J., Plasterk R.H. Dicer functions in RNA interference and in synthesis of small RNA involved in developmental timing in *C. elegans*. Genes Dev 2001; 15:2654–9.

Keyl H.G. Chromosomenevolution bei *Chironomus*. II. Chromosomenumbauten und phylogenetisce Beziehungen der Arten. Chromosoma 1962; 12:464–514.

Kidera A., Konishi Y., Oka M., Ooi T., Scheraga H. A. Statistical analysis of the physical properties of the 20 naturally occurring amino acids. J Protein Chem 1985; 4:23–54.

King D.A., Zhang L., Guarente L., Marmorstein R. Structure of a HAP1–DNA complex reveals dramatically asymmetric DNA binding by a homodimeric protein. Nature Struct Biol 1999; 2:64–71.

King R., Sternberg M. Identification and application of the concepts important for accurate and reliable protein secondary structure prediction. Prot Sci 1996; 5:2298–310.

King R.D., Sternberg M.J.E. Machine learning approach for the prediction of protein secondary structure. J Mol Biol 1990; 216:441–57.

Kinne R.W., Brauer R., Stuhlmuller B., Palombo-Kinne E., Burmester G.R. Macrophages in rheumatoid arthritis. Arthritis Res 2000; 2:189–202.

Kiyama R., Trifonov E.N. What positions nucleosomes?—A model. FEBS Lett 2002; 523:7–11.

Klaerr-Blanchard M., Chiapello H., Coward E. Detecting localized repeats in genomic sequences: A new strategy and its application to *B. subtilis* and *A. thaliana* sequences. Comput Chem 2000; 24(1):57–70.

Knight S.W., Bass B.L. A role for the RNase III enzyme DCR-1 in RNA interference and germ line development in *Caenorhabditis elegans.* Science 2001; 293:2269–71.

Knudsen S. Promoter2.0: for the recognition of PolII promoter sequences. Bioinformatics 1999; 15:356–61.

Kochetov A.V., Ischenko I.V., Vorobiev D.G., Kel A.E., Babenko V.N., Kisselev L.L., Kolchanov N.A. Eukaryotic mRNAs encoding abundant and scarce proteins are statistically dissimilar in many structural features. FEBS Lett 1998; 440:351–5.

Kochetov A.V., Ponomarenko M.P., Frolov A.S., Kisselev L.L., Kolchanov N.A. Prediction of eukaryotic mRNA translational properties. Bioinformatics 1999; 15:704–12.

Kochetov A.V., Sarai A., Vorob'ev D.G., Kolchanov N.A. The context organization of functional regions in yeast genes with high-level expression. Mol Biol (Mosk) 2002; 36:1026–34.

Kohn K.W. Molecular interaction map of the mammalian cell cycle control and DNA repair systems. Mol Biol Cell 1999; 10:2703–34.

Kohzaki H., Ito Y., Murakami Y. Context-dependent modulation of replication activity of *Saccharomyces cerevisiae* autonomously replicating sequences by transcription factors. Mol Cell Biol 1999; 19:7428–35.

Kolchanov N.A. Transcriptional regulation of eukaryotic genes: databases and computer analysis. Mol Biol (Mosk) 1997; 4:581–3.

Kolchanov N.A., Ananko E.A., Kolpakov F.A., Podkolodnaya O.A., Ignatieva E.V., Goryachkovskaya T.N., Stepanenko E.L. Gene networks. Mol Biol (Mosk) 2000; 34:533–44.

Kolchanov N.A., Ananko E.A., Likhoshvai V.A., Podkolodnaya O.A., Ignatieva E.V., Ratushny A.V., Matushkin Yu.G. "Gene networks description and modeling in the GeneNet system". In *Gene Regulation and Metabolism. Postgenomic Computational Approaches.* J. Collado-Vides J., Hofestadt R., eds. Cambridge–London: MIT Press., 2002a; 149–79.

Kolchanov N.A., Ananko E.A., Podkolodnaya O.A., Ignatieva E.V., Stepanenko I.L., Kel-Margoulis O.V., Kel A.E., Merkulova T.I., Goryachkovskaya T.N., Busygina T.V., Kolpakov F.A., Podkolodny N.L., Naumochkin A.N., Romashchenko A.G. Transcription Regulatory Regions Database (TRRD): its status in 1999. Nucleic Acids Res 1999; 27:303–6.

Kolchanov N.A., Ignatieva E.V., Ananko E.A., Podkolodnaya O.A., Stepanenko I.L., Merkulova T.I., Pozdnyakov M.A., Podkolodny N.L., Naumochkin A.N., Romashchenko A.G. Transcription Regulatory Regions Database (TRRD): its status in 2002. Nucleic Acids Res 2002b; 30:312–7.

Kolchanov N.A., Lim H.A., eds. *Computer Analysis of Genetic Macromolecules: Structure, Function and Evolution.* Singapore, a.o.: World Scientific Publ. Co., 1994.

Kolchanov N.A., Nedosekina E.A., Ananko E.A., Likhoshvai V.A., Podkolodny N.L., Ratushny A.V., Stepanenko I.L., Podkolodnaya O.A., Ignatieva E.V., Matushkin Yu.G. GeneNet database: description and modeling of gene networks. In Silico Biol 2002c; 2:97–110.

Kolchanov N.A., Podkolodnaya O.A., Ananko E.A., Ignatieva E.V., Stepanenko I.L., Kel-Margoulis O.V., Kel A.E., Merkulova T.I., Goryachkovskaya T.N., Busygina T.V., Kolpakov F.A., Podkolodny N.L., Naumochkin A.N., Korostishevskaya I.M., Romashchenko A.G., Overton G.C. Transcription Regulatory Regions Database (TRRD): its status in 2000. Nucleic Acids Res 2000; 28:298–301.

Kolchanov N.A., Ponomarenko M.P., Kel A.E., Kondrakhin Yu.V., Frolov A.S., Lopakov F.A., Kel O.V., Ananko E.A., Ignatieva E.V., Podkolodnaya O.A., Stepanenko I.L., Merkulova T.I., Babenko V.N., Vorobiev D.G., Lavryushev S.V., Ponomarenko Yu.V., Kochetov A.V., Kolesov G.B., Podkolodny N.L., Milanesi L., Wingender E., Heinemeyer T., Solovyev V.V. GeneExpress: a computer system for description, analysis and recognition of regulatory sequenecs of the eukaryotic genome. ISMB, 6:95–104. MEDLINE PMID: 9783214; UI: 98456543, 1998.

Kolpakov F.A., Ananko E.A., Kolesov G.B., Kolchanov N.A. GeneNet: a database for gene networks and its automated visualization. Bioinformatics 1998; 14:529–37.

Kondrakhin Y.V., Kel A.E., Kolchanov N.A., Romashchenko A.G., Milanesi L. Eukaryotic promoter recognition by binding sites for transcription factors. Comput Appl Biosci 1995; 11:477–88.

Kono H., Sarai A. Structure-based prediction of DNA target sites by regulatory proteins. Proteins 1999; 35:114–31.

Kornberg R.D., Lorch Y. Twenty-five years of the nucleosome, fundamental particle of the eukaryote chromosome. Cell 1999; 98:285–94.

Korotkov E.V., Korotkova M.A. DNA regions with latent periodicity in some human clones. DNA Seq 1995; 5:353–8.

Korotkov E.V., Korotkova M.A. Enlarged similarity of nucleic acid sequences. DNA Res 1996; 3:157–64.

Korotkov E.V., Korotkova M.A. MIRs: family repeats that is common for many vertebrates. Mol Biol (Mosk) 2000; 34:348–53.

Korotkov E.V., Korotkova M.A., Rudenko V.M., Skryabin K.G. Latent periodicity regions in amino acid sequences. Mol Biol 1999; 33:611–7.

Korotkov E.V., Korotkova M.A., Tulko J.S. Latent sequence periodicity of some oncogenes and DNA-binding protein genes. Comput Appl Biosci 1997; 13:37–44.

Korotkov E.V, Kudryashov N.A. Information decomposition of symbolic sequences. Phys Lett A 2003; 312:198–210.

Korotkova M.A., Korotkov E.V., Rudenko, V.M. Latent periodicity of protein sequences. J Mol Model 1999; 5:103–15.

Kozak M. Initiation of translation in prokaryotes and eukaryotes. Gene 1999; 234:187–208.

Kozak M. Pushing the limits of the scanning mechanism for initiation of translation. Gene 2002; 299:1–34.

Krawczak M., Ball E., Fenton I., Stenson P.D., Abeysinghe S., Thomas N., Cooper D.N. Human Gene Mutation Database—a biomedical information and research resource. Hum Mutation 2000a; 15:45–51.

Krawczak M., Chuzhanova N.A., Stenson P., Johansen B., Ball E., Cooper D.N. Changes in primary DNA sequence complexity influence the phenotypic consequences of mutations in human gene regulatory regions. Hum Genet 2000b; 107:362–5.

Krawczak M., Cooper D.N. Gene deletions causing human genetic disease: mechanisms of mutagenesis and the role of the local DNA sequence environment. Hum Genet 1991; 86:425–41.

Krivov S.V., Chekmarev S.F., Karplus M. Potential energy surfaces and conformational transitions in biomolecules: a successive confinement approach applied to a solvated tetrapeptide. Phys Rev Lett 2002; 88:038–101.

Krivov S.V., Karplus M. Free energy disconnectivity graphs: application to peptide models. J Chem Phys 2002; 117:10894–903.

Kruger K., Grabowski P.J., Zaug A.J. *et al.* Self-splicing RNA: autoexcision and autocyclization of the ribosomal RNA intervening sequence of *Tetrahymena.* Cell 1982; 31:147–57.

Krull M., Voss N., Choi V., Pistor S., Potapov A., Wingender E. TRANSPATH®: an integrated database on signal transduction and a tool for array analysis. Nucleic Acids Res 2003; 31:97–100.

Kumar S., Tamura K., Jakobsen I.B., Nei M. MEGA2: Molecular Evolutionary Genetics Analysis software, Arizona State University, Tempe, Arizona, USA 2001.

Kunst F. *et al.* The complete genome sequence of the gram-positive bacterium *Bacillus subtilis*. Nature 1997; 390:249–56.

Kuroda M. *et al.* Whole genome sequencing of meticillin-resistant *Staphylococcus aureus*. Lancet 2001; 357:1225–40.

Kurose K., Hata K., Hattori M., Sakaki Y. RNA polymerase III dependence of the human L1 promoter and possible participation of the RNA polymerase II factor YY1 in the RNA polymerase III transcription system. Nucleic Acids Res 1995; 23(18):3704–9.

Kutsenko A.S., Gizatullin R.Z., Al-Amin A.N., Wang F., Kvasha S.M., Podowski R.M., Matushkin Y.G., Gyanchandani A., Muravenko O.V., Levitsky V.G., Kolchanov N.A., Protopopov A.I., Kashuba V.I., Kisselev L.L., Wasserman W., Wahlestedt C., Zabarovsky E.R. NotI flanking sequences: a tool for gene discovery and verification of the human genome. Nucleic Acids Res 2002; 30:3163–70.

Kuzin A.B., Lyubomirskaya N.V., Khudaibergenova B.M., Ilyin Y.V., Kim A.I. Precise excision of the retrotransposon *gypsy* from the forked and cut loci in a genetically unstable *D. melanogaster* strain. Nucleic Acids Res 1994; 22:4641–45.

Labrador M., Corces V.G. Protein determinants of insertional specificity for the *Drosophila gypsy* retrovirus. Genetics 2001; 158:1101–10.

Labrador M., Corces V.G. Transposable elements–host interactions: Regulation of insertion and excision. Ann Rev Genet 1997; 31:381–404.

Lafay B., Lloyd A.T., McLean M.J., Devine K.M., Sharp P.M., Wolfe K.H. Proteome composition and codon usage in spirochaetes: species-specific and DNA strand-specific mutational biases. Nucleic Acids Res 1999; 27:1642–9.

Lagos-Quintana M., Rauhut R., Lendeckel W., Tuschl T. Identification of novel genes coding for small expressed RNAs. Science 2001; 294:853–8.

Lagos-Quintana M., Rauhut R., Yalcin A., Meyer J., Lendeckel W., Tuschl T. Identification of tissue-specific microRNAs from mouse. Curr Biol 2002; 12:735–9.

Laskin A.A., Korotkov E.V., Chaley M.B., Kudryashov N.A. The locally optimal method of cyclic alignment to reveal latent periodicities in genetic texts: the NAD-binding protein sites. Mol Biol 2003; 37.

Laskin, A.A., Korotkov, E.V., Kudryashov, N.A. "Detection of hidden periodicity in protein sequences and its correlation with structure and function of proteins". In *Genome Informatics 2001*, H. Matsuda, S. Miyano, T. Takagi, L. Wong, eds., 2001; 343–4.

Laskin A., Korotkov E., Kudryashov N. New method of latent periodicity detection may determine structurally related proteins and protein families. Proceedings of the Third International Conference on Bioinformatics of Genome Regulation and Structure; 2002 July 14–20; Novosibirsk: ICG, 2002; 3.

Latour S., Gish G., Helgason C.D., Humphries R.K., Pawson T., Veillette A. Regulation of SLAM-mediated signal transduction by SAP, the X-linked lymphoproliferative gene product. Nat Immunol 2001; 2:681–90.

Lau N.C., Lim L.P., Weinstein E.G., Bartel D.P. An abundant class of tiny RNAs with probable regulatory roles in *Caenorhabditis elegans*. Science 2001; 294:858–62.

Lavery R., Sklenar H. The definition of generalized helicoidal parameters and axis curvature for irregular nucleic acids. J Biomol Struct Dyn 1988; 6:63–91.

Ledermann, W., Lloyd, E. *Handbook of Applicable Mathematics*. Vol. 6. Chichester: John Wiley & Sons, 1984.

Lee S., Elenbaas B., Levine A.J., Griffith J. p53 and its 14 kDa C-terminal domain recognize primary DNA damage in the form of insertion/deletion mismatches. Cell 1995; 81:1013–20.

Lee Y., Jeon K., Lee J.T., Kim S., Kim V.N. MicroRNA maturation: stepwise processing and subcellular localization. EMBO J 2002; 21:4663–70.

Lescot M., Régnier M. *In Silico* prediction of regulatory signals on plant promoter sequences. 2003 (submitted).

Lesnik T., Solomovici J., Deana A., Ehrlich R., Reiss C. Ribosome traffic in *E. coli* and regulation of gene expression. J Theor Biol 2000; 202:175–85.
Levitsky V.G., Podkolodnaya O.A., Kolchanov N.A., Podkolodny N.L. Nucleosome formation potential of eukaryotic DNA: tools for calculation and promoters analysis. Bioinformatics 2001a; 17:998–1010.
Levitsky V.G., Podkolodnaya O.A., Kolchanov N.A., Podkolodny N.L. Nucleosome formation potential of exons, introns, and Alu repeats. Bioinformatics 2001b; 17:1062–4.
Levitsky V.G., Ponomarenko M.P., Ponomarenko J.V., Frolov A.S., Kolchanov N.A. Nucleosomal DNA property database. Bioinformatics, 1999; 15:582–92.
Li S.C., Gish G., Yang D., Coffey A.J., Forman-Kay J.D., Ernberg I., Kay L.E., Pawson T. Novel mode of ligand binding by the SH2 domain of the human XLP disease gene product SAP/SH2D1A. Curr Biol 1999; 9:1355–62.
Li T., Stark M.R., Johnson A.D., Wolberger C. Crystal structure of MATal/MATα2 homeodomain hetrodimer bound to DNA. Science 1995; 270:262–9.
Li W.-H., Luo L.J. The relation between codon usage, base correlation and gene expression level in *Escherichia coli* and yeast. J Theor Biol 1996; 181:111–24.
Liao G.C., Rehm E.J., Rubin G.M. Insertion site preferences of the P transposable element in *Drosophila melanogaster*. Proc Natl Acad Sci USA 2000; 97:3347–51.
Likhoshvai, V.A. "Rare codons: fortunity or regularity?" In *Modeling and Computer Methods in Molecular Biology and Genetics*, V.A. Ratner, N.A. Kolchanov, eds. New York: Nova Science Publishers, Inc., USA, 1992; 463–9.
Likhoshvai V.A., Matushkin Yu.G. Differentiation of single-cell organisms according to elongation stages crucial for gene expression efficacy. FEBS Lett 2002a; 516:87–92.
Likhoshvai V.A., Matushkin Yu.G. On the theory of prediction of global modes in the function of gene networks. Proceeding Thrird International Conference on Bioinformatics of Genome Regulation and Structure; 2002b July 14–20; Novosibirsk: ICG, 2002b; 2.
Likhoshvai V.A., Matushkin Yu.G, Fadeev S.I. Relationship between a gene network graph and qualitative modes of its functioning. Mol Biol (Mosk) 2001; 35:1080–7.
Likhoshvai V.A., Matushkin Yu.G., Fadeev S.I. Problems in the theory of gene network operation. Zh Indust Mat 2003; 6:64–80.
Likhoshvai V.A., Matushkin Yu.G., Vatolin Yu.N., Bazhan S.I. A generalized chemical kinetic method for simulating complex biological systems. A computer model of λ phage ontogenesis. Comput Technol 2000; 5:87–99.
Likhoshvai V.A., Nedosekina E.A., Ratushny A.V., Podkolodny N.L. Technology of using experimental data for verification of models of gene network operation dynamics. Proceedings of the Third International Conference on Bioinformatics of Genome Regulation and Structure; 2002 July 14–20; Novosibirsk: ICG, 2002.
Lim V. Algorithms for prediction of α-helical and β-structural regions in globular proteins. J Mol Biol 1974; 88:873–94.
Lim V.I., Ptitsyn O.B. On the constancy of the hydrophobic nucleus volume in molecules of myoglobins and hemoglobins. Mol Biol (USSR) 1970; 4:372–82.
Liu R., Blackwell T.W., States D.J. Conformational model for binding site recognition by *E. coli MetJ* transcription factor. Bioinformatics 2001; 17:(7):622–33.
Liu R., States D. Consensus promoter identification in the human genome utilizing expressed gene markers and gene modeling. Genome Res 2002; 12:462–9.
Llave C., Xie Z., Kasschau K.D., Carrington J.C. Cleavage of Scarecrow like mRNA targets directed by a class of *Arabidopsis* miRNA. Science 2002; 297:2053–6.
Lloyd A.T., Sharp P.M. Synonymous codon usage in *Kluyveromyces lactis*. Yeast 1993; 9:1219–28.

Lobry J.R. Asymmetric substitution patterns in the two DNA strands of bacteria. Mol Biol Evol 1996; 13:660–5.

Loktev K.A., Tkachev Yu.A., Ananko E.A., Podkolodny N.L. A system for visual modeling of gene networks' structural and functional organization. Proceedings of the Third International Conference on Bioinformatics of Genome Regulation and Structure; 2002 July 14–20; Novosibirsk: ICG, 2002; 2.

Lonardi, S. *Global detectors of unusual words: design, implementation, and applications to pattern discovery in biosequences*. PhD thesis, Department of Computer Sciences, Purdue University, 2001. 145.

Lowary P.T., Widom J. New DNA sequence rules for high affinity binding to histone octamer and sequence-directed nucleosome positioning. J Mol Biol 1998; 276:19–42.

Lowe T.M., Eddy S.R. tRNAscan-SE: a program for improved detection of transfer RNA genes in genomic sequence. Nucleic Acids Res 1997; 25:955–64.

Luger K. Structure and dynamic behavior of nucleosomes. Curr Opin Genet Devel 2003; 13:127–35.

Luger K., Mader A.W., Richmond R.K., Sargent D.F., Richmond T.J. Crystal structure of the nucleosome core particle at 2.8 Å resolution. Nature 1997; 389:251–60.

Lukas J., Bartkova J., Bartek J. Convergence of mitogenic signaling cascades from diverse classes of receptors at the cyclin D-cyclin-dependent kinase-pRb-controlled G1 checkpoint. Mol Cell Biol 1996; 16:6917–25.

Lukaszewicz M., Feuermann M., Jerouville B., Stas A. Boutry M. *In vivo* evaluation of the context sequence of the translation initiation codon in plants. Plant Sci 2000; 154:89–98.

Lundberg A.S., Weinberg R.A. Functional inactivation of the retinoblastoma protein requires sequential modification by at least two distinct cyclin-cdk complexes. Mol Cell Biol 1998; 18:753–61.

MacDonald P.M., Struhl G. A molecular gradient in early *Drosophila* embryos and its role in specifying the body pattern. Nature 1986; 324:537–45.

Mahalanobis P.C. On the generalised distance in statistics. Proc Natl Inst Sci India 1936; 12:49–55.

Malik H.S., Eickbush T.H. Modular evolution of the integrase domain in the Ty3/*Gypsy* class of LTR retrotransposons. J Virol 1999; 73(6):5186–90.

Malik H.S., Eickbush T.H. Phylogenetic analysis of ribonuclease H domains suggests a late, chimeric origin of LTR retrotransposable elements and retroviruses. Genome Res 2001; 11:1187–97.

Manning G., Whyte D.B., Martinez R., Hunter T., Sudarsanam S. The protein kinase complement of the human genome. Science 2002; 298:1912–34.

Mans R.M., Pleij C.W. *et al.* tRNA-like structures. Structure, function and evolutionary significance. Eur J Biochem 1991; 201:303–24.

Marmorstein R., Carey M., Ptashne M., Harrison S.C. DNA recognition by GAL4: structure of a protein–DNA complex. Nature 1992; 356:408–14.

Marmorstein R., Harrison S.C. Crystal structure of a PPR1–DNA complex: DNA recognition by proteins contaning a Zn_2Cys_6 binuclear cluster. Genes Dev 1994; 8:2504–12.

Marnellos, G., Mjolsness, E. "A gene network approach to modeling early neurogenesis in Drosophila." In *Pacific Symposium on Biocomputing*. R.B. Altman, A.K. Dunker, L. Hunter, T. Klein, eds. World Scientific, 1998.

Marsan L., Sagot M.F. Extracting structured motifs using a suffix tree-algorithms and application to promoter consensus identification. Proceedings RECOMB'00, Tokyo. ACM, 2000.

Marsan, L. *Inférence de motifs structurés: algorithmes et outils appliqués à la détection de sites de fixation dans des séquences génomiques*. PhD thesis, University of Marne-la-Vallée, 2002.

Martin J. Chromosomes as tools in taxonomy and phylogeny of Chironomidae. Ent Scand 1979; Suppl 10: 67–74.

Martino G., Furlan R., Poliani P.L. The pathogenic role of inflammation in multiple sclerosis. Rev Neurol 2000; 30:1213–7.

Matthews B.W. Protein–DNA interaction. No code for recognition. Nature 1988; 335:294–5.

Mayoraz E. C++ tools for logical analysis of data. Rutcor Research Report 1995; 1–95.

McCarthy J.E.G. Posttranscriptional control of gene expression in yeast. Microbiol Mol Biol Rev 1998; 62:1492–553.

McConnell K.J., Beveridge D.L. DNA structure: what's in charge? J Mol Biol 2000; 304:803–20.

McLean M.J., Wolfe K.H., Devine K.M. Base composition skews, replication orientation, and gene orientation in 12 prokaryote genomes. J Mol Evol 1998; 47:691–6.

Meierhans D., Sieber M., Allemann R.K. High affinity binding of MEF-2C correlates with DNA bending. Nucleic Acids Res 1997; 25:4537–44.

Mejlumian L., Pelisson A., Bucheton A., Terzian C. Comparative and functional studies of *Drosophila* species invasion by the gypsy endogenous retrovirus. Genetics 2002; 160:201–9.

Merika M., Thanos D. Enhanceosomes. Curr Opin Genet Devel 2001; 11:205–8.

Metropolis N., Rosenbluth A.W., Rosenbluth M.N., Teller A.N., Teller E. J Chem Phys 1953; 21:1087.

Meyerowitz E.M. Genetic control in cell division patterns in developing plants. Cell 1997; 88:299–308.

Middleton T.F., Hernandez-Rojas J., Mortenson P.N., Wales D.J. Crystals of binary Lennard–Jones solid. Phys Rev B 2001; 64:184–201.

Mikhalap S.V., Shlapatska L.M., Berdova A.G., Law C.L., Clark E.A., Sidorenko S.P. CDw150 associates with src-homology 2-containing inositol phosphatase and modulates CD95-mediated apoptosis. J Immunol 1999; 162:5719–27.

Mironov A.A., Djakonova L.P., Kister A.E. A kinetic approach to the prediction of RNA secondary structures. J Biomol Struct Dynam 1985; 2:953–62.

Mjolsness E., Sharp D.H., Reinitz J. A connectionist model of development'. J Theor Biol 1991; 152(4):429–454.

Morra M., Howie D., Grande M.S., Sayos J., Wang N., Wu C., Engel P., Terhorst C. X-linked lymphoproliferative disease: a progressive immunodeficiency. Annu Rev Immunol 2001; 19:657–82.

Morra M., Lu J., Poy F., Martin M., Sayos J., Calpe S., Gullo C., Howie D., Rietdijk S., Thompson A., Coyle A.J., Denny C., Yaffe M.B., Engel P., Eck M.J., Terhorst C. Structural basis for the interaction of the free SH2 domain EAT-2 with SLAM receptors in hematopoietic cells. Embo J 2001; 20:5840–52.

Moss E.G., Lee R.C., Ambros V. The cold shock domain protein LIN-28 controls developmental timing in *C. elegans* and is regulated by the *lin-4* RNA. Cell 1997; 88:637–46.

Mourelatos Z., Dostie J., Paushkin S., Sharma A., Charroux B., Abel L., Rappsilber J., Mann M., Dreyfuss G. miRNPs: a novel class of ribonucleoproteins containing numerous microRNAs. Genes Dev 2002; 16:720–8.

Mrázek J., Karlin S. Strand compositional asymmetry in bacterial and large viral genomes. Proc Natl Acad Sci USA 1998; 95:3720–5.

Muller H., Moroni M.C., Vigo E., Petersen B.O., Bartek J., Helin K. Induction of S-phase entry by E2F transcription factors depends on their nuclear localization. Mol Cell Biol 1997; 17:5508–20.

Muller C.W., Rey F.A., Sodeoka M., Verdine G. L., Harrison S.C. Structure of the NF-kB p50homodimer bound to DNA. Nature 1995; 373:311–7.

Muller H.P., Varmus H.E. DNA bending creates favored sites for retroviral integration: an explanation for preferred insertion sites in nucleosomes. EMBO J 1994; 13:4704–14.

Mulligan M.E., Hawley D.K., Entriken R., McClure W.R. *Escherichia coli* promoter sequences predict *in vitro* RNA polymerase selectivity. Nucleic Acids Res 1984; 12:789–800.

Murray, R.K., Granner, D.K., Mayes, P.A., Rodwell, V.W. *Harpers Biochemistry*. Norwalk, Connecticut/San Mateo, California: Appleton & Lange, 1988; 1.

Muto A., Ushida C., Himeno H. A bacterial RNA that functions as both a tRNA and an mRNA. Trends Biochem Sci 1998; 23:25–9.

Myllykallio H., Lopez P., Lopez-Garcia P., Heilig R., Saurin W., Zivanovic Y., Philippe H., Forterre P. Bacterial mode of replication with eukaryotic-like machinery in a hyperthermophilic archaeon. Science 2000; 288:2212–5.

Nadeau J.H., Taylor B.A. Length of chromosomal segments conserved since man and mouse. Proc Natl Acad Sci USA 1984; 81:814–8.

Naykova T.N., Kondrakhin Y.V., Rogozin I.B., Voevoda M.I., Yudin N.S., Romaschenko A.G. Concerted changes in the nucleotide sequences of the intragenic promoter regions of all the eukaryotic tRNA genes' types. J Mol Evol 2003 (in press).

Nedosekina E.A., Ananko E.A. Gene network of macrophage activation under the action of interferon-gamma and lipopolysaccharides. Proceedings of the Third International Conference on Bioinformatics of Genome Regulation and Structure; 2002 July 14–20; Novosibirsk: ICG, 2002.

Nedosekina E.A., Ananko E. A., Likhoshvai V.A. Construction of mathematical model of the gene network on macrophage activation under the action of IFN-γ and LPS. Proceedings of the Third International Conference on Bioinformatics of Genome Regulation and Structure; 2002 July 14–20; Novosibirsk: ICG, 2002.

Neria E., Fischer S., Karplus M. Simulation of activation free energies in molecular systems. J Chem Phys 1996; 105, 1902–21.

Neuwald A.F., Poleksic A. PSI-BLAST searches using hidden Markov models of structural repeats: prediction of unusual sliding DNA clamp and of beta-propellers in UV-damaged DNA-binding protein. Nucleic Acids Res 2000; 28:3570–80.

Newman M., Strzelecka T., Dorner L.F., Schildkraut I., Aggarwal A.K. Structure of BamHI endonuclease bound to DNA: partial folding and unfolding on DNA binding. Science 1995; 269:656–63.

Nicholls A., Sharp K.A., Honig B. Protein folding and association: insights from the interfacial and thermodynamic properties of hydrocarbons. Proteins 1991; 11:281–96.

Nicodème P. Fast approximate motif statistics. J Comput Biol 2001; 8(3):235–48.

Niepel M., Ling J., Gallie D.R. Secondary structure in the 5'-leader and 3'-untranslated region reduces protein yield but does not affect the functional interaction between the 5'-cap and the poly(A) tail. FEBS Lett 1999; 462:79–84.

Nohturfft A., Yabe D., Goldstein J.L., Brown M.S., Espenshade P.J. Regulated step in cholesterol feedback localized to budding of SCAP from ER membranes. Cell 2000; 102:315–23.

Nuel, G. *Grandes déviations et chaines de Markov pour l'étude des mots exceptionnels dans les séquences biologiques*. PhD thesis, Université René Descartes, Paris V, 2001. (defended in July, 2001).

Olsen P.H., Ambros V. The *lin-4* regulatory RNA controls developmental timing in *Caenorhabditis elegans* by blocking LIN-14 protein synthesis after the initiation of translation. Dev Biol 1999; 216:671–80.

OMG The Common Object Request Broker Architecture: 2.0/IIOP Specification, OMG Document Number 96.08.04. OMG (Object Management Group), 1996.

Orgel L.E. Evolution of the genetic apparatus. J Mol Biol 1968; 38:381–93.

Oshchepkov D.Yu., Turnaev I.I., Vityaev E.E. SITECON: a method for recognizing transcription factor binding sites basing on analysis of their conservative physicochemical and conformational properties. Proceedings of the Third International Conference on Bioinformatics of Genome Regulation and Structure; 2002 July 14–20; Novosibirsk: ICG, 2002; 1.

Oshchepkov D.Yu., Turnaev I.I., Vityaev E.E. Study of the context-dependent conformational and physicochemical properties of DNA functional sites. Proceedings of the Third International Conference on Bioinformatics of Genome Regulation and Structure; 2002 July 14–20; Novosibirsk: ICG, 2002; 1.

Overbeek R., Larsen N., Pusch G.D., D'Souza M., Selkov E. Jr., Kyrpides N., Fonstein M., Maltsev N., Selkov E. WIT: integrated system for high-throughput genome sequence analysis and metabolic reconstruction. Nucleic Acids Res 2000; 28:123–5.

Park W., Li J., Song R., Messing J., Chen X. CARPEL FACTORY, a Dicer homolog, and HEN1, a novel protein, act in microRNA metabolism in *Arabidopsis thaliana.* Curr Biol 2002; 12:1484–95.

Pavesi A. Relationships between transcriptional and translational control of gene expression in *Saccharomyces cerevisiae*: a multiple regression analysis. J Mol Evol 1999; 48:133–41.

Pennec X., Ayache N. A geometric algorithm to find small but highly similar 3D substructures in proteins. Bioinformatics 1998; 14:516–22.

Percudani R., Ottonello S. Selection at the wobble position of codons read by the same tRNA in *Saccharomyces cerevisiae.* Mol Biol Evol 1999; 16:1752–62.

Peri S., Pandey A. A reassessment of the translation initiation codon in vertebrates. Trends Genet 2001; 17:685–7.

Pevzner P.A., Borodovski M., Mironov A. Linguistic of nucleotide sequences: the significance of deviations from the mean: statistical characteristics and prediction of the frequency of occurrences of words. J Biomol Struct Dynam 1989; 6:1013–1026.

Pieler T., Hamm J., Roeder R.G. The 5S gene internal control region is composed of three distinct sequence elements, organized as two functional domains with variable spacing. Cell 1987; 48(1):91–100.

Podkolodnaya O.A., Stepanenko I.L. Mechanisms of transcriptional regulation of erythroid specific genes. Mol Biol (Mosk) 1997; 31(4):671–683.

Podkolodnaya O.A., Levitsky V.G., Podkolodnyi N.L. Locus control regions: description in the LCR-TRRDatabase. Mol Biol (Mosk) 2001; 35:802–9.

Poletaev I.A. Volterra's models predator–victim and certain generalizations using Liebig's principle. Zh Obshch Biol 1973; 34:43.

Ponomarenko J.V., Ponomarenko M.P., Frolov A.S., Vorobyev D.G., Overton G.C., Kolchanov N.A. Conformational and physicochemical DNA features specific for transcription factor binding sites. Bioinformatics 1999; 15(7/8):654–68.

Ponomarenko M.P., Ponomarenko J.V., Frolov A.S., Podkolodny N.L., Savinkova L.K., Kolchanov N.A., Overton G.C. Identification of sequence-dependent DNA features correlating to activity of DNA sites interacting with proteins. Bioinformatics 1999; 15(7/8):687–703.

Ponomarenko M.P., Ponomarenko J.V., Kel A.E., Kolchanov N.A. Search for DNA conformational features for functional sites. Investigation of the TATA box. Biocomputing: Proceedings of the 1997 Pacific Symposium; Word Sci Publ, Singapore, 1997.

Ponomarenko M.P., Ponomarenko Yu.V., Kel' A.E., Kolchanov N.A., Karas H., Wingender E., Sklenar H. Computer analysis of conformational features of the eukaryotic TATA-box DNA promoters. J Mol Biol 1997; 31:733–740.

Poy F., Yaffe M.B., Sayos J., Saxena K., Morra M., Sumegi J., Cantley L.C., Terhorst C., Eck M.J. Crystal structures of the XLP protein SAP reveal a class of SH2 domains with extended, phosphotyrosine-independent sequence recognition. Mol Cell 1999; 4:555–61.

Praz V., Périer RC., Bonnard C., Bucher P. The Eukaryotic Promoter Database, EPD: new entry types and links to gene expression data. Nucleic Acids Res 2002; 30:322–4.

Prestrige D.S. Predicting Pol II promoter sequences using transcription factor binding sites. J Mol Biol 1995; 249:923–932.

Proscura A.L., Levitsky V.G., Oshchepkov D.Yu., Pozdnyakov M.A., Ignatieva E.V. Expression of lipid metabolism genes: description in TRRD database and computer-assisted analysis. Proceedings of the Third International Conference on Bioinformatics of Genome Regulation and Structure; 2002 July 14–20; Novosibirsk: ICG, 2002; 3.

Puglisi J.D., Wyatt J.R., Tinoco I. A pseudoknotted RNA oligonucleotide. Nature 1988; 331:283–286.

Qiu D., Shenkin P.S., Hollinger F.P., Still W.C. A fast analytical method for the calculation of approximate Born radii. J Phys Chem A 1997; 101:3005–14.

Rackovsky S. Hidden sequence periodicities and protein architecture. Proc Natl Acad Sci USA 1998; 95:8580–4.

Ratner V.A. A concept of limiting genetic factors of expression, organization, and evolution. Genetika 1990; 5:789–803.

Ratner, V.A., Tchuraev, R.N. "Simplest genetic systems controlling ontogenesis: organization principle and models of their function". In *Progress in Theoretical Biology*, R. Rosen, F.M. Snell, eds. New York, San Francisco, London: Acad. Press, 1978; 5:81–127.

Ray B.K., Brandler T.G., Adya S., Daniels-McQeen S., Miller J.K., Hershey J.W.B., Grifo J.A., Merrick W.C., Thach R.E. Role of mRNA competition in regulating translation: further characterization of mRNA discriminatory initiation factors. Proc Natl Acad Sci USA 1983; 80:663–7.

Raychaudhuri S., Byers R., Upton T., Eisenberg S. Functional analysis of a replication origin from Saccharomyces cerevisiae: identification of a new replication enhancer. Nucleic Acids Res 1997; 25:5057–64.

Read T.D. *et al.* Genome sequences of *Chlamydia trachomatis* MoPn and *Chlamydia pneumoniae* AR39. Nucleic Acids Res 2000; 28:1397–406.

Reese J.C. Basal transcription factors. Curr Opin Genet Devel 2003; 13:114–8.

Régnier M. A Unified Approach to Word Occurrences Probabilities. Special issue on Computational Biology; preliminary version at RECOMB'98. Discrete Applied Mathematics 2000; 104(1):259–280.

Régnier M. Complexity of unusual words counting. Lecture Notes in Computer Science. Proceedings of JOBIM'00, Montpellier: Springer-Verlag, 2001; 2066:101–117.

Régnier M., Denise A. Rare Events on Random Strings, 2003 (submitted), http://algo.inria.fr/regnier/index.html.

Régnier M., Szpankowski W. On pattern frequency occurrences in a markovian sequence. Algorithmica 1997; 22(4): 631–649 (preliminary draft at ISIT'97).

Reinert G., Schbath S., Waterman M. Probabilistic and statistical properties of words: an overview. J Comput Biol 2000; 7(1):1–46.

Reinhart B.J., Weinstein E.G., Rhoades M.W., Bartel B., Bartel D.P. MicroRNAs in plants. Genes Dev 2002; 16:1616–26.

Reinitz J., Sharp D.H. Mechanism of eve stripe formation. Mech Dev 1995; 49:133–158.

Rhoades M.W., Reinhart B.J., Lim L.P., Burge C.B., Bartel B., Bartel D.P. Prediction of plant microRNA targets. Cell 2002; 110:513–20.

Rich A., RajBhandary U.L. Transfer RNA: molecular structure, sequence, and properties. Ann Rev Biochem 1976; 45:805–60.

Richards F.M., Kundrot C.E. Proteins: Struct Funct Genet 1988; 3:71–84.

Richmond, T.J., Widom, J. "Nucleosome and Chromatin Structure". In *Chromatin Structure and Gene Expression*, 2nd ed., J.L. Workman, S.C. Elgin, eds. Oxford: Oxford University Press, 2000.

Ripley L.S. Model for the participation of quasi-palindromic DNA sequences in frameshift mutation. Proc Natl Acad Sci USA 1982; 79:4128–4132.

Rivard N., L'Allemain G., Bartek J., Pouyssegur J. Abrogation of p27Kip1 by cDNA antisense suppresses quiescence (G0 state) in fibroblasts. J Biol Chem 1996; 271:18337–41.

Rivas E., Eddy S.R. Secondary structure alone is generally not statistically significant for the detection of noncoding RNAs. Bioinformatics 2000; 16:583–605.

Robin S., Daudin J.J. Exact distribution of word occurrences in a random sequence of letters. J Appl Prob 1999; 36:179–193.

Robin S., Schbath S. Numerical comparison of several approximations on the word count distribution in random sequences. J Comput Biol 2001; 8(4):349–59.

Rocha E.P., Danchin A. Ongoing evolution of strand composition in bacterial genomes. Mol Biol Evol 2001; 18:1789–99.

Rost B. PHD: predicting one-dimensional protein structure by profile based neural networks. Meth Enzymol 1996; 266:525–539.

Rost B., Sander C., Schneider R. Redefining the goals of protein secondary structure prediction. J Mol Biol 1994; 235:13–26.

Roulet E., Fisch I., Bucher P., Mermod N. Evaluation of computer tools for prediction of transcription factor binding sites on genomic DNA. In Silico Biol 1998; 1:21–28.

Roux B., Simonson T. Implicit solvent models. Biophys Chem 1999; 78:1–20.

Russel, P.J. *Genetics*. 4th ed. New York: Harper Collins Publishers, 1996.

Saenger, W. *Principles of Nucleic Acid Structure*. Springer-Verlag; New York Inc., 1984.

Saitou N., Nei M. The neighbor-joining method: A new method for reconstructing phylogenetic trees. Mol Biol Evol 1987; 4:406–425.

Sakai J., Nohturfft A., Cheng D., Ho Y.K., Brown M.S., Goldstein J.L. Identification of complexes between the COOH-terminal domains of sterol regulatory element-binding proteins (SREBPs) and SREBP cleavage-activating protein. J Biol Chem 1997; 272:20213–21.

Salamov A., Solovyev V. Prediction of protein secondary structure by combining nearest-neighbor algorithms and multiple sequence alignment. J Mol Biol 1995; 247:11–15.

Salgado H., Santos-Zavaleta A., Gama-Castro S., Millan-Zarate D., Diaz-Peredo E., Sanchez-Solano F., Perez-Rueda E., Bonavides-Martinez C., Collado-Vides J. RegulonDB (version 3.2): transcriptional regulation and operon organization in *Escherichia coli* K-12. Nucleic Acids Res 2001; 29(1):72–4.

Sankoff D., Nadeau J.H. Conserved synteny as a measure of genetic distance. Discrete Appl Math 1996; 71:247–257.

Savageau M.A. Mathematics of organizationally complex systems. Biomed Biochim Acta 1985; 6:839–44.

Sayos J., Wu C., Morra M., Wang N., Zhang X., Allen D., van Schaik S., Notarangelo L., Geha R., Roncarolo M.G., Oettgen H., De Vries J.E., Aversa G., Terhorst C. The X-linked lymphoproliferative-disease gene product SAP regulates signals induced through the co-receptor SLAM. Nature 1998; 395:462–9.

Schaefer M., Bartels C., Karplus M. Solution conformations and thermodynamics of structured peptides: Molecular dynamics simulation with an implicit solvation model. J Mol Biol 1996; 284:835–48.

Schaefer M., Karplus M. A comprehensive analytical treatment of continuum electrostatics. J Phys Chem 1996; 100:1578–99.

Scheres B. The role of position and lineage. Plant Physiol 2001; 125:112–114.

Scherf M., Klingenhoff A., Werner T. Highly specific localization of promoter regions in large genomic sequences by PromoterInspector: a novel context analysis approach. J Mol Biol 2000; 297:599–606.

Schomburg D., Schomburg L., Chang A., Bänsch C. BRENDA the Information System for Enzymes and metabolic Information. Proceedings of the German Conference on Bioinformatics, Germany, 1999; 226–7.

Seeman N.C. Macromolecular design, nucleic acid junctions and crystal formation. J Biomol Struct Dynam 1985; 3(1):11–34.

Selvaraj S., Kono H., Sarai A. Specificity of protein–DNA recognition revealed by structure-based potentials: symmetric/asymmetric and cognate/non-cognate binding. J Mol Biol 2002; 322:907–15.

Selvin, S. "*F* distribution". In *Encyclopedia of Biostatistics*. P. Armitage, T. Colton, eds. Vol. 2. Chichester: John Wiley & Sons, 1998; 1469–72.

Shapiro, B.E., Levchenko, A., Mjolsness, E. "Automatic model generation for signal transduction with applications to MAP-kinase pathways." In *Foundations of Systems Biology*. H. Kitano, ed. Cambridge, Massachusetts: MIT Press, 2001.

Shapiro B.E., Mjolsness E. Developmental simulations with Cellerator. Proceedings of the Second International Conference on Systems Biology; 2001 November 4–7; Pasadena, CA, 2001.

Sharp S., DeFranco D., Dingermann T., Farrell P., Soll D. Internal control regions for transcription of eukaryotic tRNA genes. Proc Natl Acad Sci USA 1981; 78:6657–61.

Sharp P.M., Devine K.M. Codon usage and gene expression level in *Dictiostelium discodideum:* highly expressed genes do "prefer" optimal codons. Nucleic Acids Res 1989; 17:5029–39.

Sharp P.M., Li W.-H. Codon usage in regulatory genes in *Escherichia coli* does not reflect selection for "rare" codons. Nucleic Acids Res 1986; 14:7737–49.

Sharp P.M., Li W.-H. The codon adaptation index—a measure of directional synonymous codon usage bias, and its potential applications. Nucleic Acids Res 1987; 15:1281–95.

Sharp K.A., Nicholls A., Fine R.F., Honig B. Reconciling the magnitude of the microscopic and macroscopic hydrophobic effects. Science 1991; 252:106–9.

Sharp S.J., Schaak J., Cooley L., Burke D.J., Soll D. Structure and transcription of eukaryotic tRNA genes. CRC Crit Rev Biochem 1985; 19:107–44.

Sheaff R.J., Groudine M., Gordon M., Roberts J.M., Clurman B.E. Cyclin E-CDK2 is a regulator of p27Kip1. Genes Dev 1997; 11:1464–78.

Sherlock G. *et al.* The Stanford microarray database. Nucleic Acids Res 2001; 29:152–155.

Shields D.C., Sharp P.M. Synonymous codon usage in *Bacillus subtilis* reflects both translational selection and mutational biases. Nucleic Acids Res 1987; 15:8023–40.

Shields D.C., Sharp P.M., Higgins D.G., Wright F. "Silent" sites in *Drosophila* genes are not neutral: evidence of selection among synonymous codons. Mol Biol Evol 1988; 5:704–16.

Shioiri C., Takahata N.J. Skew of mononucleotide frequencies, relative abundance of dinucleotides, and DNA strand asymmetry. Mol Evol 2001; 53:364–76.

Shlapatska L.M., Mikhalap S.V., Berdova A.G., Zelensky O.M., Yun T.J., Nichols K.E., Clark E.A., Sidorenko S.P. CD150 association with either the SH2-containing inositol phosphatase or the SH2-containing protein tyrosine phosphatase is regulated by the adaptor protein SH2D1A. J Immunol 2001; 166:5480–7.

Shobanov N.A., Zotov V.S. Cytogenetic aspects of the phylogeny of the genus *Chironomus* Meigen (Diptera, Chironomidae). Entomol Review 2001; 80:180–92.

Sibbald P.R., Argos P. Weighting aligned protein or nucleic acid sequences to correct for unequal representation. J Mol Biol 1990; 216:813–8.

Sidorenko S.P., Clark E.A. The dual-function CD150 receptor subfamily: the viral attraction. Nat Immunol 2003; 4:19–24.

Singer J.D., Gurian-West M., Clurman B., Roberts J.M. Cullin-3 targets cyclin E for ubiquitination and controls S phase in mammalian cells. Genes Dev 1999; 13:2375–87.

Sippl M. Calculation of conformational ensembles for potentials of mean force: an approach to the knowledge-based prediction of local structures in globular proteins. J Mol Biol 1990; 213:859–83.

Slack F.J., Basson M., Liu Z., Ambros V., Horvitz H.R., Ruvkun G. The *lin-41* RBCC gene acts in the *C. elegans* heterochronic pathway between the *let-7* regulatory RNA and the LIN-29 transcription factor. Mol Cell 2000; 5:659–69.

Smits H.A., Boven L.A., Pereira C.F., Verhoef J., Nottet H.S. Role of macrophage activation in the pathogenesis of Alzheimer's disease and human immunodeficiency virus type 1-associated dementia. Eur J Clin Invest 2000; 30:526–35.

Söll, D., RajBhandary, U.L. *tRNA: structure, biosynthesis and function.* Washington, D.C.: ASM Press, 1995.

Solovyev V., Salamov A. The Gene-Finder computer tools for analysis of human and model organism genome sequences. Proceeding of the Fifth International Conference on Intelligent Systems for Molecular Biology; 1997; 294–302.

Solovyev V., Shahmuradov I. PromH: promoters identification using orthologous genomic sequences. Nucleic Acids Res 2003; 31:3540–5.

Spana C., Corces V.G. DNA bending is a determinant of binding specificity for a *Drosophila* zinc finger protein. Genes Dev 1990; 4:1505–15.

Spellman P.T., Sherlock G., Zhang M.Q., Iyer V.R., Anders K., Eisen M.B., Brown P.O., Botstein D., Futcher B. Comprehensive identification of cell cycle-regulated genes of the yeast *Saccharomyces cerevisiae* by microarray hybridization. Mol Biol Cell 1998; 9:3273–97.

Spirin, A.S. *Molecular Biology: Ribosome Structure and Protein Biosynthesis*: Textbook for Higher Biological Education. Moscow: Vysshaya Shkola, 1986.

Spirov A.V., Borovsky M., Spirova O.A. HOX Pro DB: the functional genomics of hox ensembles. Nucleic Acids Res 2002; 30:351–3.

Sprinzl M., Horn C. *et al.* Compilation of tRNA sequences and sequences of tRNA genes. Nucleic Acids Res 1998; 26:148–53.

Starr D.B., Hoopes B.C., Hawley D.K. DNA bending is an important component of site-specific recognition by the TATA binding protein. J Mol Biol 1995; 250:434–46.

Steeves, T.A., Sussex, I.M. *Patterns in plant development.* New York: Cambridge University Press, 1989.

Stevens R., Baker P., Bechofer S. TAMBIS: Transparent Access to Multiple Bioinformatics Information Sources. Bioinformatics 2000; 16:184–5.

Still W., Tempczyk A. Hawley R., Hendrickson T. Semianalytical treatment of solvation for molecular mechanics and dynamics. J Am Chem Soc 1990; 112:6127–9.

Stoeckert C.J. jr., Salas F., Brunk B., Overton G.C. EpoDB: a prototype database for the analysis of genes expressed during vertebrate erythropoiesis. Nucleic Acids Res 1999; 27(1):200–3.

Stormo G.D., Schneider T.D., Gild L. Quantitative analysis of the relationship between nucleotide sequence and functional activity. Nucleic Acids Res 1986; 14:6661–79.

Streisinger G., Okada Y., Emrich J., Newton J., Tsugita A., Terzaghi E., Inouye M. Frameshift mutations and the genetic code. Cold Spring Harbor Symp Quant Biol 1966; 31:77–84.

Sutcliffe J.G., Milner R.J., Bloom F.E., Lerner R.A. Common 82-nucleotide sequence unique to brain RNA. Proc Natl Acad Sci USA 1982; 79(16):4942–6.

Suzuki M., Amano N., Kakinuma J., Tateno M. Use of 3D structure data for understanding sequence-dependent conformational aspects of DNA. J Mol Biol 1997; 274:421–35.

Suzuki Y., Ishihara D., Sasaki M., Makagawa H., Hata H., Tsunoda T., Watanabe M., Komatsu T., Ota T., Isogai T., Suyama A., Sugano S. Statistical analysis of the 5'-untranslated region of human mRNA using "oligo-capped" cDNA libraries. Genomics 2000; 64:286–97.

Takai-Igarashi T, Kaminuma T. A pathway finding system for the cell signaling networks database. In Silico Biol 1999; 1:129–46.

Tamayo P., Slonim D., Mesirov J., Zhu Q., Kitareewan S., Dmitrovsky E., Lander E.S., Golub T.R. Interpreting patterns of gene expression with self-organizing maps: methods and application to hematopoietic differentiation. Proc Natl Acad Sci USA 1999; 96:2907–12.

Tang G., Reinhart B.J., Bartel D.P., Zamore P.D. A biochemical framework for RNA silencing in plants. Genes Dev 2003; 17:49–63.

Tatsuo H., Ono N., Tanaka K., Yanagi Y. SLAM (CDw150) is a cellular receptor for measles virus. Nature 2000; 406:893–7.

Tautz D. Regulation of the *Drosophila* segmentation gene hunchback by two maternal morphogenetic centres. Nature 1988; 332:281–4.

Tavazoie S., Hughes J.D., Campbell M.J., Cho R.J., Church G.M. Systematic determination of genetic network architecture. Nat Genet 1999; 22:281–5.

Taylor S.S., Knighton D.R., Zheng J., Ten Eyck L.F., Sowadski J.M. Structural framework for the protein kinase family. Annu Rev Cell Biol 1992; 8:429–62.

Tchuraev R.N. A new method for the analysis of the dynamics of the molecular genetic control systems. I. Description of the method of generalized threshold models. J Theor Biol 1991; 151:71–87.

Tchuraev R.N., Galimzyanov A.V. The solution of the problems on parametric stability for ontogenesis control gene networks. Proceedings of 3rd Dagstuhl Seminar for Information and Simulation Systems for the Analysis of Gene Regulation and Metabolic Pathways; 2001 Juny 24–29; Dagstuhl: IBFI Gem. GmbH Schloss Dagstuhl, 2001a.

Tchuraev R.N., Galimzyanov A.V. Modeling real eukaryotic control gene subnetworks based on generalized threshold models. Mol Biol (Mosk) 2001b; 35:1088–94.

Tchuraev, R.N., Ratner, V.A. "Modeling of dynamics of the l phage development control system". In *Researches on Mathematical Genetics*, V.A. Ratner, ed. Novosibirsk: ICG, 1975.

Thåström A., Lowary P.T., Widlund H.R., Cao H., Kubista M., Widom J. Sequence motifs and free energies of selected natural and non-natural nucleosome positioning DNA sequences. J Mol Biol 1999; 288:213–29.

The Arabidopsis Genome Initiative. Analysis of the genome sequence of the flowering plant Arabidopsis thaliana. Nature 2000; 408:796–815.

The Object Database Standard: ODMG-93, Release 2.0. R. Cattell, D.K. Barry, eds. San Francisco, CA: Morgan Kauffmann Publishers, 1997.

Thomas R., Thieffry D., Kaufman M. Dynamical behavior of biological regulatory networks I. Biological role of feedback loops and practical use of the concept of the loop-characteristic state. Bull Math Biol 1995; 57:247–76.

Titov I.I., Vorobiev D.G., Ivanisenko V.A., Kolchanov N.A. A fast genetic algorithm for RNA secondary structure analysis. Chemistry of Natural Compounds and Bioorganic Chemistry 2002; 07:1135–44.

Todorovic V., Falaschi A., Giacca M. Replication origins of mammalian chromosomes: the happy few. Front Biosci 1999; 4:D859–68.

Tomasi J., Persico M. Molecular interactions in solution: An overview of methods based on continuum distribution of the solvent. Chem Rev 1994; 94:2027–94.

Tomii K., Kanehisa M. Analysis of amino acid indices and mutation matrices for sequence comparison and structure prediction of proteins. Protein Eng 1996; 9:27–36.

Trifonov E.N. Genetic level of DNA sequences is determined by superposition of many codes Mol Biol (Mosk). 1997; 31(4):759–67.

Tsai S.F., Jang C.C., Prikhod'ko G.G., Bessarab D.A., Tang C.Y., Pflugfelder G.O., Sun Y.H. *gypsy* retrotransposon as a tool for the in vivo analysis of the regulatory region of the optomotor-blind gene in Drosophila. Proc Natl Acad Sci USA 1997; 94:3837–41.

Tsurimoto, T., Matsubara, K. "Replication of bacteriophage λ DNA". In *Cold Spring Harbor Symposia on Quantitative Biology*. Cold Spring Harbor, New York, 1983; 47:681–691.

Turner D.H., Sugimoto N., Freier S.M. RNA structure prediction. Ann Rev Biophys Biophys Chem 1988; 17:167–92.

Udomkit A., Forbes S., Dalgleish G., Finnegan D.J. BS, a novel LINE-like element in *Drosophila melanogaster*. Nucleic Acids Res 1995; 23:1354–8.

URL: AMBER6 Home page: http://www.amber.ucsf.edu/amber/index.html, 2001.

URL: GROMOS96 Home page: http://www.igc.ethz.ch/gromos/ 2000.

Vandenbogaert M. Recherche de signaux fonctionnels dans les zones de régulation. Thèse de 3ieme cycle, Université de Bordeaux I. 2003.

Vandenbogaert M., Makeev V. Analysis of bacterial RM-systems through genome-scale analysis and related taxonomy issues. Proceedings of the Third International Conference on Bioinformatics of Genome Regulation and Structure; 2002 July 14–20; Novosibirsk: ICG, 2002; 2.

Viadiu H., Aggarwal A.K. Structure of BamHI bound to non-specific DNA: a model for DNA sliding. Mol Cell 2000; 5:889–95.

Vigdal T., Kaufman C., Izsvak Z., Voytas D., Ivics Z. Common physical properties of DNA affecting target site selection of Sleeping Beauty and other Tc1/mariner transposable elements. J Mol Biol 2002; 323:441–52.

Vinogradov D.V., Mironov A.A. SITEPROB: Yet Another Algorithm to Find Regulatory Signals in Nucleotide Sequences. Proceedings of the Third International Conference on Bioinformatics of Genome Regulation and Structure; 2002 July 14–20; Novosibirsk: ICG, 2002; 1.

Vishnevsky O.V., Vityaev E.E. Analysis and recognition of erythroid-specific gene promoters basing on degenerate oligonucleotide motifs. Mol Biol (Mosk) 2001; 35:1–9.

Vorobjev Y.N., Scheraga H.A. A fast adaptive multigrid boundary element method for macromolecular electrostatics in a solvent. J Comp Chem 1997; 18:569–83.

Vorobjev Yu., Hermans J. SIMS, computation of a smooth invariant molecular surface. J Biophys 1997; 72:722–32.

Vorobjev Yu.N., Almagro J.C., Hermans J. Discrimination between native and intentionally misfolded conformations of proteins: ES/IS, a new method for calculating conformational free energy. Proteins Struct Funct Gen 1998; 32:399–413.

Vorobjev Yu.N., Hermans J. ES/IS: estimation of conformational free energy by combining dynamics simulations with explicit solvent continuum model. Biophys Chem 1999; 78:195–205.

Vorobjev Yu.N., Hermans J. Free energies of protein decoys provide insight into determinants of protein stability. Protein Sci 2001; 10:2498–506.

Wallace A.C., Laskowski R.A., Thornton J.M. Derivation of 3D coordinate templates for searching structural databases: application to the Ser-His-Asp catalytic triads of the serine proteinases and lipases. Protein Sci 1996; 5:1001–13.

Waterman, M. *Introduction to Computational Biology*. London: Chapman and Hall Press, 1995.

Waterman, M.S. "Map Sequences and Genomes". In *Introduction to Computational Biology*. London: Chapman and Hall Press, 1995.

Waterston R.H., Lindblad-Toh K., Birney E. *et al.* Initial sequencing and comparative analysis of the mouse genome. Nature 2002; 420:520–62.

Waugh M. Pathdb helps researchers analyze metabolism. Technical Report 1, National Center for Genome Resources, 2000.

Wells J., Boyd K.E., Fry C.J., Bartley S.M., Farnham P.J. Target gene specificity of E2F and pocket protein family members in living cells. Mol Cell Biol 2000; 20:5797–807.

Whalen J.H., Grigliatti T.A. Molecular characterization of a retrotransposon in *Drosophila melanogaster*, nomad, and its relationship to other retrovirus-like mobile elements. Mol Gen Genet 1998; 260:401-9.

Widlund H.R., Cao H., Simonsson S., Magnusson E., Simonsson T., Nielsen P.E., Kahn J.D., Crothers D.M., Kubista M. Identification and characterization of genomic nucleosome-positioning sequences. J Mol Biol 1997; 267:807–17.

Widom J. Structure, dynamics, and function of chromatin *in vitro*. Ann Rev Biophys 1998; 27:285–327.

Wightman B., Ha I., Ruvkun G. Posttranscriptional regulation of the heterochronic gene *lin-14* by *lin-4* mediates temporal pattern formation in *C. elegans*. Cell 1993; 75:855–62.

Willis I.M. RNA polymerase III. Genes, factors and transcriptional specificity. Eur J Biochem 1993; 212:1–11.

Wingender E., Chen X., Fricke E., Geffers R., Hehl R., Liebich I., Krull M., Matys V., Michael H., Ohnhauser R., Pruss M., Schacherer F., Thiele S., Urbach S. The TRANSFAC system on gene expression regulation. Nucleic Acids Res 2001; 29:281–3.

Wingender E., Dietze P., Karas H., Knuppel R. TRANSFAC: a database on transcription factors and their DNA binding sites. Nucleic Acids Res 1996; 24(1):238–41.

Woese, C. "The evolution of the genetic code". In *The Genetic Code*. New York: Harper & Row, 1967; 179–95.

Wolberger C., Vershon A.K., Liu B., Johnson A., Pabo C.O. Crystal structure of a MATα2 homeodomain–operator complex suggests a general model for homeodomain–DNA interactions. Cell 1991; 67:517–28.

Wolffe A.P., Matzke M.A. Epigenetics: regulation through repression. Science 1999; 286(5439):481–6.

Wülker W., Dévai G., Dévai I. Computer assisted studies of chromosome evolution in the genus *Chironomus* (Dipt.). Comparative and integrated analysis of chromosome arms A, E and F. Acta Biol Debr Oecol Hung 1989; 2: 373–87.

Wyrick J.J., Aparicio J.G., Chen T., Barnett J.D., Jennings E.G., Young R.A., Bell S.P., Aparicio O.M. Genome-wide distribution of ORC and MCM proteins in *S. cerevisiae*: high-resolution mapping of replication origins. Science 2001; 294:2357–60.

Xu M., Sheppard K.A., Peng C.Y., Yee A.S., Piwnica-Worms H. Cyclin A/CDK2 binds directly to E2F-1 and inhibits the DNA-binding activity of E2F-1/DP-1 by phosphorylation. Mol Cell Biol 1994; 14:8420–31.

Yamao F., Andachi Y., Muto A., Ikemura T., Osawa S. Levels of tRNAs in bacterial cells as affected by amino acid usage in proteins. Nucleic Acids Res 1991; 19:6119–22.

Yang W., Lee H.-W., Hellinga H., Yang J.J. Structural analysis, identification, and design of calcium-binding sites in proteins. Proteins 2002; 47:344–56.

Yang X.-J., Seto E. Collaborative spirit of histone deacetylases in regulating chromatin structure and gene expression. Curr Opin Genet Devel 2003; 13:143–53.

Yoda K., Ando S., Okuda A., Kikuchi A., Okazaki T. *In vitro* assembly of the CENP-B/alpha-satellite DNA/core histone complex: CENP-B causes nucleosome positioning. Genes Cells 1998; 3:533–48.

Yun D.F., Laz T.M., Clements J.M., Sherman F. mRNA sequences influencing translation and the selection of AUG initiator codons in the yeast *Saccharomyces cerevisiae*. Mol Microbiol 1996; 19:1225–39.

Zakharov I.A., Nikiforov V.S., Stepanyuk E.V. Homology and evolution of the gene orders: simulation and reconstruction of the evolutionary process. Genetika 1997; 33:31–9.

Zawel L., Reinberg D. Advances in RNA polymerase II transcription. Curr Opin Cell Biol 1992; 4(3):488–95.

Zemla A., Venclovas C., Fidelis K., Rost B. A modified definition of SOV, a segment based measure for protein secondary structure prediction assessment. Proteins: Struct Funct Genet 1999; 34:220–3.

Zeng Y., Wagner E.J., Cullen B.R. Both natural and designed micro RNAs can inhibit the expression of cognate mRNAs when expressed in human cells. Mol Cell 2002; 9:1327–33.

Zhang M.Q. Identification of human gene core promoters *in silico*. Genome Res 1998; 8:319–26.

Zhang X., Mesirov J.P., Waltz D.L. Hybrid system for protein secondary structure prediction. J Mol Biol 1992; 225:1049–63.

Zheng J., Knighton D.R., Ten Eyck L.F., Karlsson R., Xuong N., Taylor S.S., Sowadski J.M. Crystal structure of the catalytic subunit of cAMP-dependent protein kinase complexed with MgATP and peptide inhibitor. Biochemistry 1993; 32:2154–61.

Zheng N., Fraenkel E., Pabo C.O., Pavletich N.P. Structural basis of DNA recognition by the heterodimeric cell cycle transcription factor E2F-DP. Genes & Development 1999; 13:666–74.

Index

E

F

G

H

I

K

L

M

N

O

P

R

S

T

V

W

Y